recent advances in phytochemistry

volume 31

Functionality of Food Phytochemicals

RECENT ADVANCES IN PHYTOCHEMISTRY

Proceedings of the Phytochemical Society of North America
General Editor: John T. Romeo, *University of South Florida, Tampa, Florida*

Recent Volumes in the Series:

A Continuation Order Plan is available for this series. A continuation order will bring delivery of each new volume immediately upon publication. Volumes are billed only upon actual shipment. For further information please contact the publisher.

recent advances in phytochemistry

volume 31

Functionality of Food Phytochemicals

Edited by

Timothy Johns
McGill University
Ste. Anne de Bellevue
Quebec, Canada

and

John T. Romeo
University of South Florida
Tampa, Florida

PLENUM PRESS • NEW YORK AND LONDON

Library of Congress Cataloging-in-Publication Data

Functionality of food phytochemicals / edited by Timothy Johns and John T. Romeo.
p. cm. -- (Recent advances in phytochemistry ; v. 31)
"Proceedings of the Thirty-six Annual Meeting of the Phytochemical Society of North America on Functionality of Food Phytochemicals: Flavors, Stimulants, and Health Promoters, held August 10-14, 1996, in New Orleans, Louisiana"--T.p. verso.
Includes bibliographical references and index.
ISBN 0-306-45691-5
1. Nutrition--Congresses. 2. Food--Composition--Congresses. 3. Botanical chemistry--Congresses. I. Johns, Timothy, 1950- II. Romeo, John T. III. Meeting of the Phytochemical Society of North America on Functionality of Food Phytochemicals: Flavors, Stimulants, and Health Promoters (1996 : New Orleans, La.) IV. Series.
[DNLM: 1. Plants, Edible--chemistry--congresses. 2. Food--congresses. W1 RE105Y v.31 1997 / QK 861 F979 1997]
QK861.R38 vol.31
[QP141.A1]
572'.2 s--dc21
[612.3]
DNLM/DLC
for Library of Congress 97-15679
CIP

Proceedings of the Thirty-Sixth Annual Meeting of the Phytochemical Society of North America on Functionality of Food Phytochemicals: Flavors, Stimulants, and Health Promoters, held August 10–14, 1996, in New Orleans, Louisiana

ISBN 0-306-45691-5

A Division of Plenum Publishing Corporation
233 Spring Street, New York, N. Y. 10013

http://www.plenum.com

10 9 8 7 6 5 4 3 2 1

Printed in the United States of America

PREFACE

Phytochemists are aware that their focus of interest is receiving attention from a wider segment of society and from a greater diversity of disciplines within the scientific community than ever before. Nonetheless, they were bemused to learn three years ago that "until recently scientists didn't even know phytochemicals existed" (*Newsweek*, April 24, 1994). Changing public perception of the positive contributions of phytochemicals to human well-being has foundations in scientific advances. With popular reports emphasizing the important implications of phytochemicals in the daily lives of people, there is a pressing need for those working in this area to explain their diverse scientific activities to the public.

Chemicals from plant foods are linked through epidemiological and experimental studies with reduced incidence of chronic degenerative diseases. Phytomedicines, standardized according to particular constituents, are making increasing contributions to health care. Naturally occurring constituents of plants are recognized as fundamental to the appeal, quality, and marketability of food products. In light of such developments, perceptions by phytochemists of their own discipline and its applications are expanding. Until recently, food phytochemistry largely implied food toxicants. Food plants were familiar, but seldom the source of novel economically important compounds. Increasingly sophisticated methods of analysis, however, have opened new opportunities for understanding the nature and functions of food constituents, and for manipulating them to improve the quality, acceptability, and value of food products.

As a reflection of the active nature of current studies of the functional significance of food constituents, new terminology is emerging. *Functional foods*, while consumed as part of a usual diet, have demonstrated physiological and health benefits beyond basic nutritional functions. *Nutraceuticals* are produced from foods and have similar benefits, but are sold in pills, powders, and other medicinal forms not generally associated with food. Refined ingredients, produced from foods, may be added to enhance the benefits of the ingestion of other products, as well as enhance their color, flavor, texture, and preservation.

The papers in this volume represent the emerging scope of the field of food phytochemistry. They were presented at the annual meeting of the Phytochemical

Society of North America, held in New Orleans, August 10–14, 1996. The symposium considered why phytochemicals are receiving increasing attention, as well as the new methodologies that make it possible to exploit their potential. The increasing recognition of food phytochemicals as health promoters is addressed most directly in the papers by Block et al. and Montanari et al. on the chemistry of garlic and citrus, respectively, and by Stavric's discussion of chemopreventive agents.

The paper by Sotelo, in considering both nutrient and toxicological aspects of food plants, illustrates the continued relevance of this classic approach in phytochemistry. Indigenous food plants have great potential as food sources in developing regions of the world and, as links are made globally between the need to preserve biodiversity while meeting the subsistence needs of a growing human population, analysis of their properties is needed more than ever. Flores and Flores consider the importance of indigenous root and tuber crops and emphasize how with advanced analytical techniques they can be useful models for understanding basic biochemical mechanisms.

In looking at behavioral effects of food and herbal constituents, Johns provides a perspective on both the physiological and psychosocial factors that help explain why and when humans ingest phytochemicals. One of the motivators for seeking such chemicals from plant products is flavor, as is illustrated directly in the paper by Ebeler on the sensory properties of wine. Understanding of mechanisms of flavor are extended in the paper by Crouzet. Each of these papers illustrates situations where phytochemical characteristics can be manipulated for advantage.

Cormier and Voelker extend this theme to consideration of the application of biotechnology in the production of useful food phytochemical products, the former on colorants from cell-culture and the latter on transgenic manipulations of edible oilseeds. As fatty acid composition has been linked with health, the work of Voelker and colleagues has clear ties with the aforementioned papers on health promoters. Similar links can be made with the paper by Fulcher et al. on quantitative imagining of cereal carbohydrates. The importance of technological advances to food research is emphasized in this and other papers.

We acknowledge the contribution of Dr. Nicholas Fischer who hosted the meeting in New Orleans and Dr. Francois Cormier who played a major role in organizing the symposium. T.J. acknowledges the assistance of V. Yaylayan. J.T.R. expresses gratitude to Dawn McGowan for her technical expertise and commitment to the project.

Timothy Johns
MacDonald Campus, McGill University
John T. Romeo
University of South Florida

CONTENTS

recent advances in phytochemistry

volume 31

Functionality of Food Phytochemicals

Chapter One

PEELING THE ONION*

Organosulfur and -selenium Phytochemicals in Genus *Allium* Plants

Eric Block,[1] Elizabeth M. Calvey,[2] Charles W. Gillies,[3] Jennifer Z. Gillies,[4] and Peter Uden[5]

[1] Department of Chemistry, State University of New York at Albany
Albany, New York 12222

[2] Center for Food Safety and Applied Nutrition, Food and Drug Administration
Washington, DC 20204

[3] Department of Chemistry, Rensselaer Polytechnic Institute
Troy, New York 12180

[4] Department of Chemistry, Siena College
Loudonville, New York 12211

[5] Department of Chemistry, University of Massachusetts
Amherst, Massachusetts 01003

* Dedicated with warm regards to Professor Dieter Seebach on the occasion of his 60th birthday.

Functionality of Food Phytochemicals
edited by Johns and Romeo, Plenum Press, New York, 1997

INTRODUCTION

There are more than 600 known species in the genus *Allium*. While some are little more than botanical curiosities, others are attractive ornamental plants of diverse size and hue (*e.g.*, *A. moly* L., *A. giganteum* Regel, *A. flavum* L., *A. pulchellum*, *A. roseum*) or economically important spices and vegetables (*e.g.*, onion, garlic, leek, shallot, chive, and scallion, respectively *A. cepa*, *A. sativum*, *A. porrum* L., *A. ascalonicum* auct., *A. schoenoprasum* L., and *A. fistulosum* L.).[1] The antibiotic, anticancer, antithrombotic, cholesterol-lowering, and other beneficial health effects associated with consumption of genus *Allium* plants are widely touted in the popular and scientific/medical press.[2,3] Typical culinary usage of these plants involves cutting or crushing them so as to maximize flavor and aroma release. Cutting or crushing results in disruption of plant tissue with ensuing enzymatic and chemical reactions generating the actual flavorants and aroma compounds.[4] The flavorants and aroma compounds probably serve as protective agents for the plant against attack by predators and infectious microorganisms.[4] At the same time several insect pests, such as the leek moth or onion maggot, key in on these compounds to locate their next meal or egg-laying site.[5]

The health benefits of *Allium* species are variously attributed to the biological activity of: compounds found in the intact plants, flavorants formed on cutting or crushing the plants, substances derived from further reactions of these flavorants, or metabolic degradation products of these three types of compounds.[2–4] The majority of the *Allium* species-derived compounds of interest from either a health or flavor standpoint contain sulfur, often in forms rarely found elsewhere in nature. The focus of our research has been the identification and chemical study of such compounds — *e.g.*, "*Allium* chemistry" — with the overall goal of better understanding both the flavor and medicinal chemistry of these plants.

This chapter summarizes recent work on the characterization of organosulfur and organoselenium compounds formed upon disruption of *Allium* tissue. The organosulfur compounds are discussed in order of increasing number of carbon atoms present, *e.g.* the one carbon (C1) and C3 intermediates, the C3 lachrymatory factor (LF), the C2, C4, and C6 thiosulfinates, the C6 LF dimer, zwiebelanes, and bissulfine, and the C9 ajoene and cepaenes, followed by the non-protein sulfur and selenium amino acids and volatile selenium compounds. Classes of unusual sulfur compounds to be discussed below include, among others, sulfenic acids, thiosulfinates, and sulfines, *e.g.* RSOH, RS(O)SR', and $EtCH{=}S^{+}\text{-}O^{-}$, where R/R' are methyl (Me), *n*-propyl (*n*-Pr), 1-propenyl

(MeCH=CH) or 2-propenyl (allyl; CH_2=$CHCH_2$). In these compounds sulfur is bonded to both oxygen and carbon, and sometimes to a second sulfur atom.

Major concerns of our research include: 1) efficient and rapid isolation of flavorants from the plant matrix or plant homogenates; 2) characterization of rapidly formed and short lived flavorants; 3) identification and quantification of longer lasting primary flavorants including those formed in trace amounts; 4) elucidation of mechanisms of formation of flavorants from stable precursors and identification of such precursors; 5) avoidance of artifact formation problems; 6) synthesis of flavorants; 7) extension of analyses to less well known Alliums. The characteristics of *Allium* flavorants make them particularly useful in demonstrating the power of a variety of modern instrumental methods.

ONION LACHRYMATORY FACTOR; *ALLIUM* FLAVORANT INTERMEDIATES

The unique ability of the onion to bring tears to those that would cut it has been widely recognized since earliest times. Thus, Shakespeare writes:[6] "And if the boy have not a woman's gift to rain a shower of commanded tears, an onion will do well for such a shift, which in a napkin being close conveyed, shall in despite enforce a watery eye." When an onion, or other *Allium* spp., is cut, alliinase enzymes commingle with *S*-alk(en)yl-L-cysteine *S*-oxides (**1a-d**, Fig. 1) forming the respective C1 and C3 sulfenic acid intermediates (**2a-d,** Fig. 1). While **2a-d** cannot be directly observed because of their very short lifetimes, they are thought to be the precursors of thiosulfinates ("sulfenic anhydrides") (**3**, Fig. 1), and other flavorant compounds discussed below (Fig. 2). Methanesulfenic acid (**2a**), the C1 intermediate, when independently generated by flash vacuum pyrolysis (FVP) of *tert*-butyl methyl sulfoxide (**4a**, Fig. 3), was found by absorption microwave (MW) spectroscopy,[7] as well as other techniques,[8] to have the structure CH_3S-O-H and a gas phase lifetime of *ca.* one minute at 0.1 Torr and 25° C. When condensed at -196° C, **2a** could not be recovered on warming in vacuum, instead forming C2 thiosulfinate (**3a**, MeS(O)SMe). In the presence of D_2O, MeSOH undergoes ready exchange in the waveguide of the

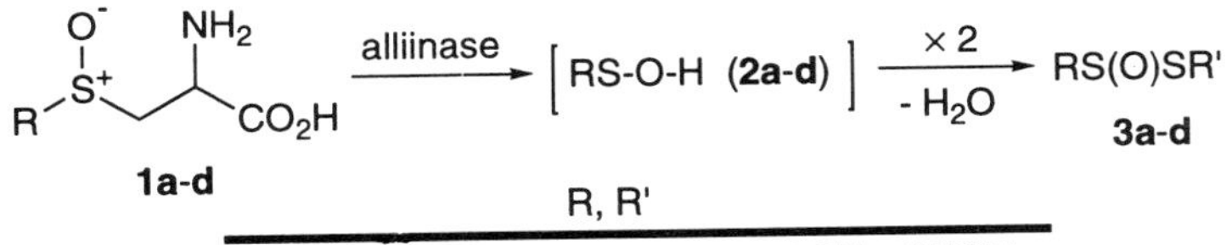

1a-d —alliinase→ [RS-O-H (**2a-d**)] —×2, - H_2O→ RS(O)SR' **3a-d**

R, R': **a** Me; **b** (*E*)-MeCH=CH; **c** *n*-Pr; **d** CH_2=$CHCH_2$

Figure 1. Alliinase-induced formation of intermediates from *S*-alk(en)yl-L-cysteine *S*-oxides.

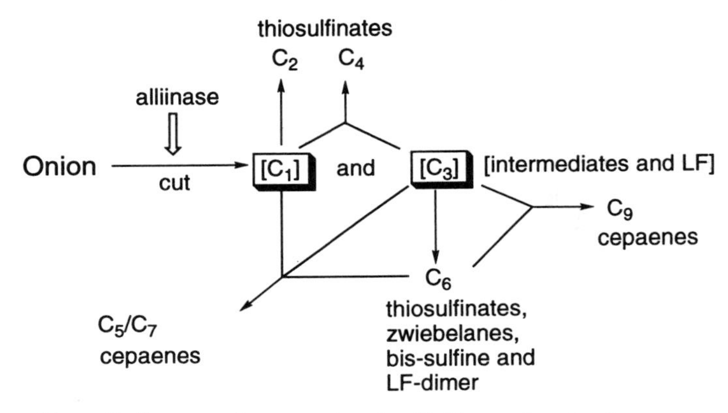

Figure 2. Flavorants from cut onion classified by number of carbon atoms.

MW spectrometer giving MeSOD.[7] The technique of MW spectroscopy is particularly well suited for the detection and characterization of reactive small molecules such as MeSOH and related compounds discussed below.

It was shown in 1961 by Virtanen that **1b** (*ca.* 0.2% by weight in onion) is the precursor of the onion lachrymatory factor (LF), molecular formula C_3H_6OS.[9] Virtanen also found that alliinases convert **1a**, **1c** (*ca.* 0.005–0.02% by weight in onion) and **1d** (*ca.* 0.5–1.4% by weight in garlic), to respective thiosulfinates **3a,c,d**.[10] Virtanen suggested that sulfenic acids, formulated by him as MeS(O)H, *n*-PrS(O)H and CH_2=$CHCH_2S(O)H$, are the immediate precursors of the thiosulfinates **3a,c,d** and that the onion LF is the sulfenic acid formulated as MeCH=CHS(O)H. However, using absorption MW,[11] pulsed-beam Fourier-transform microwave (FT-MW)[12] as well as NMR spectroscopy,[12,13] we found that the onion LF is in fact (*E*,*Z*)-propanethial *S*-oxide (EtCH=S^+-O^-, **5b**). The proposed mechanism for its formation is shown (Fig.

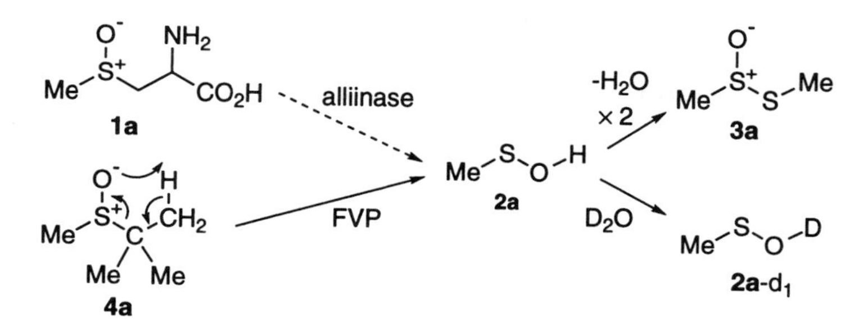

Figure 3. Formation and reactions of methanesulfenic acid.

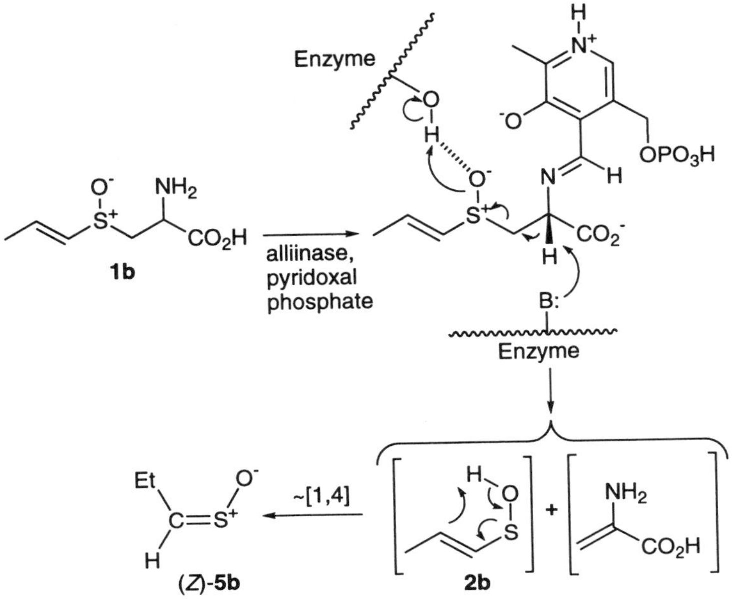

Figure 4. Proposed mechanism for formation of LF from cut onion.

4), and is supported by the observation that from onions macerated in the presence of D_2O, (*Z*)-**5b**-d_1 is detected by FT-MW spectroscopy. The latter compound is also detected by FT-MW spectroscopy, along with (*Z*)-**5b**, when *tert*-butyl (*E*)-1-propenyl sulfoxide (**4b**) is subjected to flow pyrolysis in the presence of D_2O (Fig. 5).[12] Theoretical calculations (HF/6–31G*)[14] and studies of (*E*,*Z*)-ethanethial *S*-oxide ($MeCH{=}S^{+}{-}O^{-}$; the lower homolog of **5b**),[12,15] also support the mechanism of Fig. 4. Quantification of the LF can be accomplished by gas chromatography (GC).[16]

LACHRYMATORY FACTOR DIMER

Upon standing, onion LF **5b** spontaneously dimerizes to afford *trans*-3,4-diethyl-1,2-dithietane 1,1-dioxide, $C_6H_{12}SO_2$ (**6**), a C6 compound presumably formed from the LF by a [3+2] cycloaddition process followed by rearrangement, as shown in Figure 6.[12,17] The structure of dimer **6** was established by chemical means as shown (Fig. 6). Theoretical calculations (HF/6–31G*; unpublished results, R. Sustmann and W. Sicking) support the [3+2] cycloaddition step.

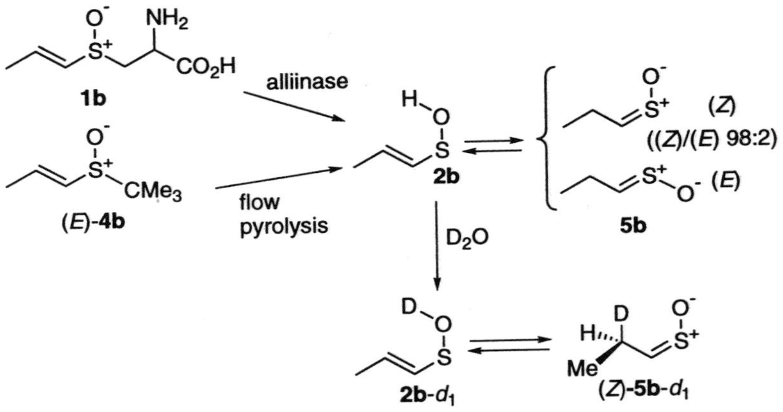

Figure 5. Generation of onion LF in the presence of D_2O.

DETECTION OF THIOSULFINATES IN *ALLIUM* HOMOGENATES; AVOIDANCE OF ARTIFACT FORMATION DURING ANALYSIS

In 1944, Cavallito et al. described the isolation and characterization of allicin (**3b**; CH_2=$CHCH_2S(O)SCH_2CH$=CH_2), an odoriferous, unstable antibacterial substance obtained from cut garlic cloves by extraction with ethanol.[18,19] Decomposition of allicin afforded diallyl disulfide as a major breakdown product. Cavallito found that peracid oxidation of diallyl disulfide represents the most

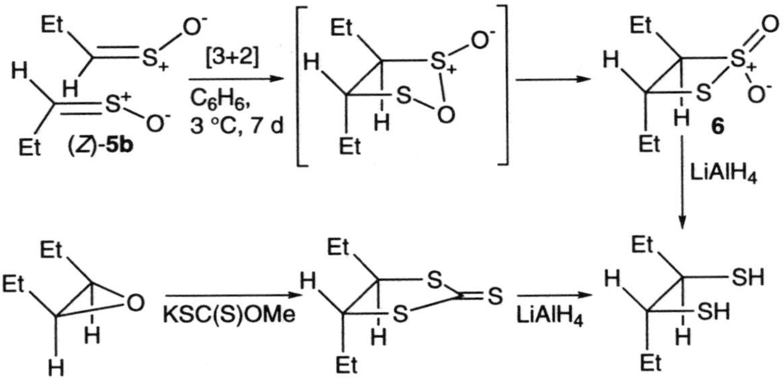

Figure 6. Dimerization of onion LF.

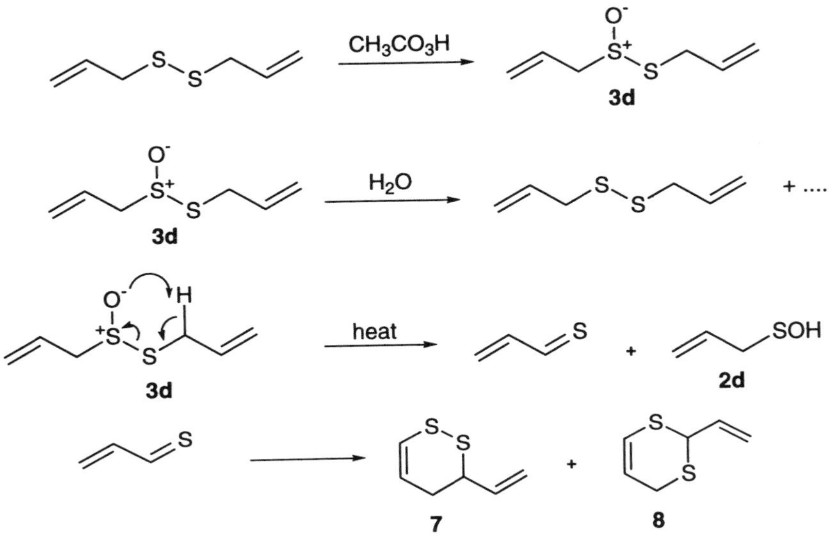

Figure 7. Synthesis and thermal decomposition of allicin.

direct method of synthesizing allicin (Fig. 7).[20] On injection into a GC, allicin decomposes giving two $C_6H_8S_2$ isomers identified by us as thioacrolein dimers 3-vinyl-3,4-dihydro-1,2-dithiin (**7,** Fig. 7) and 2-vinyl-2,4-dihydro-1,3-dithiin (**8,** Fig. 7).[21] Subsequent to Cavallito's work, numerous papers have appeared on the detection of complex mixtures of polysulfides from extracts, volatiles and distilled oils of various *Allium* species. At the same time, analysis by mild methods such as TLC suggest that mixtures of thiosulfinates **3** (RS(O)SR') are present in room temperature extracts of Alliums.

Once the analyst recognizes that thiosulfinates are an especially labile and reactive class of compounds, three major problems must be faced: 1) rapid, efficient isolation of the compounds from homogenates before substantial decomposition has occurred; 2) qualitative and quantitative analysis of thiosulfinates and related initial flavorants under conditions that avoid thermally and hydrolytically induced artifact formation; 3) unambiguous differentiation among regio- and stereoisomeric structures for the naturally-produced thiosulfinates and related compounds. Isolation procedures typically involve trimming plants to remove dried outer leaves and roots and "juicing" with a commercial juicer (*e.g.* for an onion) or homogenization with a tissue homogenizer (for denser plants such as garlic and shallots). Thiosulfinates can be directly extracted from the plant juice or homogenates with organic solvents such as methylene chloride or ether. Direct extraction methods work reasonably well with garlic homogenates, which contain relatively high levels of flavorants. In the case of onion, where levels of flavorants are lower and emulsion problems more severe, direct solvent extraction is used for rapid isolation and GC determination of the onion LF.[16] Emulsion problems can be reduced if the plant homogenates are subjected to "room-temperature steam distillation", *e.g.* vacuum distillation of the aqueous homogenate at room temperature, prior to extraction.[22] Supercritical fluid extraction (SFE) with carbon dioxide has proven to be an especially rapid and efficient procedure, resulting in higher yields of thiosulfinates from garlic homogenates than achieved with organic solvents.[23] SFE using carbon dioxide has the additional advantage of avoiding the safety hazards and costs associated with the use and disposal of conventional organic solvents.

Allium extracts can be analyzed by GC-MS if care is taken to employ mild analyte introduction conditions, *e.g.* cryogenically cooled injection port and column,[22] and to verify the stability of thiosulfinates and the LF with authentic standards,[24] as illustrated with an onion extract (Fig. 8). However, even under the mildest GC conditions tested, some thiosulfinates and many higher molecular

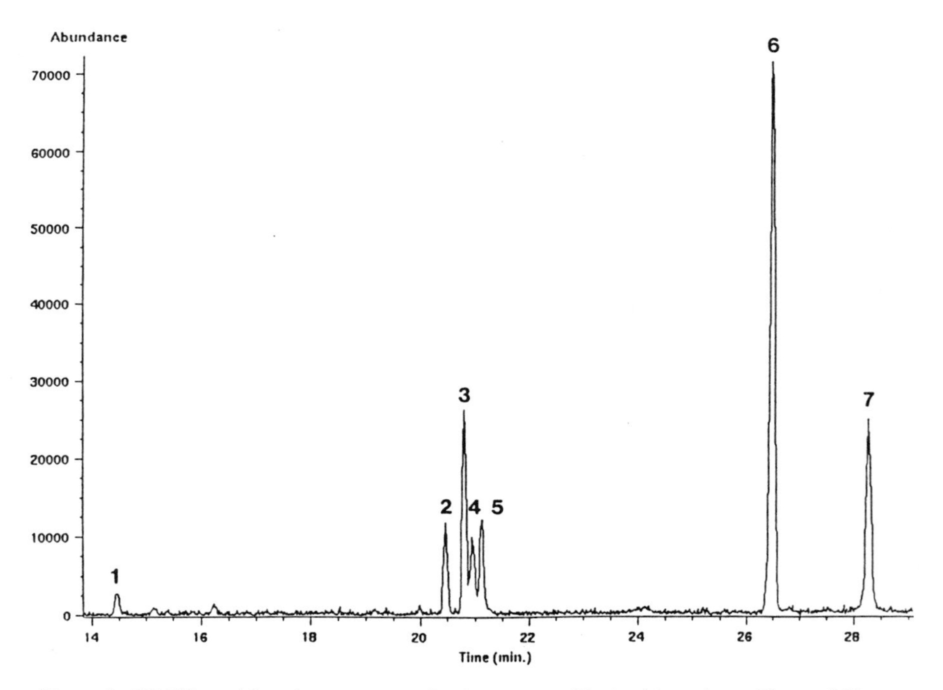

Figure 8. GC-MS total ion chromatogram of onion extract. Obtained by using a 30 m × 0.53 mm methyl silicone gum capillary column, injector and column programmed from 0 to 200 °C. Identification of components: 1. MeS(O)SMe; 2. MeS(O)SPr-*n*; 3. MeCH=CHSS(O)Me; 4. MeSS(O)Pr-*n*; 5. MeCH=CHS(O)SMe; 6. MeCH=CHSS(O)Pr-*n* + *trans*-zwiebelane; 7. *cis*-zwiebelane.

weight *Allium* flavorants decompose, which leads to artifact formation.[25] For this reason, we have explored the use of room temperature analytical methods such as supercritical fluid chromatography-MS (SFC-MS),[26] liquid chromatography (LC),[27] and particularly LC-MS with atmospheric pressure chemical ionization (APCI) and tandem MS (MS-MS). Analysis of SF extracts of garlic by these methods indicated the presence of allicin (>50% of total thiosulfinates), allyl/methyl and 1-propenyl/methyl isomers (>25%), allyl/1-propenyl isomers (>19%), traces of allyl/*n*-propyl isomers, and ajoene (see below). A typical LC-APCI-MS reconstructed ion chromatogram for garlic is shown (Fig. 9). When an acetonitrile/water LC mobile phase is used in these analyses, the mass spectra of the thiosulfinates have a base peak associated with the protonated molecular ion (MH^+). The fragmentation patterns of the MS/MS spectra of each peak compare well with the corresponding patterns of synthetic standards (see below). Analysis by LC-APCI-MS of an SF extract of onion shows methyl/

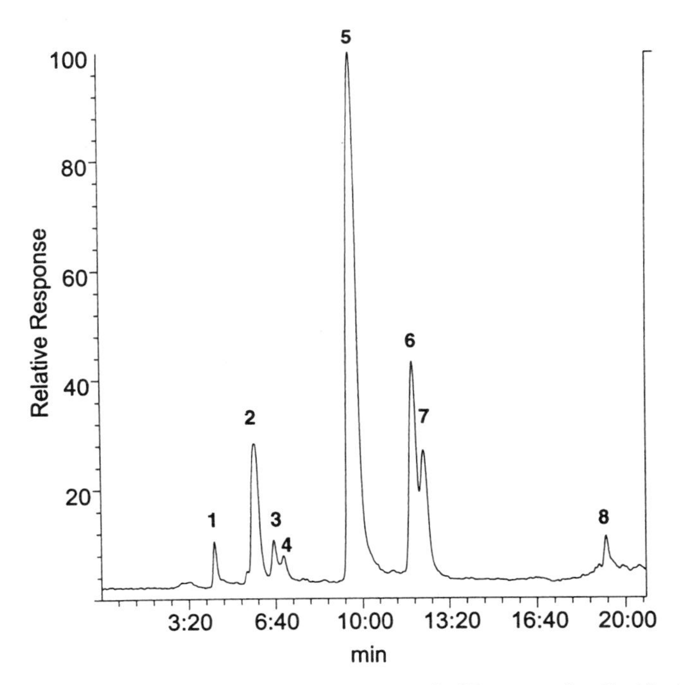

Figure 9. LC-APCI-MS reconstructed ion chromatogram of a SF extract of garlic. Identification of components: 1. MeS(O)SMe; 2. AllS(O)SMe/AllSS(O)Me; 3. MeS(O)SCH = CHMe; 4. MeSS(O)CH = CHMe; 5. $CH_2 = CHCH_2S(O)SCH_2CH = CH_2$ (allicin); 6. $CH_2 = CHCH_2S(O)SCH = CHMe$; 7. $CH_2 = CHCH_2SS(O)CH = CHMe$; 8. $CH_2 = CHCH_2S(O)CH_2CH = CHSSCH_2CH = CH_2$ (*E*,*Z*-ajoene).

methyl, methyl/1-propenyl, methyl/propyl, 1-propenyl/propyl, and propyl/propyl thiosulfinates along with other compounds, discussed below.

SYNTHESIS OF THIOSULFINATE STANDARDS

In order to fully characterize *Allium* flavorants, it is necessary to synthesize all possible C4 and C6 thiosulfinates that could result from condensation of sulfenic acids (**2a-d**). While thiosulfinates with identical alkyl groups (**3**, R = R'), such as MeS(O)SMe (**3a**), *n*-PrS(O)SPr-*n* (**3c**) and allicin (**3d**) are known, thiosulfinates (see Fig. 1) with different alkyl groups (**3**, R ≠ R') were mostly unknown. When groups R and R' in thiosulfinates RS(O)SR' are methyl, propyl, or allyl, specific compounds can be prepared by pyridine-catalyzed condensation of the appropriate sulfinyl chloride RS(O)Cl with thiol R'SH.[21] Proton and carbon NMR spectroscopy confirmed that compounds with a MeS(O) group have signals at 2.98–3.04 (^{1}H) and 42.1–42.7 (^{13}C) ppm while those with a MeS group have signals at 2.60–2.69 (^{1}H) and 14.5–17.3 (^{13}C) ppm. Regioisomers

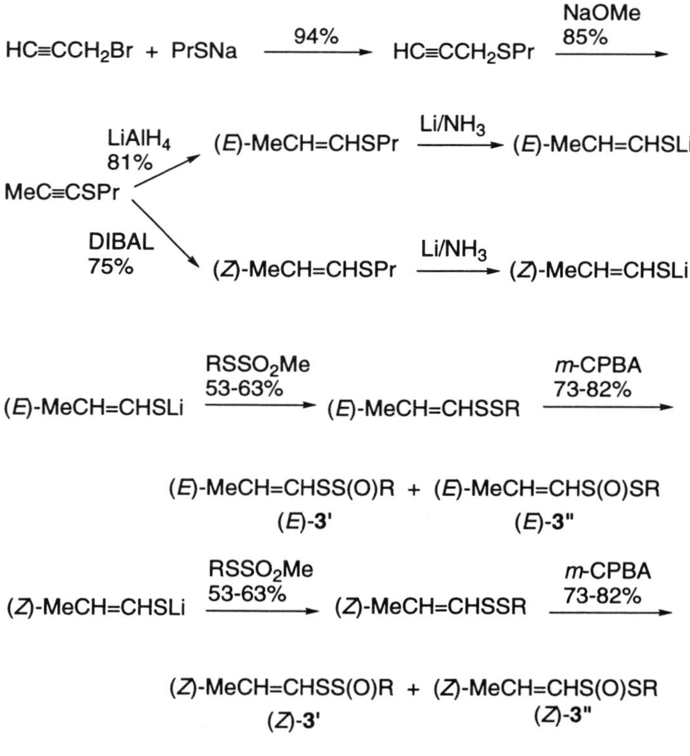

$$HC{\equiv}CCH_2Br + PrSNa \xrightarrow{94\%} HC{\equiv}CCH_2SPr \xrightarrow[85\%]{NaOMe}$$

$$MeC{\equiv}CSPr \xrightarrow[81\%]{LiAlH_4} (E)\text{-}MeCH{=}CHSPr \xrightarrow{Li/NH_3} (E)\text{-}MeCH{=}CHSLi$$

$$MeC{\equiv}CSPr \xrightarrow[75\%]{DIBAL} (Z)\text{-}MeCH{=}CHSPr \xrightarrow{Li/NH_3} (Z)\text{-}MeCH{=}CHSLi$$

$$(E)\text{-}MeCH{=}CHSLi \xrightarrow[53\text{-}63\%]{RSSO_2Me} (E)\text{-}MeCH{=}CHSSR \xrightarrow[73\text{-}82\%]{m\text{-}CPBA}$$

$$\underset{(E)\text{-}\mathbf{3'}}{(E)\text{-}MeCH{=}CHSS(O)R} + \underset{(E)\text{-}\mathbf{3''}}{(E)\text{-}MeCH{=}CHS(O)SR}$$

$$(Z)\text{-}MeCH{=}CHSLi \xrightarrow[53\text{-}63\%]{RSSO_2Me} (Z)\text{-}MeCH{=}CHSSR \xrightarrow[73\text{-}82\%]{m\text{-}CPBA}$$

$$\underset{(Z)\text{-}\mathbf{3'}}{(Z)\text{-}MeCH{=}CHSS(O)R} + \underset{(Z)\text{-}\mathbf{3''}}{(Z)\text{-}MeCH{=}CHS(O)SR}$$

Figure 10. Synthesis of isomeric alk(en)yl/1-propenylthiosulfinates (R = Me, *n*-Pr or $CH_2{=}CHCH_2$).

such as MeS(O)SCH_2CH=CH_2 and MeSS(O)CH_2CH=CH_2 can also be distinguished by characteristic MS fragmentation patterns. When R or R' are 1-propenyl, the above synthesis can not be used because of the unavailability of the corresponding sulfinyl chloride (MeCH=CHS(O)Cl) and thiol (MeCH=CHSH). The lengthier procedure outlined in Figure 10, which also controls the double bond stereochemistry, was therefore employed.[24] By the above methods, the thiosulfinates shown in Figure 11 were synthesized. Retention times, molecular weights (listed below each compound in Figure 11) and fragmentation patterns from MS analysis were matched with compounds found in *Allium* extracts. With the exception of MeCH=CHS(O)SCH=CHMe (in dashed box; see below), all of the thiosulfinates in Figure 11 have now been found by LC-APCI-MS methods in *Allium* extracts, although some only in trace amounts, *e.g.* allyl/propyl thiosulfinates in garlic (E. Calvey, E. Block and coworkers, unpublished results). Related results have been reported for "*Allium* odors" (volatiles).[28,29]

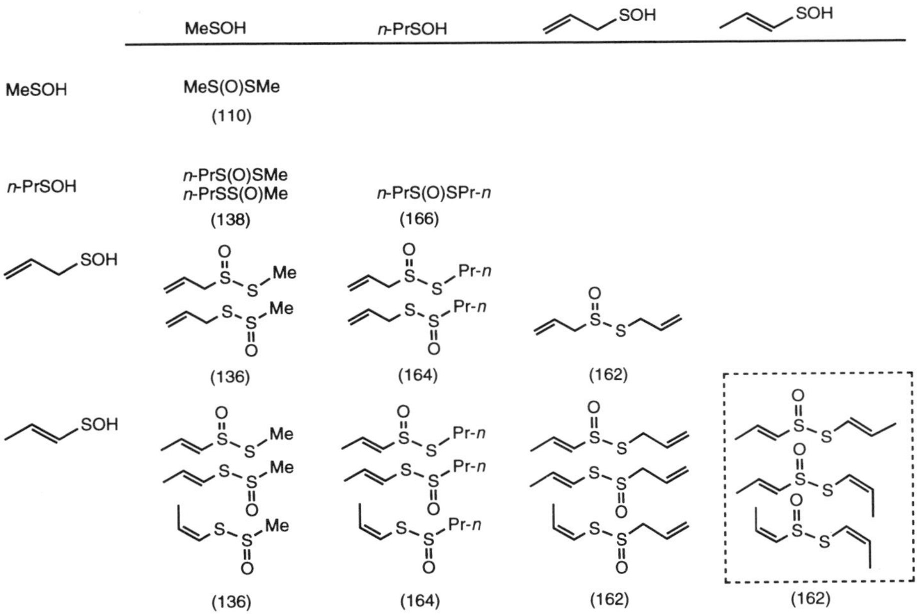

	MeSOH	*n*-PrSOH	SOH	SOH
MeSOH	MeS(O)SMe (110)			
n-PrSOH	*n*-PrS(O)SMe *n*-PrSS(O)Me (138)	*n*-PrS(O)SPr-*n* (166)		
SOH	Me Me (136)	Pr-*n* Pr-*n* (164)	(162)	
SOH	Me Me Me (136)	Pr-*n* Pr-*n* Pr-*n* (164)	(162)	(162)

Figure 11. Thiosulfinates (MW in parentheses) from pair-wise condensation of MeSOH, *n*-PrSOH, CH_2=$CHCH_2SOH$, and MeCH=CHSOH.

ZWIEBELANES AND BISSULFINE

Figure 11 suggests that stereoisomeric thiosulfinates of type MeCH=CHS(O)SCH=CHMe should be formed in *Allium* homogenates from self-condensation of 1-propenesulfenic acid, MeCH=CHSOH. Initial efforts to prepare authentic samples of these thiosulfinates by peracid oxidation of stereoisomers of bis(1-propenyl) disulfide, MeCH=CHSSCH=CHMe (**9**), led instead to a pair of novel, isomeric $C_6H_{10}OS_2$ sulfoxides (**10a** and **b**, Fig. 12). Compounds **10a**,**b** were isolated from extracts of onion homogenates and were named "zwiebelanes" (from the German word for onion, "zwiebel").[30,31] Full stereochemical characterization of *trans*-zwiebelane (**10a**) was achieved by X-ray crystallographic analysis of oxidation product **12**, prepared by oxidation of symmetrical bis-sulfoxide **11**(Fig. 12).[31] Detailed mechanistic studies involving

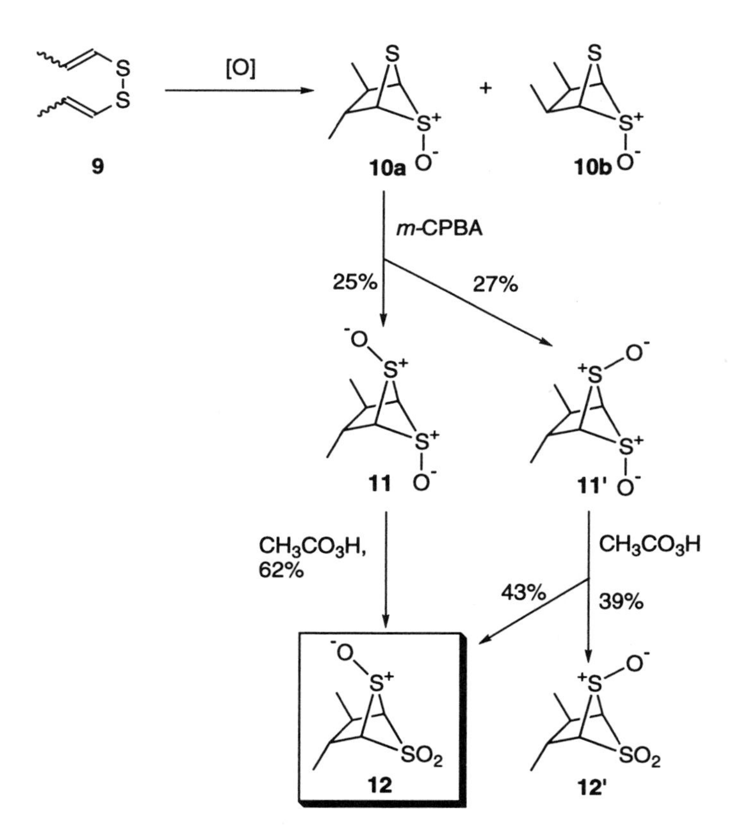

Figure 12. Synthesis of zwiebelanes, **10a**,**b**; proof of stereochemistry in **10a** by X-ray crystallography of derivative **12**.

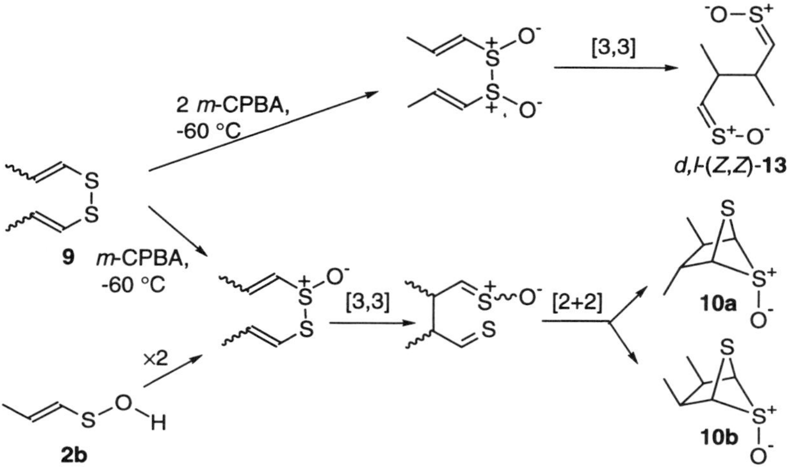

Figure 13. Proposed mechanism for formation of zwiebelanes, **10a,b**, from 1-propenesulfenic acid, **2b**. Synthesis of bissulfine, **13**.

low temperature peracid oxidation of individual isomers of **9**, whereby the corresponding isomers of MeCH=CHS(O)SCH=CHMe could each be directly observed, led us to propose a mechanism for formation of zwiebelanes in onion (Fig. 13).[24] Further study of the low temperature oxidation of MeCH=CHS(O)SCH=CHMe isomers combined with careful analysis of onion extracts led to identification of the $C_6H_{10}O_2S_2$ compound $^-$O-S$^+$=CHCHMeCH-MeCH=S$^+$-O$^-$ (**13**), another structurally unique C6 flavorant which contains two sulfine (C=S$^+$-O$^-$) functions.[24,32] Compound **13** could be synthesized as shown (Fig. 13). Compounds **10a,b** and **13** have been identified in onion extracts using LC-APCI-MS/MS methods (see below).

Acyclic thiosulfinates RS(O)SR' as well as *trans*-zwiebelane (**10a**) are chiral and can therefore exist as pairs of enantiomers or, from natural sources, as individual enantiomers. Through the use of chiral γ-cyclodextrin GC columns operating at 90–120 °C we find that enantiomers of MeS(O)SMe, MeS(O)SPr-*n* and MeSS(O)Pr-*n* as well as **10a** can be resolved and that individual enantiomers of the thiosulfinates are stable under the analytical conditions.[25] Analysis of an onion extract on the chiral column showed that thiosulfinates and **10a** were all present as racemic mixtures. This observation suggests that asymmetric induction is not involved in the formation of thiosulfinates and **10a** from achiral sulfenic acids (**2**).

AJOENE AND CEPAENES

Guided by assays of antithrombotic (anticoagulant) activity, Jain and Apitz-Castro identified an unusual compound from garlic extracts of formula $C_9H_{14}OS_3$ that showed particularly high activity. Collaborative research ultimately led to the characterization of this new compound as CH_2=$CHCH_2S(O)CH_2CH$=$CHSSCH_2C$=CH_2 (**14**), dubbed "ajoene" ("ajo" is Spanish for garlic).[21,33] Ajoene, a γ-sulfinyldisulfide, could be prepared by decomposition of allicin (**3b**) in various solvents. It is presumably formed as shown in Figure 14. Ajoene is found in garlic extracts, along with various thiosulfinates, by LC-MS (Fig. 9). Compounds related to ajoene, but with seven or five carbon atoms, *e.g.* $MeS(O)CH_2CH$=$CHSSCH_2CH$=CH_2 and $MeS(O)CH_2CH$=$CHSSMe$, isolated from *Allium ursinum*,[34] have been synthesized from unsymmetrical thiosulfinate $MeS(O)SCH_2CH$=CH_2.[21]

Similar biological assay-guided methods led Wagner,[35,36] and independently, Kawakishi and Morimitsu,[37,38] to both discover α-sulfinyl disulfides related to ajoene in onion extracts. These compounds, *e.g.* MeCH=CHS(O)CHEtSSCH=CHMe, $C_9H_{16}OS_3$, are termed "cepaenes" and are thought to be formed through interaction of the onion LF with sulfenic acids (Fig. 15).[37,38] Cepaenes can be identified in onion extracts using LC-MS methods, even when they are incompletely separated from other compounds. Thus, the identification of the two diastereomers of MeS(O)CHEtSSMe (MH^+ *m/z* 185), which cannot be baseline resolved from the two zwiebelanes (MH^+ *m/z* 163), is achieved by monitoring the individual protonated parent ions as well as employing tandem MS (MS/MS) techniques (Fig. 16), calibrating with synthetic standards.[39]

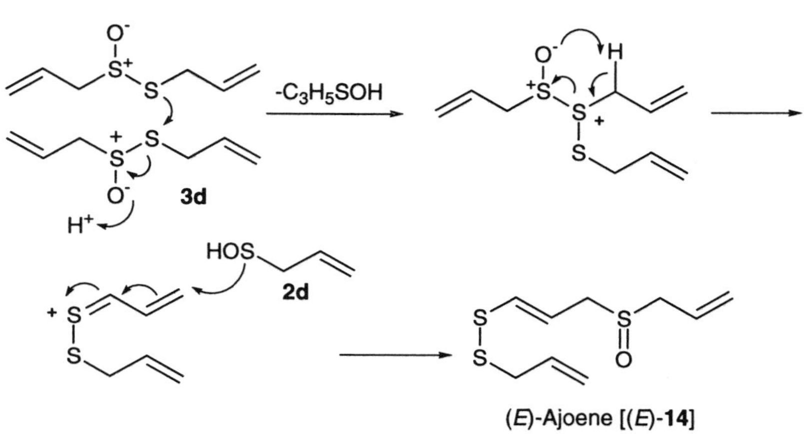

Figure 14. Proposed formation of (*E*)-ajoene ((*E*)-**14**) from allicin, **3d**.

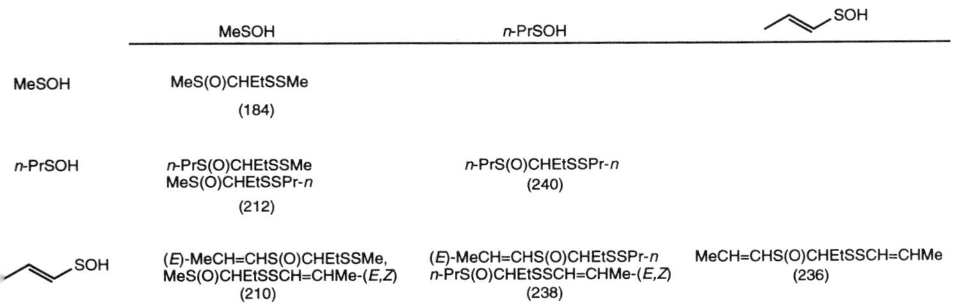

	MeSOH	n-PrSOH	SOH
MeSOH	MeS(O)CHEtSSMe (184)		
n-PrSOH	n-PrS(O)CHEtSSMe MeS(O)CHEtSSPr-n (212)	n-PrS(O)CHEtSSPr-n (240)	
SOH	(E)-MeCH=CHS(O)CHEtSSMe, MeS(O)CHEtSSCH=CHMe-(E,Z) (210)	(E)-MeCH=CHS(O)CHEtSSPr-n n-PrS(O)CHEtSSCH=CHMe-(E,Z) (238)	MeCH=CHS(O)CHEtSSCH=CHMe (236)

Figure 15. Cepaenes (MW in parentheses) from condensation of MeSOH, *n*-PrSOH, and MeCH=CHSOH with onion LF.

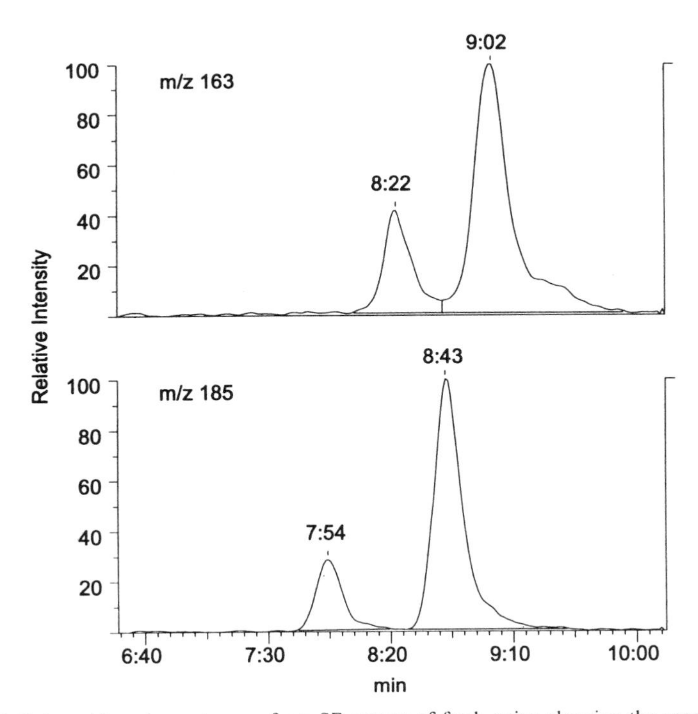

Figure 16. Selected ion chromatogram from SF extract of fresh onion showing the spectrometric resolution of zwiebelanes (**10a**,**b**, retention times 8:22 and 9:02 min, respectively; MH^+ *m/z* 163) and diastereomeric methyl/methyl cepaenes (MeS(O)CHEtSSMe, retention times 7:54 and 8:43 min; MH^+ *m/z* 185).

ALLIUM ORGANOSELENIUM COMPOUNDS

In 1964, Finnish Nobel Laureate A.I. Virtanen, who did much of the pioneering research on onion sulfur biochemistry, reported on the basis of radioisotope studies that the selenoamino acids selenocystine ($(HOOCCH(NH_2)CH_2Se)_2$) and selenomethionine ($HOOCCH(NH_2)CH_2CH_2SeMe$) were present in onion.[40] Virtanen's results suggested that there might be a selenium (Se) based flavor chemistry in *Allium* spp. parallel to that based on sulfur (Fig. 17), *e.g.* originating from soil selenate (SeO_4^{-2}) or selenite (SeO_3^{-2})(Fig. 18) rather than sulfate. Garlic and onion contain 0.28 and 0.015 μg Se per g fresh weight, respectively,[41] compared to 3.3 and 0.84 mg S, *e.g.* 12,000 to 56,000 times higher levels of S than Se are present in these *Allium* spp. Nonetheless, because the odor threshhold for low molecular weight organoselenium compounds is lower than for the corresponding sulfur compounds, trace amounts of Se may contribute as flavorants. Certain regions of California are major sources of domestic garlic, and are known to contain higher than average levels of Se in the soil. Furthermore, studies suggest that Se-enriched garlic[42] as well as yeast[43] possess cancer preventative properties. We obtained information on the nature and amounts of organoselenium compounds in garlic and related *Allium* species as well as information on how the body handles the Se consumed from these plants.

Analysis of Se-flavorants and their precursors from *Allium* spp. plants is especially challenging because these compounds would be expected to have

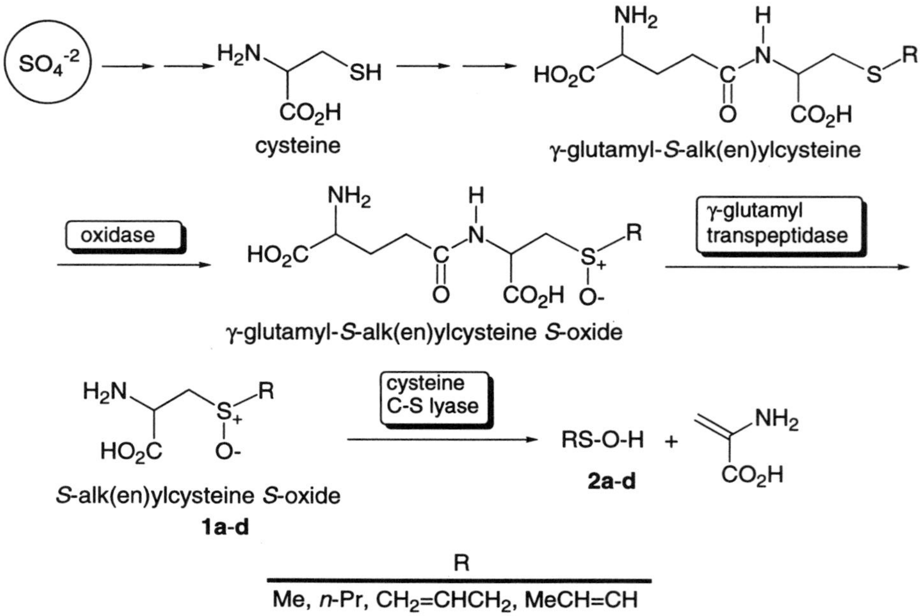

SO_4^{-2} → → cysteine → → γ-glutamyl-*S*-alk(en)ylcysteine

oxidase → γ-glutamyl-*S*-alk(en)ylcysteine *S*-oxide → γ-glutamyl transpeptidase →

S-alk(en)ylcysteine *S*-oxide **1a-d** → cysteine C-S lyase → RS-O-H **2a-d** + $CH_2=C(NH_2)CO_2H$

R
Me, *n*-Pr, $CH_2=CHCH_2$, MeCH=CH

Figure 17. Proposed biosynthetic route to *S*-alk(en)ylcysteine *S*-oxides from sulfate.

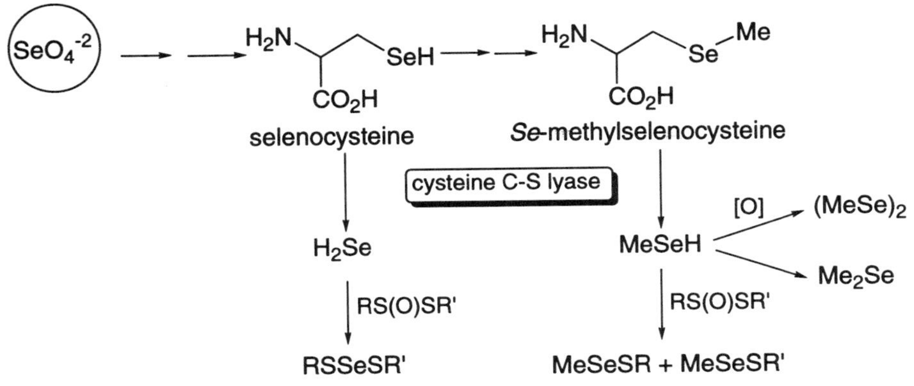

Figure 18. Proposed biosynthetic route to selenocysteine and *Se*-methylselenocysteine from selenate.

physical properties quite similar to those of the far more abundant homologs containing sulfur. In our work, we have used GC with atomic plasma spectral emission detection (GC-AED) for element specific measurement of natural abundance organoselenium compounds in *Allium* spp. plants, plant extracts and volatiles, and human exhaled breath following plant consumption. The technique of GC-AED (Fig. 19) has the important advantages of elemental selectivity, high sensitivity, and the possibility of simultaneous multi-element analysis.[44] The AED response can flag compounds in the GC effluent that contain specific elements, even though these compounds may be present in very small amounts or may coelute with other components. We have used the Se emission line at 196 nm to identify organoselenium species while concurrently monitoring S and C by lines at 181 and 193 nm, respectively; assignments were confirmed by GC-MS. The above techniques were used to analyze the headspace (HS) above homogenized garlic, elephant garlic, onion, Chinese chive and broccoli as well as the headspace above garlic and onion grown in a Se-fertilized medium (Se-enriched plants) or plant homogenates augmented by addition of synthetic selenoamino acids.[45,46] Augmentation of Se was useful in raising the levels of Se compounds to the point where HS-GC-MS could be used for compound identification.

In the case of the HS-GC-AED analysis of volatiles from natural garlic (Fig. 20), there are no Se-containing peaks observed in the C or S channels because at those signal levels they are so small as to be lost within the background signals. The S channel shows MeS_nMe, MeS_nAll, and $AllS_nAll$ ($n = 1–3$, All = allyl), typical of garlic and garlic-like Alliums. The Se channel shows seven peaks: dimethylselenide (MeSeMe), methanesulfenoselenoic acid methylester (MeSeSMe), dimethyl diselenide (MeSeSeMe), bis(methylthio)selenide ($(MeS)_2Se$), allylmethylselenide (MeSeAll), 2-propenesulfenoselenoic acid

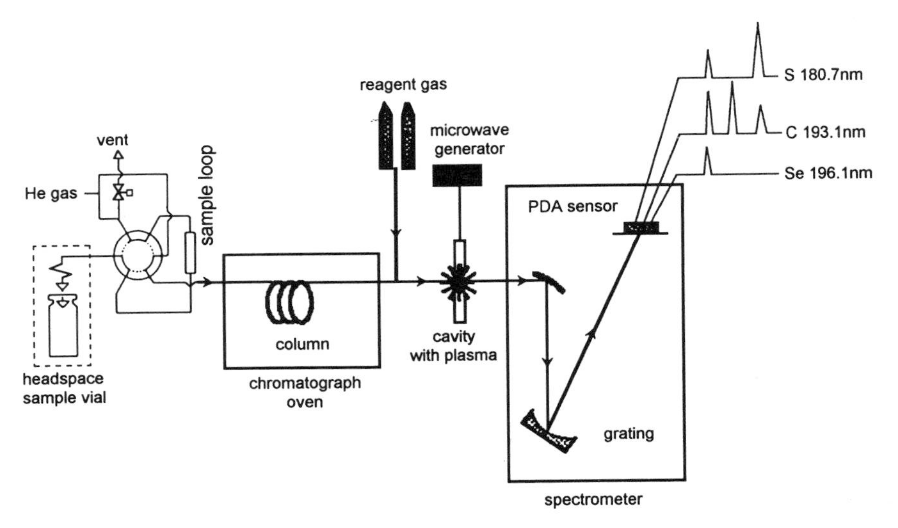

Figure 19. Gas chromatograph with atomic emission detector and headspace attachment.

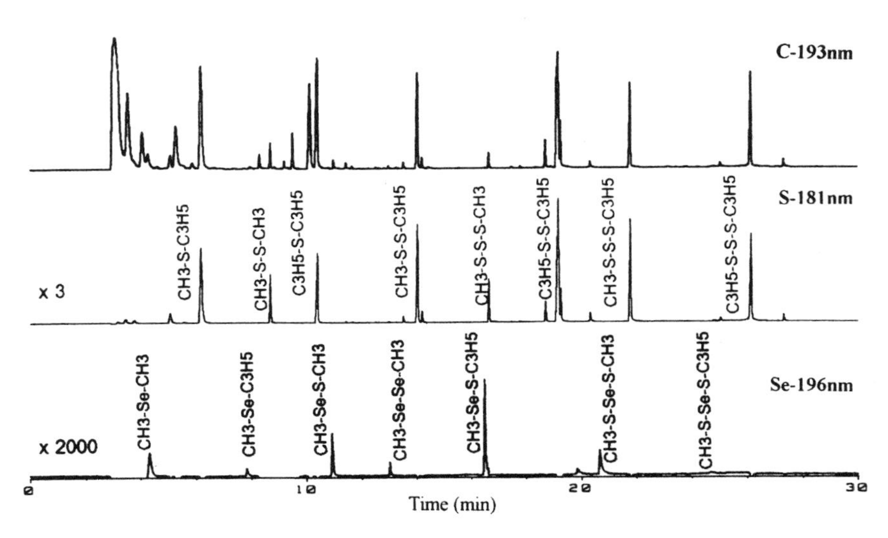

Figure 20. Headspace-GC-AED analysis of sulfur and selenium volatiles from garlic.

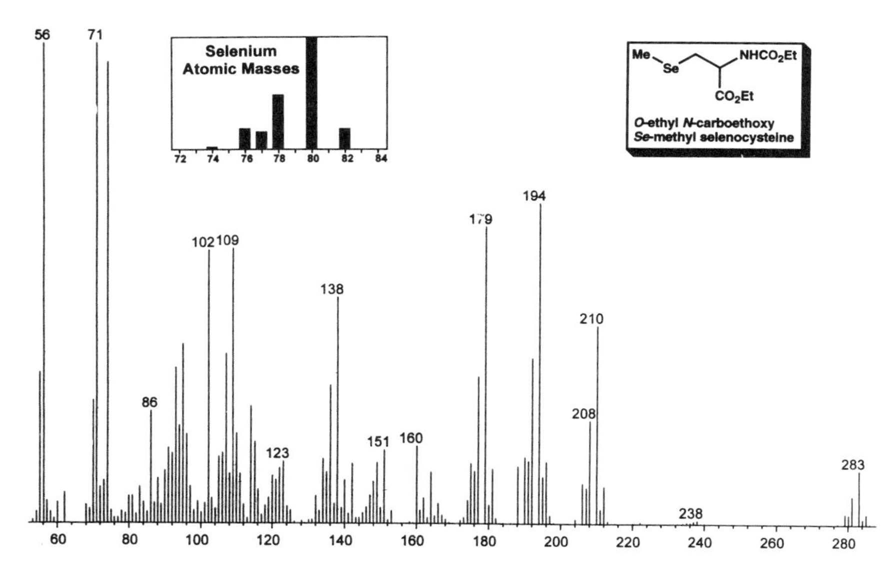

Figure 21. Mass spectral fragmentation pattern of ethylchloroformate-protected *Se*-methylselenocysteine.

methylester (MeSeSAll), and (allylthio)(methylthio)selenide (MeSSeSAll). Structures were established by GC-MS through comparison with spectra of synthetic samples. The headspace above chopped onion contained methylpropylselenide, MeSePr.

Lyophilized normal garlic (0.02 ppm Se) or moderately Se-enriched (68 ppm Se) garlic was derivatized with ethylchloroformate to volatize the selenoamino acids, likely precursors of the headspace Se compounds. Analysis by GC-AED showed selenocysteine, identified by comparison with the mass spectral fragmentation (Fig. 21) and the retention time of an authentic standard (Fig. 22). In garlic, more heavily Se-enriched (1355 ppm Se), *Se*-methylselenocysteine was the major selenoamino acid found along with minor amounts of selenocysteine and traces of *Se*-methionine; the S channel showed 2:1 allylcysteine and allylcysteine *S*-oxide along with minor amounts of methionine (Fig. 23).[47] There were only minor changes in the ratios of the sulfur amino acids as the level of Se was varied from 0.02 to 1355 ppm. Similar analysis of Se-enriched onion (96 ppm Se) revealed the presence of equal amounts of *Se*-methylselenocysteine and selenocysteine in the Se channel.

High performance ion chromatography as well as C18 LC with inductively coupled plasma mass spectrometry (ICP-MS) detection has been used to analyze the selenium species in garlic and onion without the application of heat.[48] These analyses show that *Se*-methylselenocysteine is the major component along with

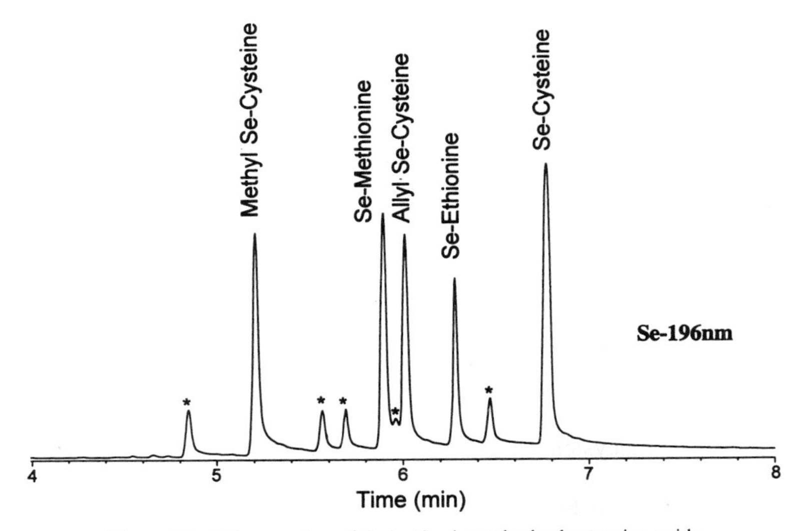

Figure 22. GC separation of derivatized standard selenoamino acids.

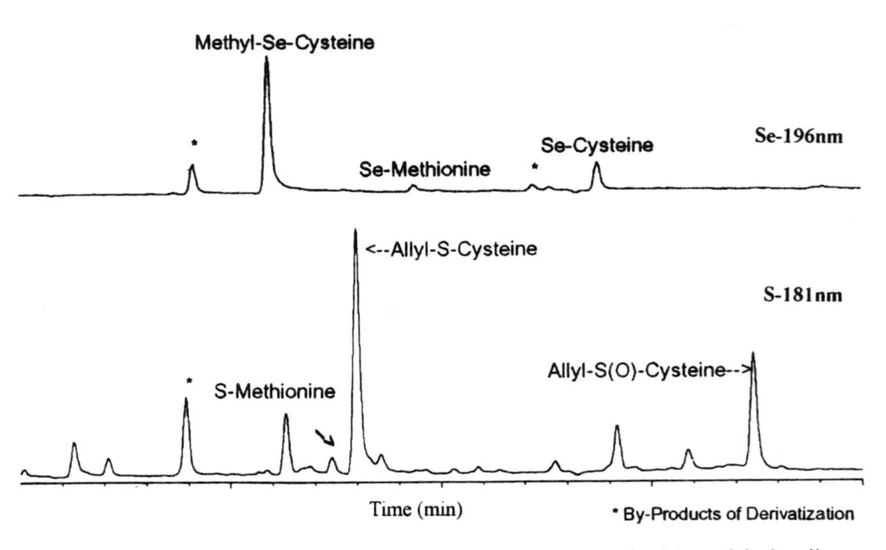

Figure 23. Derivatized sulfur and selenoamino acids from lyophilized Se-enriched garlic.

lesser amounts of *Se*-methionine, selenocystine, and selenate and selenite salts. It is likely that when *Allium* spp. are presented with high levels of inorganic Se fertilizer, excess selenocysteine formed is *Se*-methylated to give the major selenoamino acid, *Se*-methylselenocysteine, as commonly seen with other plants upon exposure to selenium.[49] The presence of *Se*-methylselenocysteine as the major source of Se in Se-enriched *Allium* spp. is significant since *Se*-methylselenocysteine is reported to exhibit cancer chemopreventative activity.[42] *Se*-Methyl-L-selenocysteine is superior to *S*-methyl-L-cysteine as a substrate for L-methionine γ-lyases and *S*-alkylcysteine α,β-lyases in bacteria, due to the superior leaving group ability of $MeSe^-$ compared to MeS^-.[50,51] Related enzymes in *Allium* spp. probably cleave this selenoamino acid to MeSeH (or $MeSe^-$), which reacts with RSS(O)R' formed when the plant is cut to give MeSeSR and MeSeSR', or forms MeSeSeMe on oxidation. Analogous reaction of H_2Se, released from selenocysteine, could afford RSSeSR', while Se-methylation of *Se*-methyl-L-selenocysteine followed by enzymatic cleavage would afford MeSeMe (Fig. 24).

Analysis of the headspace above Se-enriched garlic by HS-GC-AED showed a similar profile of compounds whether or not synthetic *Se*-methylselenocysteine was added, although all of the peaks were enhanced by the addition of the synthetic selenoamino acid. Addition of synthetic *Se*-allylselenocysteine to Se-enriched garlic showed a profile quite different from that of Se-enriched garlic by HS-GC-AED, with the major peaks being diallylselenide (not seen in normal garlic; small peak in Se-enriched garlic) and AllSSeSAll (or isomer;

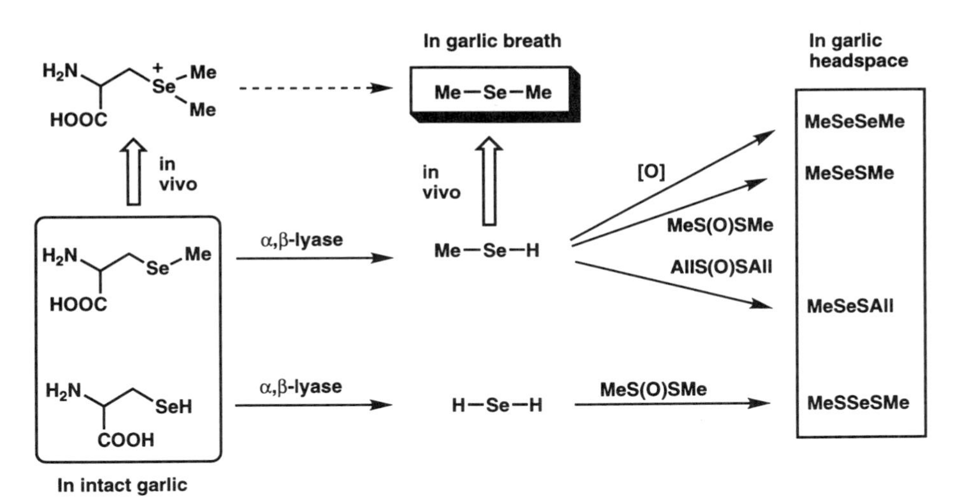

Figure 24. Proposed enzymatic cleavage of selenoamino acids to give volatile, low-molecular weight organoselenium compounds.

small peak in Se-enriched garlic). We conclude that *Se*-allylselenocysteine is not present in our lyophilized garlic samples. *Se*-Allylselenocysteine is presumably cleaved in garlic homogenates to CH_2=$CHCH_2SeH$, which is oxidized to thermally unstable AllSeSeAll, which in turn loses Se affording AllSeAll. It seems likely that when garlic is presented with high levels of inorganic Se fertilizer, the excess selenocysteine formed is *Se*-methylated to give the major selenoamino acid, *Se*-methylselenocysteine.

TRANSFORMATION OF GARLIC FLAVORANTS FOLLOWING COOKING AND INGESTION

Laboratory reactions of garlic flavorants and their precursors have been summarized above. How are these compounds transformed by cooking? Alliinase is completely inactivated when unpeeled whole garlic cloves are boiled for 15–20 minutes.[2] However, before this process is complete, up to 1% of the precursor alliin is converted into allicin, perhaps by mechanical abrasion of the cloves during boiling.[2] Boiling rapidly converts any allicin formed into diallyltrisulfide and related polysulfides, which are detected in breath following ingestion of boiled garlic. At the same time, peptide precursors of alliin (*e.g.* γ-glutamylcysteines) are reported to undergo hydrolysis to *S*-allylcysteine and *S*-1-propenylcysteine, which undergo further breakdown on continued boiling.[2] If garlic is diced or crushed prior to boiling, most alliin is converted to allicin and other thiosulfinates, and these, after boiling for *ca.* 20 minutes, are completely transformed into polysulfides. Boiling cut garlic in water in an open container leads to loss of 97% of volatile sulfides by evaporation. If milk is used instead of water, up to 70% of the volatile sulfides remain even after 40 minutes, presumably due to the protective action of milk fat.[2] Even at the high temperatures (*ca.* 180 °C) involved in stir-frying chopped garlic cloves in hot soybean oil for one minute, 16% of the sulfides are retained in the oil.[2] Denaturization of alliinase occurs particularly rapidly on microwaving; complete inactivation of alliinase in individual 5–6 gram garlic cloves occurs within 15–30 seconds (650 watts microwave power).[2]

What happens to garlic flavorants upon ingestion? Our analysis of human garlic breath by GC-AED (Fig. 25; the subject consumed, with brief chewing, 3 g of fresh garlic with small pieces of white bread, followed by 50 mL of cold water) showed in the Se channel dimethylselenide (MeSeMe) as the major Se component along with one-tenth to one-fortieth the amount of $MeSeC_3H_5$, MeSeSMe and $MeSeSC_3H_5$; the S channel showed AllSH, MeSAll and AllSSAll with lesser amounts of MeSSMe, $MeSSC_3H_5$, an isomer of AllSSAll (presumably MeCH=CHSSAll), $C_3H_5SC_3H_5$ and $C_3H_5SSSC_3H_5$.[52,53] In this same study, we also examined the composition of the Se and S compounds in garlic breath as a function of time. After four hours, the levels of MeSeMe, AllSSAll, AllSAll and

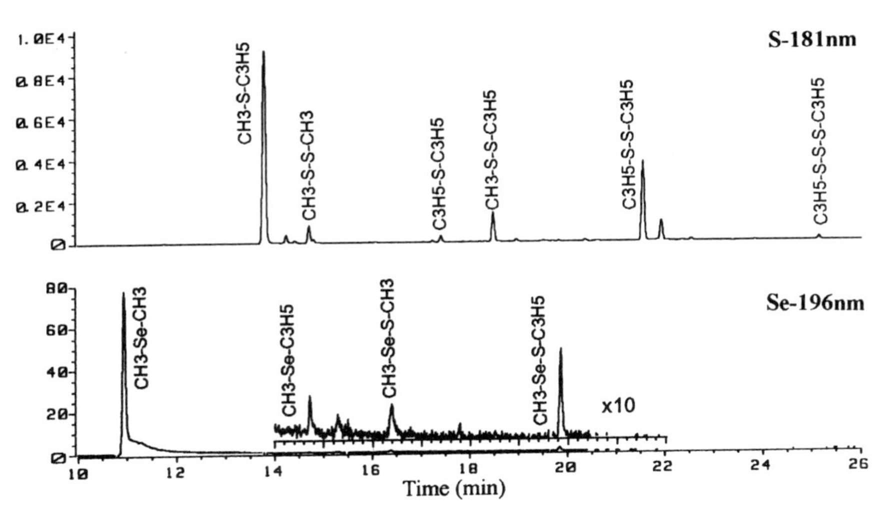

Figure 25. GC-AED trace of S and Se compounds in human garlic breath following ingestion of garlic.

MeSSMe were reduced by 75% from the initial levels of 0.45 ng/L (MeSeMe), 45 ng/L (AllSSAll), 6.5 ng/mL (AllSAll), and 1.8 ng/L (MeSSMe)(Fig. 26). The AllSH could only be found in breath immediately after ingestion of garlic. In view of the reported very low threshhold detection level for low molecular weight organoselenium compounds,[54] it is likely that compounds such as MeSeMe contribute to the overall odor associated with garlic breath. It has been previously reported that MeSeMe, which has a garlic-like odor, is found in the breath air of animals fed inorganic Se compounds[55] and humans who have accidentally ingested Se compounds.[56] Studies involving consumption of larger quantities of garlic (38 g) indicate persistence of elevated levels of sulfur compounds (as high as 900 ppb) in the subject's breath for more than 32 hours.[57] The presence of elevated levels of acetone in the subject's breath were attributed to enhanced metabolism of blood lipids.[57]

ORGANOLEPTIC CHARACTERISTICS OF *ALLIUM* FLAVORANTS

S-Alk(en)ylcysteine *S*-oxides, when acted on by alliinase enzymes, are the source of *Allium* flavors. For maximum yield of flavorants, the plants should be chopped under conditions of neutral or slightly acidic pH and

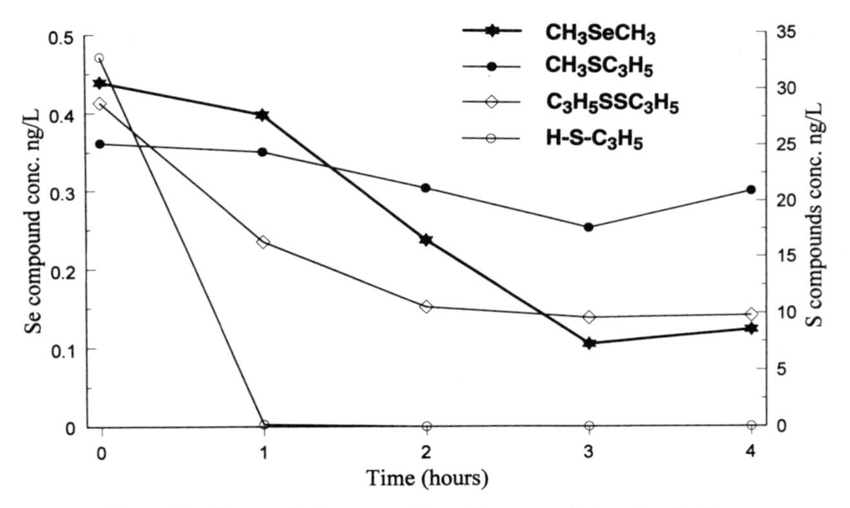

Figure 26. Changes in the composition of human garlic breath with time.

moderate temperature. Once formed, many of the flavorants are tolerant of elevated temperatures and lower pH (*e.g.* acetic acid). Prolonged heating degrades *Allium* flavorants into strong smelling, low detection threshold polysulfides. While it was originally thought that such polysulfides represent the actual *Allium* flavorants, it is now clear that they are merely artifacts of overly harsh analytical conditions. However, such polysulfides are the principal components of distilled oils of *Allium* species and are likely to be important as flavorants when *Allium* species are subjected to elevated temperatures during cooking.[4,58] The principal flavorants from onion include the pungent tear-producing LF (Fig. 4) and lesser amounts of α,β-unsaturated thiosulfinates, zwiebelanes and bissulfine (Fig. 11 and 13), all of which are quite unstable. In the case of garlic, the β,γ-unsaturated thiosulfinate allicin (**3d**, Fig. 1 and 7) is the dominant flavorant. With heat or even at room temperature, allicin undergoes a variety of reactions giving ajoene, dithiins, and polysulfides (Fig. 7 and 14). A variety of pure synthetic samples of *Allium* flavorants have been evaluated by "expert flavorists". Their description of odor and taste (organoleptic) characteristics of these compounds is summarized in Table 1. While the detection threshold for the majority of the compounds listed ranges from 10–100 ppb, several compounds could be detected at levels as low as 0.001–0.01 ppb. In view of the similar, low threshold levels for many of the compounds, it is apparent that *Allium* aromas and flavors are made up of a kaleidoscope of "chemical colors". As the ratios of these compounds change on processing the plants, the aroma and flavor can undergo significant alteration.

Table 1. Flavor descriptions and thresholds for various *Allium* flavorants and related compounds

Compound[a]	Threshold (ppb)[b]	Flavor Description[b]
(*E*)-MeCH=CHSSCH$_2$CH=CH$_2$	10-50	sulfur, fruity, creamy garlic (10-50 ppb) pungent garlic, metallic, sulfury, buttery (0.1-ppm)
(*Z*)-MeCH=CHSSMe	50	cabbage, sulfur, rotten vegetable, onion-like
(*Z,Z*)-MeCH=CHSSCH=CHMe	1-10 (0.01)	sulfur, meaty, livery, rubbery, green slight onion
(*E,E*)-MeCH=CHSCHEtSSCH=CHMe	50	green, fruity sulfur, slight rubbery
(*E,E*)-MeCH=CHS(O)CHEtSSCH=CHMe	10	fruity, fresh onion, sulfur, green, melon-like
MeS(O)SMe	100	cabbage, sulfury, slight sweet creamy, broccoli, green-onion, cauliflower
MeS(O)SPr-*n*/MeSS(O)Pr-*n*	500	creamy cabbage, sulfur, slight green, meaty, liver
n-PrS(O)SPr-*n*	500	fresh onion, chive, green onion
MeCH=CHSS(O)Me	10	sulfur, cabbage, onion, shallot-like
MeCH=CHS(O)SMe	10	cabbage, radish, meaty, sulfur
CH$_2$=CHCH$_2$S(O)SCH$_2$CH=CH$_2$	100	raw garlic, sulfur, rubbery, green
MeCH=CHSS(O)Pr-*n*	100	green, raw, fresh onion
MeCH=CHSS(O)CH$_2$CH=CH$_2$	50	slight garlic, sulfur, creamy, metallic (50 ppb); very fresh garlic, creamy, pungent (1 ppm)
MeCH=CHSSO$_2$Pr-*n*	10 (0.001)	oniony, sulfur, solvent, rubber-like, burnt hair (not fresh onion)
(Z)-EtCH=S$^+$–O$^-$	10-100	onion, rubbery, livery, metallic, tongue burn/heat, sulfur, gaseous
cis-zwiebelane	100	metallic, liver, onion, green, sauteed, sweet brown onion
trans-zwiebelane	100	onion, sulfur, sweet, green, melon-like
MeSSR (R = 3,4-dimethyl-2-thienyl)	5-10	nutty
n-PrSSR (R = 3,4-dimethyl-2-thienyl)	10-20	cabbage, green, vegetative
MeCH=CHSSR (R = 3,4-dimethyl-2-thienyl)	10-20	nutty, green, slightly sulfury

[a]*Chemical Abstracts* names for compounds in this table can be found in references 27 and 58.
[b]Data from references 27, 39, 58, and 59.

SUMMARY OF *ALLIUM* BIOLOGICAL ACTIVITY

In laboratory studies, allicin and related thiosulfinates show antibacterial, antifungal and antitumor activity.[2,3] Allicin also displays lipid biosynthesis inhibitory and antithrombotic activity.[2,3] Both ajoene and cepaenes possess antithrombotic activity; ajoene also displays antifungal activity.[2,3] All of the *Allium* thiosulfinates (as well as ajoene) react readily with thiols such as cysteine and undergo hydrolysis. These facts severely limit the *in vivo* lifetimes of ajoene and *Allium* thiosulfinates. Thus, metabolites of these compounds are more likely to be the active agents *in vivo*. It has been reported that the allicin precursor γ-glutamyl-*S*-allylcysteine (Fig. 17) inhibits the blood pressure-regulating angiotensin converting enzymes while the MeS(O)SMe (**3a**, Fig. 3) precursor *S*-methylcysteine *S*-oxide (**1a**) inhibits the formation of benzo[a]pyrene-induced micronucleated polychromatic erythrocytes (an indicator for genetoxicity).[3] In other laboratory studies, allylic sulfides and selenides are reported to inhibit growth of tumor cells.[3] A recent epidemiological study suggests a correlation between consumption of fresh garlic and decreased risk of gastrointestinal cancer.[60]

One of us wrote: "ideally, all garlic products sold for their health benefit should be validated using randomized, double-blind, placebocontrolled clinical trials with well-defined biological endpoints. Such studies ... should employ analyzed ingredients together with marker compounds, along with evaluation of product safety. Statistical analysis in these trials should be based on "between group comparisons" and differences should be presented with confidence intervals ... Health claims based on *in vitro* data need to be tempered by considering the effects of realistic human doses expected for the consumption of fresh garlic (estimated U.S. per capita consumption of garlic is only 0.5 kg/year!) and various garlic-derived compounds, rather than larger doses given to laboratory animals."[3]

FUTURE RESEARCH DIRECTIONS

Many questions in the field of *Allium* chemistry remain unanswered. After *Allium* plants are digested, what are the biologically most active and important forms of the derived sulfur and selenium compounds? Does 2-propenethiol in the blood play a role as a nitric oxide carrier (a role generally ascribed to cysteine), and, if so, can this explain a portion of the biological activity of garlic? Specifically how does the onion LF trigger lachrymation, and how is the LF converted to dipropyldisulfide, the major component of distilled onion oil? How do *Se*-methylselenocysteine and related compounds in selenium enriched garlic and yeast inhibit tumor formation? Hopefully, research in progress should provide answers to most of these questions.

ACKNOWLEDGMENTS

Financial support from NSF, NIH-NCI the NRI Competitive Grants Program/USDA (Awards No. 92–37500–8068 and 96–35500–3351), and McCormick & Company is gratefully acknowledged. The authors acknowledge with appreciation the efforts of our enthusiastic and dedicated coworkers whose names appear in the publications from our laboratories. We thank Dr. Chester Cavallito for discussions concerning the possible involvement of 2-propenethiol in the nitric oxide transport process.

REFERENCES

1. DAVIES, D. 1992. Alliums - The Ornamental Onions. Timber Press, Portland, OR, 168 pp.
2. KOCH, H.P., LAWSON, L.D. (eds.) 1996. Garlic. The Science and Therapeutic Application of *Allium sativum* L. and Related Species. 2nd Ed. Williams & Wilkins, Baltimore, 329 pp.
3. BLOCK, E. 1996. The chemistry and health benefits of organosulfur and organoselenium compounds in garlic (*Allium sativum*): recent findings. In: Hypernutritious Foods. (J.W. Finley, D.J. Armstrong, S. Nagy, and S.F. Robinson, eds.), Agscience, Inc., Auburndale FL, pp. 261–292.
4. BLOCK, E. 1992. The organosulfur chemistry of the genus *Allium* - implications for organic sulfur chemistry. Angew. Chem., Int. Ed. Engl. 31:1135–1178.
5. AUGER, J., LECOMTE, C., THIBOUT, E. 1989. Leek odor analysis by gas chromatography and identification of the most active substance for the leek moth, *Accrolepiopsis assectella*. J. Chem. Ecol. 15: 1847–1854.
6. SHAKESPEARE, W. 1623. The Taming of the Shrew, Introduction, Scene I.
7. PENN, R.E., BLOCK, E., REVELLE, L.K. 1978. Methanesulfenic acid. J. Am. Chem. Soc. 100:3622–3623.
8. LACOMBE, S., LOUDET, M., BANCHEREAU, E., SIMON, M., PFISTER-GUILLOUZO, G. 1996. Sulfenic acids in the gas phase: A photoelectron study. J. Am. Chem. Soc. 118:1131–1138.
9. VIRTANEN, A.I. 1962. Some organic sulfur compounds in vegetables and fodder plants and their significance in human nutrition. Angew. Chem., Int. Ed. Engl. 1:299–306.
10. VIRTANEN, A.I. 1965. Studies on organic sulphur compounds and other labile substances in plants. Phytochemistry 4:207–228.
11. BLOCK, E., PENN, R.E., REVELLE, L.K. 1979. Structure and origin of the onion lachrymatory factor. A microwave study. J. Am. Chem. Soc. 101:2200–2201.
12. BLOCK, E., GILLIES, J.Z., GILLIES, C.W., BAZZI, A.A., PUTMAN, D., REVELLE, L.K., WALL, A., WANG, D., ZHANG, X. 1996. *Allium* chemistry: Microwave spectroscopic identification, mechanism of formation, synthesis, and reactions of (*E,Z*)-propanethial *S*-oxide, the lachrymatory factor of the onion (*Allium cepa*). J. Am. Chem. Soc. 118:7492–7501.
13. BLOCK, E., REVELLE, L.K., BAZZI, A.A. 1980. The lachrymatory factor of the onion: An nmr study. Tetrahedron Lett. 21:1277–1280.
14. TURECEK, F., MCLAFFERTY, F.W., SMITH, B.J., RADOM, L. 1990. Neutralization-reionization study of the $CH_2{=}CHSOH \rightarrow CH_3CH{=}SO$ rearrangement. Int. J. Mass Spectrom. Ion Processes 101:283–300.
15. GILLIES, J.Z., GILLIES, C.W., GRABOW, J.-U., HARTWIG, H., BLOCK, E. 1996. Microwave investigation of (*Z*)- and (*E*)-ethanethial *S*-oxide. J. Phys. Chem. 100:18708–18717.

16. SCHMIDT, N.E., SANTIAGO, L.M., EASON, H.D., DAFFORD, K.A., GROOMS, C.A., LINK, T.E., MANNING, D.T., COOPER, S.D., KEITH, R.C., CHANCE, W.O., III, WALLA, M.D., COTHAM, W.E. 1996. A rapid extraction method of quantitating the lachrymatory factor of onion using gas chromatography. J. Agric. Food Chem. 44:2690–2693.
17. BLOCK, E., BAZZI, A.A., REVELLE, L.K. 1980. The dimer of the onion lachrymatory factor: The first stable 1,2-dithietane derivative. J. Am. Chem. Soc. 102: 2490–2492.
18. CAVALLITO, C.J., BAILEY, J.H. 1944. Allicin, the antibacterial principle of *Allium sativum*. 1. Isolation, physical properties and antibacterial action. J. Am. Chem. Soc. 66:1950–1951.
19. CAVALLITO, C.J., BUCK, J.S., SUTER, C.M. 1944. Allicin, the antibacterial principle of *Allium sativum*. 2. Determination of the chemical structure. J. Am. Chem. Soc. 66:1952–1954.
20. SMALL, L.D., BAILEY, J.H., CAVALLITO, C.J. 1947. Alkyl thiolsulfinates. J. Am. Chem. Soc. 69:1710–1713.
21. BLOCK, E., AHMAD, S., CATALFAMO, J., JAIN, M.K., APITZ-CASTRO, R. 1986. Antithrombotic organosulfur compounds from garlic: Structural, mechanistic and synthetic studies. J. Am. Chem. Soc. 108:7045–7055.
22. BLOCK, E., PUTMAN, D., ZHAO, S.-H. 1992. *Allium* chemistry: GC-MS Analysis of thiosulfinates and related compounds from onion, leek, scallion, shallot, chive and Chinese chive. J. Agric. Food Chem. 40:2431–2438.
23. CALVEY, E.M., BETZ, J.M., MATUSIK, J.E., WHITE, K.D., BLOCK, E., LITTLEJOHN, M.H., NAGANATHAN, S., PUTMAN, D. 1994. Off-line supercritical fluid extraction of thiosulfinates from garlic and onion. J. Agric. Food Chem. 42:1335–1341.
24. BLOCK, E., BAYER, T., NAGANATHAN, S., ZHAO, S.-H. 1996. *Allium* chemistry: Synthesis and sigmatropic rearrangements of alk(en)yl 1-propenyl disulfide *S*-oxides from cut onion and garlic. J. Am. Chem. Soc. 118:2799–2810.
25. BLOCK, E., CALVEY, E.M. 1994. Facts and artifacts in *Allium* chemistry. In Sulfur Compounds in Foods, ACS Symposium Series 564 (C.J. Mussinan and M.E. Keelan, eds.), American Chemical Society, Washington DC, pp. 63–79.
26. CALVEY, E.M., ROACH, J.A.G., BLOCK, E. 1994. Supercritical fluid chromatography of garlic (*Allium sativum*) extracts with mass spectrometric identification of allicin. J. Chromatogr. Sci. 32:93–96.
27. BLOCK, E., NAGANATHAN, S., PUTMAN, D., ZHAO, S.-H. 1992. *Allium* chemistry: HPLC quantitation of thiosulfinates from onion, garlic, wild garlic, leek, scallions, shallots, elephant (great-headed) garlic, chives and Chinese chives. Uniquely high allyl to methyl ratios in some garlic samples. J. Agric. Food Chem. 40:2418–2430.
28. FERARY, S., THIBOUT, E., AUGER, J. 1996. Direct analysis of odors emitted by freshly cut *Allium* using combined high-performance liquid chromatography and mass spectrometry. Rapid Commun. Mass Spectrom. 10:1327–1332.
29. FERARY, S., AUGER, J. 1996. What is the true odour of cut Allium? Complementarity of various hyphenated methods: Gas chromatography-mass spectrometry and high-performance liquid chromatography-mass spectrometry with particle beam and atmospheric pressure ionization interfaces in sulphenic acids rearrangement components discrimination. J. Chromatog. A 750:63–74.
30. BAYER, T., WAGNER, H., BLOCK, E., GRISONI, S., ZHAO, S.H., NESZMELYI, A. 1989. Zwiebelanes: novel 2,3-dimethyl-5,6-dithiabicyclo[2.1.1]hexanes from onion. J. Am. Chem. Soc. 111:3085–3086.
31. BLOCK, E., THIRUVAZHI, M., TOSCANO, P.J., BAYER, T., GRISONI, S., ZHAO, S.-H. 1996. *Allium* chemistry: Structure, synthesis, natural occurrence in onion (*Allium cepa*), and reactions of 2,3-dimethyl-5,6-dithiabicyclo[2.1.1]hexane *S*-oxides. J. Am. Chem. Soc. 118:2790–2798.
32. BLOCK, E., BAYER, T. 1990. (*Z,Z*)-*d,l*-2,3-Dimethyl-1,4-butanedithial 1,4-dioxide: a novel biologically active organosulfur compound from onion. Formation of *vic*-disulfoxides in onion extracts. J. Am. Chem. Soc. 112:4584–4585.

33. BLOCK, E., AHMAD, S., JAIN, M.K., CRECELY, R.W., APITZ-CASTRO, R.. CRUZ, M.R. 1984. (*E,Z*)-Ajoene: A potent antithrombotic agent from garlic, J. Am. Chem. Soc. 106:8295–8296.
34. SENDL, A., WAGNER, H. 1991. Isolation and identification of homologues of ajoene and alliin from bulb-extracts of *Allium ursinum*. Planta Med. 57:361–362.
35. BAYER, T., WAGNER, H., WRAY, V., DORSCH, W. 1988. Inhibitors of cyclooxygenase and lipoxygenase in onions. Lancet 906.
36. BAYER, T., BREU, W., SELIGMANN, O., WRAY, V., WAGNER, H. 1989. Biologically active thiosulphinates and α-sulphinyldisulfides from *Allium cepa*. Phytochemistry 28:2373–2377.
37. KAWAKISHI. S., MORIMITSU, Y. 1988. New inhibitor of platelet aggregation in onion oil. Lancet 330.
38. MORIMITSU, Y., KAWAKISHI. S. 1990. Inhibitors of platelet aggregation from onion. Phytochemistry 29:3435–3439.
39. BLOCK, E., ZHAO, S.-H. 1992. *Allium* chemistry: Simple syntheses of antithrombotic "cepaenes" from onion and "deoxycepaenes" from oil of shallot by reaction of 1-propene-thiolate with sulfonyl halides. J. Org. Chem. 57:5815–5817.
40. SPÅRE, C.-G., VIRTANEN, A.I. 1964. On the occurrence of free selenium-containing amino acids in onion (*Allium cepa*). Acta Chem. Scand. 18:280–282.
41. MORRIS, V.C., LEVANDER, O.A. 1970. Selenium content of foods. J. Nutr. 100:1383–1388.
42. IP, C., LISK, D.J., STOEWSAND, G.S. 1992. Mammary cancer prevention by regular garlic and selenium-enriched garlic. Nutr. Cancer 17:279–286.
43. CLARK, L.C., COMBS, G.F., TURNBULL, B.W., SLATE, E.H., CHALKER, D.K., CHOW, J., DAVIS, L.S., GLOVER, R.A., GRAHAM, G.F., GROSS, E.G., KRONGRAD, A., LESHER, J.L., PARK, H.K., SANDERS, B.B., SMITH, C.L., TAYLOR, J.R . 1996. Effects of selenium supplementation for cancer prevention in patients with carcinoma of the skin. J. Amer. Med. Assoc. 276:1957–1963.
44. UDEN, P.C. 1992. Atomic specific chromatographic detection. An overview. In: Element-Specific Chromatographic Detection by Atomic Emission Spectroscopy. ACS Symposium Series 479 (P. Uden, ed.), American Chemical Society, Washington DC, pp. 1–24.
45. CAI, X.-J., UDEN, P.C., SULLIVAN, J.J., QUIMBY, B.D., BLOCK, E. 1994. Headspace/gas chromatography with atomic emission and mass selective detection for the determination of organoselenium compounds in elephant garlic. Anal. Proc. Including Anal. Commun. 31:325–327.
46. CAI, X.-J., UDEN, P.C., BLOCK, E., ZHANG, X., QUIMBY, B.D, SULLIVAN, J.J. 1994. *Allium* chemistry: Identification of natural abundance organoselenium volatiles from garlic, elephant garlic, onion, and Chinese chive using headspace gas chromatography with atomic emission detection. J. Agric. Food Chem. 42:2081–2084.
47. CAI, X.-J., BLOCK, E., UDEN, P.C., ZHANG, X., QUIMBY, B.D, SULLIVAN, J.J. 1995. *Allium* chemistry: Identification of selenoamino acids in ordinary and selenium-enriched garlic, onion, and broccoli using gas chromatography with atomic emission detection. J. Agric. Food Chem. 43:1754–1757.
48. GE, H., TYSON, J.F., UDEN, P.C., CAI, X.-J., DENOYER, E.R., BLOCK, E. 1996. Identification of selenium species in selenium-enriched garlic, onion, and broccoli using high-performance ion chromatography with inductively coupled plasma mass spectrometry detection. Anal. Commun. 33:279–281.
49. NEUHIERL, B., BÖCK, A. 1996. On the mechanism of selenium tolerance in selenium-accumulating plants. Purification and characterization of a specific selenocysteine methyl-transferase from cultured cells of *Astragalus bisculatus*. Eur. J. Biochem. 239:235–238.
50. TAKADA, H., ESAKI, N., TANAKA, H., SODA, K. 1988. The C_3-N bond cleavage of 2-amino-3-(*N*-substituted-amino)propionic acids catalyzed by L-methionine γ-lyase. Agric. Biol. Chem. 52:2897–2901.

51. KAMITANI, H., ESAKI, N., TANAKA, H., SODA, K. 1990. Thermostable *S*-alkyl-cysteine α,β-lyase from a thermophile: Purification and properties. Agric. Biol. Chem. 54:2069–2076.
52. CAI, X.-J., BLOCK, E., UDEN, P.C., QUIMBY, B.D, SULLIVAN, J.J. 1995. *Allium* chemistry: Identification of natural abundance organoselenium compounds in human breath after ingestion of garlic using gas chromatography with atomic emission detection. J. Agric. Food Chem. 43:1751–1753.
53. BLOCK, E., CAI, X.-J., UDEN, P.C., ZHANG, X., QUIMBY, B.D, SULLIVAN, J.J. 1996. *Allium* chemistry: Natural abundance organoselenium compounds from garlic, onion, and related plants and in human garlic breath. Pure Appl. Chem. 68:937–944.
54. RUTH, J.H. 1986. Odor thresholds and irritation levels of several chemical substances: A review. Am. Ind. Hyg. Assoc. J. 47:142–151.
55. OYAMADA, N., KIKUCHI, M., ISHIZAKI, M. 1987. Determination of dimethyl selenide in breath air of mice by gas chromatography. Anal. Sci. 3:373–376.
56. BUCHAN, R.F. 1974. Garlic Breath Odor. J. Am. Med. Assoc. 227:559–560.
57. TAUCHER, J., HANSEL, A., JORDAN, A., LINDINGER, W. 1996. Analysis of compounds in human breath after ingestion of garlic using proton-transfer-reaction mass spectrometry. J. Agric. Food Chem. 44:3778–3782.
58. BLOCK, E., THIRUVAZHI, M. 1993. *Allium* chemistry: synthesis of alk(en)yl 3,4-di-methyl-2-thienyl disulfides, components of distilled oils and extracts of *Allium* species. J. Agric. Food Chem. 41:2235–2237.
59. RANDLE, W.M., BLOCK, E., LITTLEJOHN, M.H., PUTMAN, D., BUSSARD, M. 1994. Onion Thiosulfinates Respond to Increasing Sulfur Fertility. J. Agric. Food Chem. 42:2085–2088.
60. STEIMETZ, K.A., KUSHI, L.H., BOSTICK, R.M., FOLSOM, A.R., POTTER, J.D. 1994. Vegetables, fruit, and colon cancer in the Iowa Women's Health Study. Am. J. Epidemiol. 139:1–15.

Chapter Two

HEALTH PROMOTING PHYTOCHEMICALS IN CITRUS FRUIT AND JUICE PRODUCTS

Antonio Montanari, Wilbur Widmer, and Steven Nagy

Florida Department of Citrus
Scientific Research Department
Citrus Research and Education Center
700 Experiment Station Road, Lake Alfred, Florida 33850

INTRODUCTION

Citrus fruit and juice have long been traditional health-promoting foods. The origin of citrus is commonly thought to be in the southwest region of China, and citrus is at the very center of all Chinese traditional herbal medicine. All parts of the fruit are considered medicinal and wholesome, and this reputation has spread with citrus throughout the world. As modern science studies citrus

Functionality of Food Phytochemicals
edited by Johns and Romeo, Plenum Press, New York, 1997

more carefully, it seems that traditional beliefs may be founded in fact. Orange and grapefruit are excellent sources of Vitamin C, a good source of folic acid, and a fair source of niacin, and thiamine. Although the benefits of these constituents are too numerous to list in this publication, nutrition alone would be a compelling reason to consume fresh citrus or drink juice. The vitamins are only part of the story, however. Virtually every class of phytochemical known can be found in relatively high concentration in citrus including flavonoids, triterpenes (sterols), hydroxycinnamic acids, polysaccharides (fiber), hexaric acids, monoterpenes, flavor and aroma molecules. In fact, when one drinks a glass of orange or grapefruit juice, there are more phytochemicals ingested by weight than vitamins. This review article is about these phytochemicals and their impact on mammalian and perhaps human health.

FLAVANONES

Flavonoids are common in fruits and vegetables and many function as natural antioxidants.[1] Antioxidants can act as free radical scavengers and may also chelate proxidant metals, reducing their capacity to produce free radicals.

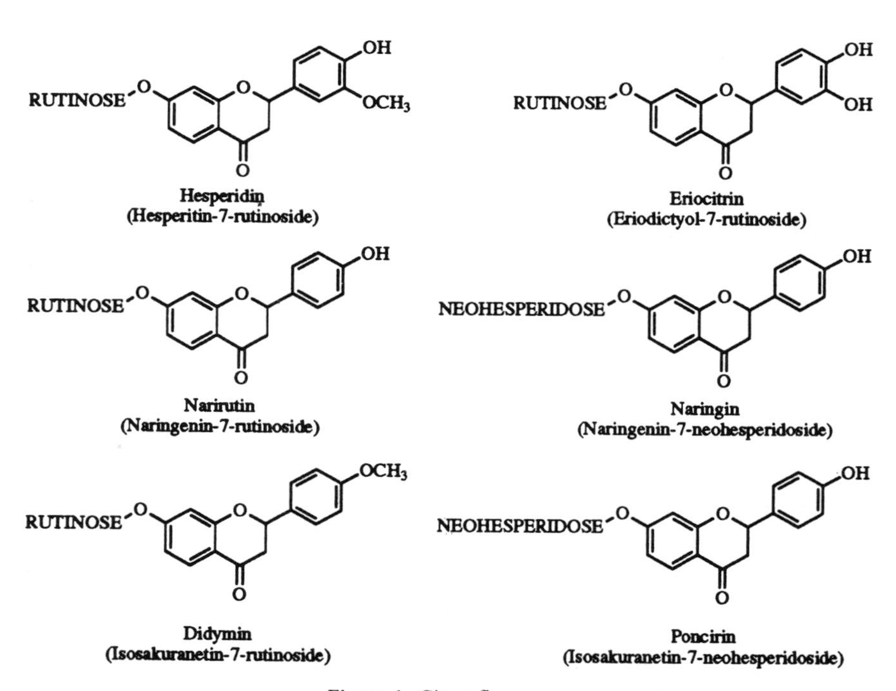

Figure 1. Citrus flavanones.

Human consumption of flavonoids has been estimated to be on the order of 1 gram per day.[2] In citrus, the predominate flavonoids are flavanones bound as glycosides (Fig. 1). Naringin and narirutin, glycosides of naringenin, are the most abundant in grapefruit, whereas hesperidin, a glycoside of hesperetin, is predominant in orange and tangerine. Eriocitrin, a glycoside of eriodictyol, and hesperidin are the predominate flavonoids in lemon and lime. The health attributes associated with flavanones have been somewhat controversial. While it has been recognized for some time that citrus flavanones are not essential vitamins, they are known to have therapeutic effects.[3] They were first reported in 1964 to decrease capillary fragility and improve blood,[4] but further research failed to substantiate that abnormal capillary permeability was due to a deficiency of bioflavonoids. Increased capillary permeability is a result, not a cause of a disease. In 1950, the Federation of American Societies for Experimental Biology recommended that the term "Vitamin P", which had been applied to flavanoids, be discontinued.[5]

The use of crude mixtures in early studies very likely was a contributing factor to inconsistencies among experimental results. The term 'bioflavonoid' refers to crude extracts which may be obtained from any of several citrus varieties including lemon, orange, mandarin or grapefruit. Crude extracts from the peel of orange and mandarin varieties contain various methoxylated flavones as minor components. Extracts from the peel of grapefruit, lemon and limes do not contain any or very few methoxylated flavones but may contain other flavones such as rutin and diosmin. Thus, in the broad definition of 'bioflavonoid', wide variations in composition of major and minor components occur.

Recent research has focused on the biological activities of individual citrus flavonoids purified from crude extracts. Vlietinck (private communication) did not find hesperidin or naringin to have any antiviral activity against herpes simplex or polio myelitis type viruses, however, hesperetin, the aglycone of hesperidin, was found by Kaul, Middleton, and Ogra[6] to actively inhibit the replication but not the infectivity of herpes simplex, polio and parainfluenza type viruses. Robbins et al.[7] found naringin affected red blood cell aggregation *in vitro*. Ingested grapefruit juice was also found to have a beneficial affect in human subjects, lowering elevated hemocrits and regulating blood viscosity *in vivo*. These findings suggested naringin was the active component in the grapefruit juice.

Intact flavanone glycosides do not seem to be absorbed by humans or other mammals. Intestinal flora in the gut, however, can metabolize them to cleave the bound sugar and release the aglycones. This was first demonstrated by Booth and coworkers for hesperidin[8] and naringin[9] where only the aglycones were detected as bound glucuronides in human urine. Naringenin glucuronides were found by Fuhr and Kummert[10] in blood plasma and urine after ingestion of naringin by human subjects. Small amounts of free naringenin were also detected in urine samples amounting to less than 0.5% of the amount of naringin consumed. Total amounts of naringenin excreated as glucuronides varied from 5 - 60% of the total amount consumed. Research of Weintraub and coworkers[11] showed similar results. Formation of naringenin glucuronides appears to take place quickly leaving little free naringenin available. Analysis of feces incubated with naringin showed a wide variation in the ability of intestinal microflora in different individuals to metabolize naringin to naringenin. Differences in intestinal microflora in individuals indicates that the amount of flavonoid aglycone available for absorption will vary considerably, and this may account for the variation in effects observed among individuals.

Several human feeding studies have demonstrated an interaction between grapefruit juice and an increase in absorption efficiency of several drugs regulated by cytochrome P450 enzymes with the suggestion that naringin is the causative agent.[12,13,14] While *in vitro* studies showed naringenin is a potent inhibitor of this enzyme,[12] *in vivo* ingestion of naringin alone failed to increase either nisoldipine[13] or felodipine[14] absorption compared to ingestion of grapefruit juice containing comparable amounts of naringin. This indicates some other component in grapefruit juice may facilitate the metabolism of naringin or itself be largely responsible for the increased absorption effect.

Anticancer Activity

Menon and coworkers[15] found naringin and naringenin to have a minor protective effect against B16F10 melanoma cells when fed to mice at the rate of 200 nmole/kg body weight. There were 26% fewer lung tumor nodules formed, and survival was reported to be 27% longer than in the control group. Catechin, epicatechin, and rutin had a great protective effect whereas ellagic acid, quercetin, and morin were inactive.

Naringenin and hesperetin were found by Carroll and coworkers[16] to be more effective than genistein, an isoflavone in soybeans, in the inhibition of MDA-MB-435 human breast cell cancer proliferation *in vitro*. A subsequent *in vivo* study in rats compared the effects of grapefruit juice, naringin, naringenin, and orange juice against induced mammary tumorigenesis by dimethylbenzanthrancene.[17] Tumor development was delayed in rats given orange juice, resulting in fewer and smaller tumors than in control animals. The group fed orange juice also had a better weight gain compared to any other test groups. Grapefruit juice and naringenin supplemented diets had no protective effect. Animals fed the naringin supplemented diet also showed a delay in tumor development and also had less weight gain than the control group. Any contributing effect of hesperidin was not tested.

Tanaka and coworkers[18] demonstrated the chemopreventive effects of hesperidin, curcumin, and ß-carotene against oral carcinogenesis induced by 4-nitroquinoline 1-oxide (4NQ) in rats. Modifying effects of test compounds were determined during both the initiation and post initiation phases. Test compounds were administered at a level of 0.05% in the diet during or after treatment with 4NQ. Protection during initiation was nearly complete for curcumin. The group fed hesperidin showed approximately one half the number of animals with oral carcinomas and hyperplastic lesions compared to the control group, and this was only slightly less effective than ß-carotene. When protective treatments were started one week after exposure to 4NQ, hesperidin exhibited a slight but not a significant protective effect. Both ß-carotene and curcumin provided a significant protective effect with the post initiation treatments.

Martin and coworkers[19] recently demonstrated that naringin was effective in protecting rats against gastric mucosal lesions induced by ethanol. Protection was dose related, and 400 mg/kg body weight ingested an hour before ingestion of ethanol resulted in a 70% reduction in ulceration. Naringin activity was accomplished by increasing the gel viscosity of mucus in the stomach.

FLAVONES

Methoxylated flavones are present as minor components in crude bioflavonoid extracts prepared from waste citrus peel. Flavones also occur as minor components in citrus fruit and juices, the most abundant flavones being fully methoxylated (Fig. 2). They are more hydrophobic than hydroxylated flavones. Many exhibit a high level of biological activity. Methoxyflavones are associated exclusively with the oil glands located in the outer layer of peel or flavedo of citrus fruit. They become incorporated into citrus juices during extraction or with addition of peel oils for flavor enhancement. Methoxylated flavones are most abundant in tangerines and oranges and comprise 0.1–0.5% of the peel on a dry weight basis (Table 1). In commercial tangerine peel oils, they are present at levels of 0.5–1%. Tangeretin and nobiletin are most abundant, with smaller

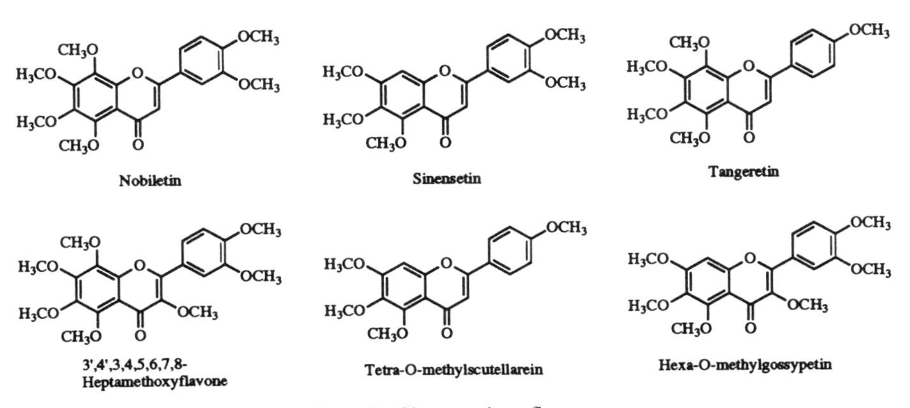

Figure 2. Citrus methoxyflavones.

Table 1. Methoxyflavone content in dried peel for selected varieties of citrus (mg/Kg)

Variety	Sin	TMIS	HexQ	Nob	Scut	Hept	Tang	Total
Ambersweet	-.-	-.-	-.-	-.-	-.-	-.-	1.2	1.2
Pineapple Orange	301.5	-.-	42.8	381.0	189.1	191.3	90.6	1196.3
Clementine Tangerine	31.3	10.4	37.4	144.1	54.1	252.8	44.6	574.7
Orlando Tangelo	35.9	44.3	3.2	424.6	19.1	149.4	218.0	894.5
Dancy Tangerine	32.5	340.4	1.5	2904.9	177.7	128.8	1617.5	5496.3
Duncan Grapefruit	-.-	-.-	-.-	60.8	1.6	35.4	34.0	131.8

Sin = sinensetin; TMIS Tetramethyl-O-isoscutellarein; HexQ = hexamethyl-O-quercetagetin; Nob = nobiletin; Scut = tetramethyl-O-scutellarein; Hept = heptamethoxyflavone; Tan = tangereretin

amounts of sinensetin, heptamethoxyflavone, tetra-O-methyl scutellarein, and hexa-O-methyl gossypetin also occurring. Commercial orange peel oils contain nobiletin and heptamethoxyflavone as the most abundant methoxyflavones. The total flavone content in orange oil ranges from 0.2–0.4%.

Robbins[20] found that isolated methoxylated flavones had greater anti-adhesive effects on red blood cells and platelets than flavanone glycosides. Reduction of clumping of blood cells may be important in the prevention of coronary thrombosis. Middleton and coworkers demonstrated that methoxylated flavones have anti-inflammatory properties and reduce allergic reactions by inhibiting histamine release.[21,22] The methoxyflavones, again, were found to be much more active than naringin, hesperidin, or their flavanone aglycones. Of 17 flavonoids evaluated for antimutagenic activity, Wall and coworkers[23] found that the glycosylated flavonoids, naringin, hesperidin, and rutin, were only weakly active, while tangeretin and nobiletin were considerably more active.

Anticancer Activity

Of considerable interest are the protective effects that methoxylated flavones have in cancer prevention. Tumors are cell populations with abnormal and uncontrolled growth characteristics and may be classified as either benign or malignant. Benign tumors are of less concern because the growing cell mass does not invade surrounding tissue and usually they are easily removed. Malignant tumors are more insidious because they grow into and invade surrounding tissues. When malignant cells gain access to the lymph and blood circulatory systems, metastatic cascade begins, spreading cancer cells throughout the body to organs distant from the primary tumor.[24]

According to Wattenberg, control of cancer may be accomplished in several ways with suppressing, blocking, or transforming agents all leading to inhibition.[24,25] Suppression agents prevent procarcinogen formation from precursor materials. Blocking agents prevent carcinogenic components from reaching critical reactive sites in tissues and cells. Transformation agents act to transform carcinogenic components into less toxic materials or prevent expression of the carcinogen. Cancer preventation may also be accomplished by inhibiting the invasion of tumor cells into surrounding tissue or by inhibiting cell mobility. Abnormal cancerous cells are thus stopped from spreading and invading other sights in the body, inhibiting onset of metastatic cascade.[24,25]

Bracke and coworkers[26] tested tangeretin, nobiletin, hesperidin, naringin, (+) catechin, and (+) epicatechin for anti-invasive activity. Healthy chicken heart tissue was exposed to MO4 malignant mouse tumor cells and varying concentrations of test materials. Tangeretin inhibited invasion by greater than 50% at a concentration of 10 μM, both tangeretin and nobiletin showed significant anti-invasive activity when present at 100 μM, while (+) catechin and epicatechin were not effective until present at a concentration of 500 μM. Hesperidin and naringin were inactive. In a subsequent study, Bracke and coworkers[27] found tangeretin also was effective in inhibiting invasion of MCF-7 human breast cancer cells into chick heart tissue when present at both 10 μM and 100 μM concentrations. Nobiletin, hesperidin, and naringin also were tested and were inactive. Measurements of the flavonoid amounts in cultures after 4 days, and tests to measure binding of each flavonoid with extracellular matrix materials (ECM), showed catechin and epicatechin (or their metabolites) underwent a high degree of binding to the ECM. In contrast, tangeretin and nobiletin amounts remained constant in the culture media and showed a poor affinity for binding with the ECM. This suggested tangeretin and nobiletin exert their anti-invasive protection differently from catechin.

Approximately 80% or more of all human cancers are derived from epithelial cells which are connected to each other by E-cadherin, a calcium dependent cell-cell adhesive glycoprotein molecule.[28,29] Recent research indicates epithelial cell invasion can result from a breakdown in the function of E-cadherin between adjacent cells.[30,31] In a series of experiments, Bracke and coworkers[27] demonstrated that tangeretin can restore the function of E-cadherin and restore cell-cell adhesion between cells. Protein synthesis was not required for the restoration of function. Some other mechanism, such as an enzymatic, glycosylation, or phosphorylation reaction, may be responsible. Restoration of tissue integrity would help prevent metastatic cascade. Cell motility is inhibited and cells are not able to enter the blood and lymph systems to spread to remote areas of the body.

Kandaswami et al.[32] found both nobiletin and tangeretin inhibited the proliferation of human squamous carcinoma HTB43 cells *in vitro*. Cell growth was inhibited by 70% or greater at concentrations of 20 μM tangeretin or nobiletin when compared to controls. Concentrations of 5 μM provided some inhibition but were less effective. Carroll and coworkers[33] also found nobiletin and tangeretin were active as growth inhibitors of MDA-MB-435 estrogen receptor negative (ER-) and MCF-7 estrogen receptor positive (ER+) human breast cancer cells *in vitro*. Concentrations of 0.5 μg/ml tangeretin or nobiletin inhibited growth by 50% in ER- cells. Similar results were found for ER+ cells. Of most interest in this study, however, were the synergistic effects seen for tangeretin or nobiletin in combination with tamoxifin and/or tocotrienols. When the methoxylated flavones were tested as 1:1 mixtures with various tocotrienols, tangeretin and γ-tocotrienol in combination had an IC_{50} of 0.05 mg/ml in ER- cells and 0.02 mg/ml in ER+ cells. Without tangeretin, γ-tocotrienol had an IC_{50} in excess of 30 mg/ml in ER- cells and 2 mg/ml in ER+ cells. Tangeretin:γ-tocotrienol: tamoxifin (1:1:1) had a lower IC_{50} of 0.01 mg/ml in ER- cells but did not affect the IC_{50} of ER+ cells. The lowest IC_{50} of 0.001 mg/ml in ER+ cells reported was for nobiletin:δ-tocotrienol:tamoxifin (compared to 0.8 mg/ml for nobiletin alone and ER+ cells). Further research is needed to investigate methoxylated flavones in combination with other components as possible treatments for breast cancer.

Methoxylated flavones also have been shown to inhibit growth of leukemia cell lines but not the growth of normal cells *in vitro*. Sugiyama et al.[34] tested the activity of 27 highly methoxylated flavones isolated from citrus against M1 and HL-60 leukemic cell lines. In the isolation of these products from mandarin peel, tangeretin accounted for 22%, and nobiletin 64% of the material isolated. Heptamethoxyflavone (30%), nobiletin (30%), tangeretin (15%), and sinensetin (15%) accounted for 90% of the material from orange peel. Isolated components included those with hydroxyl groups along with two flavone glycosides. All the isolated flavones, except the two flavone glycosides, were effective in the inhibition of growth of the M1 cell line. Inhibition ranged from 30–80% at the test concentration of 50 μM. Tangeretin, nobiletin, heptamethoxyflavone, sinensetin, and 11 other flavones were also effective in inducing differtiation of the M1 and HL-60 cells. Differentiated cells do not undergo further division. Tangeretin and heptamethoxyflavone inhibited growth by greater than 50% and induced cell differentiation at greater than 25% in both the M1 and HL-60 cell lines. Nobiletin and sinensetin inhibited growth by 78% and 38% respectively, but were less effective in inducing cell differentiation at the 50 μM concentration. The effect of substitution of hydroxyl and methoxyl groups on growth inhibition and induced differentiation appeares to be complex and not easily explained. Concentrations as low as 5 μM level are effective in growth inhibition and cell differentiation, with the effect being dose dependent.

In a similar study by Hirano et al.,[35] the effectiveness of tangeretin in inhibiting growth of HL-60 leukemic cells compared to normal lymphocytes was confirmed. HL-60 leukemic cells cultured with 27 μM tangeretin *in vitro* were significantly inhibited. Treatments at lower doses also were effective but inhibition was less. A test of the MOLT-4 cell line showed tangeretin was not effective as a growth inhibitor for this line. When tested on the growth and viability of normal healthy cells, no adverse effects were seen. Analysis of the DNA of treated cells showed no fragmentation in healthy cells, but DNA fragmentation increased in HL-60 cells treated with as little as 2.7 μM tangeretin. The amount of fragmentation observed was dependent on the dose.

LIMONOIDS

Limonoids are abundant in citrus fruit with the highest concentrations in the seeds. Grapefruit seeds are the richest source with limonoid glucosides and limonoid aglycones comprising 0.7% and 2.4% of the seed respectively, by weight.[36] Valencia orange seeds contain approximately 0.8% limonoids as glucosides and 1.5% limonoids as aglycones. Limonin, nomilin and nomilin-17-ß-d-glucoside are the predominate ones present (Fig. 3). Limonin and nomilin can easily be extracted from seeds and isolated by crystallization. Unfortunately, citrus seeds are no longer collected during processing as they once were for extraction of citrus seed oil.

Limonin and nomilin are the predominate limonoids present in the peel, membranes, and juice, and, in mature fruit, occur mostly in a bound form as glucosides. Limonin and nomilin in the aglycone form are not present in fruit tissues outside the seeds, but occur as precursors where the D lactone ring is opened to form the corresponding monolactone acid. Young fruit contain high levels of these aglycone precursors. As the fruit matures, the open ring limonoids are converted to glucosides so very little precursor is available in the mature fruit to be converted to the bitter aglycones during processing.

Commercial citrus juice contain high concentrations of limonoid glucosides.[37] Orange juice contains 320 ppm on average, while grapefruit contains approximately 190 ppm. Mandarin juices in Japan have been found to contain an average of 225 ppm total limonoid glucosides. The predominant glucoside is limonin making up approximately one half the total present. Recently, a commercial citrus juice has become available in Japan called LG1000 that is fortified with limonoid glucosides.

Fruit waste peel, seeds, and by-products such as citrus molasses are good sources of limonoid glucosides.[38,39] A process for recovery of limonoid glucosides from citrus molasses has recently been described.[40] A patent also has recently been issued in the United States for the use of limonoids as a central nervous system activator.[41] The limonoid is extracted from any of several plant

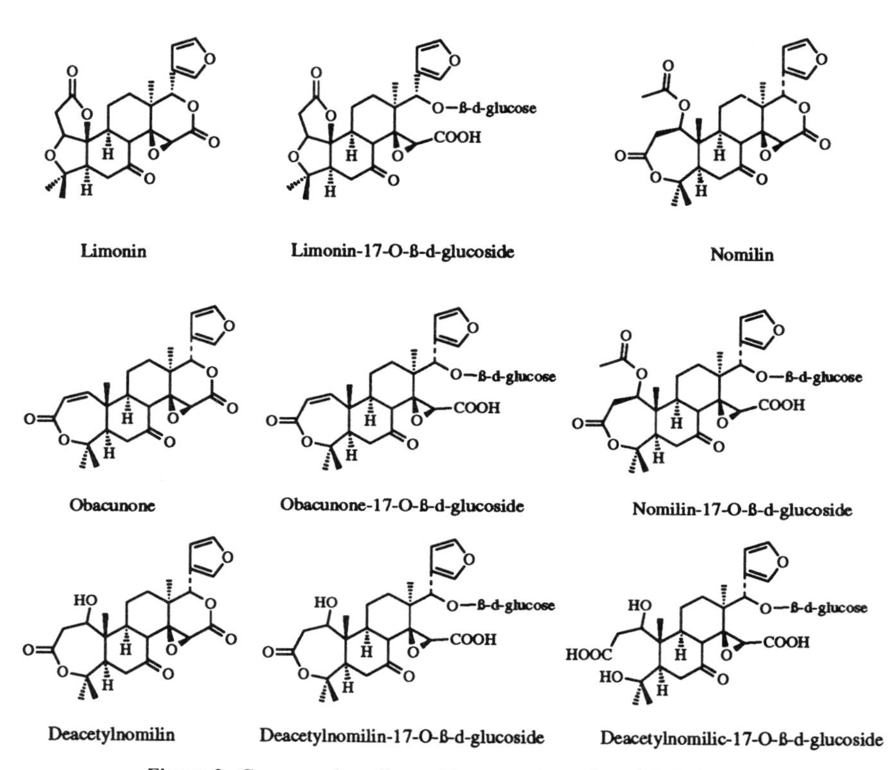

Figure 3. Common citrus limonoids present in seeds and fruit tissues.

varieties belonging of the Rutaceae family, with citrus being specifically mentioned. With further development, limonoid products may eventually become common food supplements.

Anticancer Activity

The discovery that citrus limonoids possess chemopreventive activity lends credence to the view of citrus as a healthy food, and the goodness of citrus that goes beyond the known nutritive value recognized by the nutritionists. The limonoids isolated from citrus products also possess strong anticancer activity. Inhibition of chemically induced carcinogenesis associated with the limonoid aglycones has been summarized by Lam et al.[36] Many natural products containing a furan ring will stimulate the glutathione S-transferase (GST) enzyme detoxifying system. This system catalyzes the conjugation of glutathione to activated carcinogens making them more water soluble, less reactive, and easier to excrete. Limonin has been found to be a weak stimulator of GST activity in intestinal cells but not in the forestomach or liver of mice. In contrast, nomilin at a dose level of 5 mg stimulates GST activity 3 fold in the small intestinal mucosa, 2.5 times in the liver, but has no effect in the forestomach.[42]

In a follow-up study to test whether the GST inducing activity of limonin and nomilin would provide effective protection against an induced carcinogen, female ICR mice were fed limonoids at two dose levels before and during exposure to benzopyrene. After 18 weeks, animals fed nomilin at 10 mg/dose were found to have 40% fewer tumors in those mice bearing tumors, and there were 30% fewer mice with any tumors larger than 0.5 mm when compared to the control group. Overall, there was a 50% reduction in tumors/mouse in the group fed 10 mg nomilin/day.[43] The protection afforded by the limonoids seemed to follow their ability to induce GST activity.

Lam tested the effectiveness of nomilin and limonin for protection against benzopyrone induced lung tumors. In this experiment, limonoids were mixed in the diet at several dose levels. After 18 weeks, the groups fed a diet containing 0.27% or 0.135% nomilin had a 75% and 52% reduction in the number of tumors per animal respectively. Groups fed 0.5% or 0.25% limonin had a 50% and 32% reduction respectively.

Limonoids also have been found to be effective inhibitors of skin cancer induced by dimethylbenzylanthracene. They are more effective when applied prior to exposure of the promoter TPA. When limonoid treatments of 1 mg were applied prior to induction by dimethylbenzylanthracene, there were no significant differences from the control group with either limonin or nomilin. However, when the application of limonoids was made prior to exposure of the TPA promoter, limonin at 1 or 0.25 mg doses provided 44% and 32% inhibition respectively in the number of tumors/mouse, and results were statistically different from the control group. Nomilin at 1 or 0.25 mg doses did not differ significantly from the control. Applications of limonin and nomilin were also found by Miller and coworkers[44] to inhibit the formation of DMBA induced carcinomas in hamster buccal pouches. Limonin was more effective than nomilin. Recently, it was found that commercial citrus juices contain abundant amounts of limonoids bound as glucosides.[45] Miller and coworkers found that limonin and nomilin bound as glucosides are potent inhibitors of cancer.[46]

DIETARY FIBER AND PECTIN

Dietary fiber is the skeletal remains of plant cells that are resistant to digestion by human enzymes. The role of dietary fiber in human health was

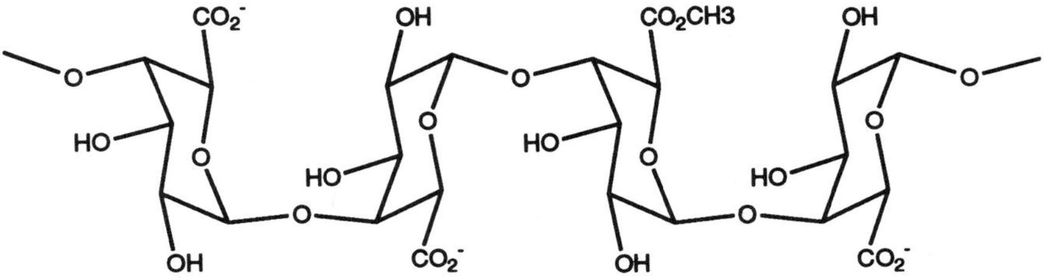

Figure 4. Pectin backbone structure.

stimulated by the work of Burkitt[47] and Painter.[48] Those investigators suggested that the relatively low level of plant fiber in the diets of Western societies predisposed those populations to diseases and disorders that differed from those in less developed societies.

Citrus fruit dietary fiber is made up of pectin, cellulose, hemicellulose, and trace amounts of lignin.[49] Lignin is a noncarbohydrate component and appears to possess no known role in human nutrition. However, cellulose, hemicellulose, and pectin are polysaccharides that possess beneficial health effects, namely, they affect large bowel function by increasing stool weight and decreasing transit time, influence glucose and lipid metabolism, influence mineral absorption, and play a role in diverticular disease, colon cancer, and coronary heart disease.[50,51]

The most studied component of citrus dietary fiber has been pectin (Fig. 4). Citrus fruits are a rich source, and it occurs both in the edible portions of the fruit and the inedible residues such as rag, core, and peel. Pectin is a polyanionic heterogeneous mixture of complex polysaccharides of high molecular weights and is predominantly composed of linear (1–5) linked galactopyranosyl uronic acid residues. Pectin, especially pectin derived from citrus, has demonstrated health-promoting activities.[52]

Coronary Heart Disease

Coronary Heart Disease (CHD) is the number one cause of disability and death in the United States. The etiology of CHD is not completely understood, however, epidemiological data suggest that elevated levels of plasma cholesterol and low density lipoprotein are two primary risk factors. Reduction of plasma cholesterol through diet, exercise, and consumption of dietary fiber has been proposed.[53,54] Research has shown that water-soluble dietary fiber, especially pectin, possesses hypocholesterolemic activity.[55,56]

Research by Cerda and co-workers with grapefruit pectin confirmed that when pectin was supplemented in the diets of laboratory animals and/or human volunteers, a significant reduction in serum cholesterol was achieved.[57–60] In rats fed high cholesterol diets, both cholestyramine (anion exchange resin) and grapefruit pectin caused a lowering in serum cholesterol and cholesterol-associated low density lipoprotein, but an increase in hepatic 3-hydroxy 3-methylglutaryl coenzyme A reductase (HMGCoA) activity.[61,62] HMGCoA catalyzes the formation of mevalonate, which is the committed step in cholesterol biosynthesis. Increased bile acid excretion resulting from pectin consumption causes an increase in cholesterol metabolism,[55] resulting in lower serum cholesterol values. Cerda and co-workers[59] also noted that the feeding of grapefruit pectin to rats caused a significant decrease in the formation of 7-α-hydroxycholesterol.

The effects of feeding grapefruit pectin to miniature pigs on their total cholesterol, low-density lipoprotein cholesterol, and low-density lipoprotein cholesterol/high-density lipoprotein cholesterol (LDLC:HDLC) ratio were investigated by Backey.[57] Grapefruit pectin: 1) significantly lowered plasma cholesterol by about 30%; 2) decreased the LDLC:HDLC ratio by about 31%; 3) resulted in an 85% decrease in plaque formation on the surface of the aorta; and 4) caused an 88% decrease in the narrowing of the coronary arteries. These last two observations suggest that grapefruit pectin may have direct beneficial effects on atherosclerosis by a mechanism independent of cholesterol levels.

Subsequently, Cerda studied the effects of grapefruit pectin on human subjects with varying levels of hypercholesterolemia.[63] Twenty-seven subjects (9 males and 18 females, ages 27 to 69) with plasma cholesterol values ranging from 208 to 420 mg per deciliter were given 15 grams grapefruit pectin per day for a period of four weeks. Measurement of the mean plasma cholesterol level after this period showed a decrease of about 7.6%. A second experiment with seven patients (utilizing a more palatable form of pectin - grapefruit pectin and eggwhite protein—Pro-Fiber™) showed more dramatic results (Table 2).[64] After 30 days, plasma cholesterol levels decreased 17%, LDLC decreased 25%, and the LDLC:HDLC ratio decreased 25%. No changes were detected on HDLC. These results are encouraging in light of many experimental findings that suggest a reduction of plasma cholesterol (by diet or medication) slows the progression rate of coronary artery disease. The NIH Consensus Development Conference on lowering blood cholesterol to prevent heart disease indicated that each 1% reduction of blood cholesterol would yield about a 2% reduction in risk for coronary heart disease.[65]

ESSENTIAL OILS

Anticancer Activity

Citrus essential oils belong to a group of non-nutrient compounds that exert inhibitory effects in experimental carcinogenesis models.[66–69] Orange peel oil, which mainly contains d-limonene (Fig. 5), is known to inhibit forestomach, lung

Table 2. Effect of Profiber™ on serum lipids[58]

	Baseline					Day 30				% Decrease	
Patient	**TC**	**LDLC**	**HDLC**	**Ratio**	**TC**	**LDLC**	**HDLC**	**Ratio**	**TC**	**LDLC**	**Ratio**
C.S.	241	154	41	3.8	202	125	36	3.5	16	19	8
M.V.	244	172	39	4.4	205	141	38	3.7	16	18	16
C.C.	293	161	29	5.5	243	114	27	4.2	17	29	24
S.E.	291	219	36	6.1	255	176	43	4.1	12	20	33
F.R.	263	124	24	5.2	219	76	27	2.8	17	39	46
C.B.	233	166	28	5.9	172	117	27	4.3	26	29	27
J.S.	276	183	33	5.5	229	139	32	4.3	17	24	22
Avg	263	168	33	5.2	218[1]	127[2]	33	3.8[3]	17	25	25

PROFIBER™ = grapefruit pectin product
TC = Total Cholesterol
LDLC = Low Density Lipoprotein Cholesterol
HDLC = High Density Lipoprotein Cholesterol

[1] $p < 0.009$
[2] $p = 0.008$
[3] $p = 0.003$

and mammary tumors.[70,71] D-limonene and citrus fruit oils (orange and lemon) also inhibit NNK-induced neoplasia (pulmonary adenoma and forestomach tumors) in female A/J mice.[69]

The mechanisms of inhibition of carcinogenesis by d-limonene have not been adequately defined. Preliminary evidence suggests that it and other essential oil components function as both blocking and suppressing agents.[72,73] Blocking agents prevent the occurrence of neoplasia by increasing the detoxification of carcinogens. When d-limonene was administered to female A/J mice one hour before administration of N-nitrosodiethyl-amine (NDEA), carcinogenesis of the forestomach and lung was inhibited.[74] Limonene and several plant essential oils induce an increase in the activity of Phase 2 enzymes that perform conjugation reactions that detoxify many carcinogenic agents. Of particular importance are the activities of glutathione S-transferase (GST), UDP-glucuronosyl transferase,

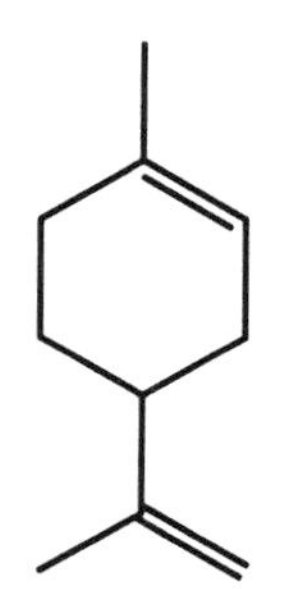

Figure 5. Limonene.

epoxide hydrolase, and NAD (P) H-quinone reductase. GST-inducing activity has been noted for the following citrus essential oil components, namely, d-limonene, β-caryophyllene, α-humulene, carvone, geraniol, p-mentha-2,8-dien-1-ol and p-mentha-8(9)-en-1,2-diol.[72]

Suppressing agents characteristically inhibit carcinogenesis when given after the complete course of carcinogen administration.[73] Orange oil, given one week after carcinogen administration (7,12-dimethylbenz (a) anthracene) (DMBA), was found to inhibit mammary tumor formation in rats. Evidence strongly indicates that d-limonene was the suppressing agent.[70] Limonene is extensively metabolized by rats and humans. Urinary metabolites of limonene include carveol and uroterpenol.[75] In tests of the chemopreventive activities of limonene, urinary metabolites (carveol and uroterpenol) were used against DMBA-induced rat mammary cancer by Crowell and coworkers.[76] The hydroxylated metabolites were more effective than limonene in decreasing tumor yield and significantly prolonged tumor latency. While carveol and uroterpenol are two primary urinary metabolites of limonene, the major circulating (plasma) metabolites are perillic acid, dihydroperillic acid and their methyl esters.[76]

The mechanisms by which limonene and/or its metabolites inhibit carcinogenesis are currently evolving. Limonene inhibits the isoprenylation of small G-proteins in the 21–26 k Da range, including ras -p21.[77] Such inhibition has the potential to alter differentiation and inhibit cell cycling. This probably involves changes in the activity levels of farnesyl-protein and geranylgeranyl-protein transferase. The evidence for the requirement of an isoprenoid product of mevalonic acid for human cell growth and division suggests that the mechanisms responsible for the chemopreventive activity of limonene might be related to alternations in isoprene metabolism.

GLUCARIC ACID

Glucaric acid (GA) is part of the class of chemicals called hexaric acids (Fig. 6). Other names for glucaric acid are saccharic acid, D-glucosaccharic acid, d-tetrahydroxyadipic acid, and glucarate. Glucarate is a common hexaric acid found in plants and animals as part of the metabolism of glucuronic acid.[78] Levels of GA in peeled citrus have been determined to be 1.29 g/kg for orange and 3.60 g/kg for grapefruit. The GA level for grapefruit reported was the highest for any fruit or vegetable measured, and the level for orange also can be considered high for a food source.[79]

Coronary Heart Disease

The biological activity of GA includes the ability to lower serum cholesterol. When female Sprague-Dawley rats were fed the AIN-76A diet (simulates

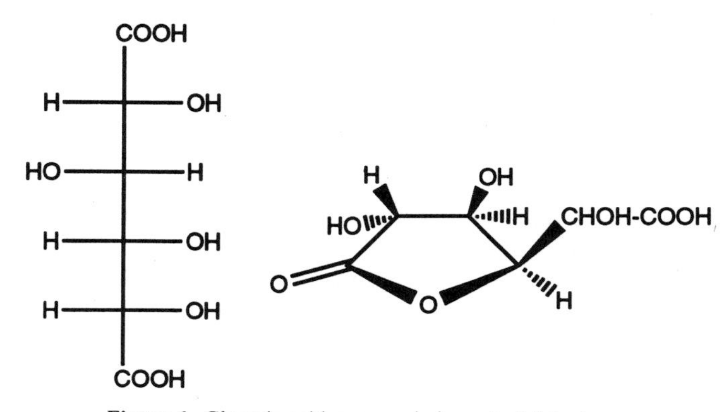

Figure 6. Glucaric acid, open and glucarate-1,4-lactone.

non-fat diet) with the experimental groups getting 17.5 and 35 mmol/kg of GA, the serum cholesterol dropped 10% and 14%, respectively.[79] Serum levels of GA increased 68% and 184% (17.5 and 35 mmol/kg of GA respectively) compared to a control group. The higher dosage caused a drop in the LDL-cholesterol serum levels of 30%. In a second experiment, Female Sprague-Dawley rats were fed AIN-76A diets supplemented with 5% and 20% of the total calories coming from fat, simulating low and high fat diets. The experimental groups had 17.5 and 35 mmol/kg of GA added to these diets. Addition of glucarate at 35 mmol/kg ranges with the high fat diet reduced levels of cholesterol by 12% and low density lipoproteins type C (LDL-C) by 35%. In all cases, the weights of the laboratory animals were not affected.[79]

Anticancer Activity

The prevention and treatment of cancer may be another area where GA can be beneficial to human health. In laboratory animals, GA prevents the formation of breast cancer induced in female Sprague-Dawley rats.[80] Glucaric acid can form glucaric acid 1,4 lactone by treatment with dilute acid. This lactone is a potent inhibitor of ß-glucuronidase which is the enzyme responsible for the removal of glucuronic acid from a conjugated chemical.[81] If glucuronic acid is not removed from conjugated carcinogens, they are excreted more effectively. This may be one mechanism of action of GA against breast cancer.

In an effort to discover the mechanism of action for GA, many known enzyme detoxification steps relative to both cancer preventing enzyme systems and cancer promoting enzyme systems were studied systematically in Sprague-Dawley rats.[82] At 35 mmol/kg food, GA did not affect the following enzymic functions: N-demethylase (N-nitroso), UDP GTPase (3-OH steroid, androsterones), bile acid sulfo-tase, catalase (important in free radical oxygen scavenging), glutathione reductase, glucose-6-P-dehydrogenase, xanthine oxidase, ß-glucuronidase, or ornithine decarboxylase. Glucaric acid inhibited protein kinase C and increased levels of cyclic AMP. As in previous studies, the effects on protein kinase C and cyclic AMP levels depended on whether GA was administered with a retinoic acid derivative. If GA and a retinoic acid are administered together, then only half the maximum dose of each is necessary to obtain optimum response.[82] The increase in c-AMP in rat liver was confirmed in another experiment which used radiolabelled GA to determine the GA pharmacokinetics.[83] Webb et al. discovered that when GA was administered orally, levels increased steadily over time and reached a peak at 24 hours. Glucaric acid was not fully metabolized until more than 5 hours later. The liver and intestinal mucosa concentrated the glucaric acid at three times the level found in the blood. This is important because ß-glucuronidase enzymatic activity is high in these organs, and GA-1,4-lactone can inhibit this activity, and excretion of conjugated carcinogens is thus increased.

The GA dose response relationship demonstrates that the compound works on the tumor itself and not on other external factors. Glucaric acid also works independantly of retinoids when fed to rats with mammary tumors. A 75% decrease in protein kinase C activity, relative to controls, was demonstrated in mammary tumors extracted from Sprague Dawley rats fed either a combination of glucaric acid and retinoic acid, or a high dose of glucaric acid. The levels of c-AMP also increased in the rat tumor with either glucarate feeding or a combination of lower dose glucarate and retinoid feeding. Both c-AMP levels and protein kinase C are thought to control the proliferation and growth of cancer cells. The levels of glucaric acid necessary to carry this out are 128mmol/kg of food, or approximately 16g/kg of food.[84]

Inhibition and treatment of cancer was not limited to Spraque Dawley rats. When GA was tested against MCF-7 human mammary tumor cells *in vitro*, it affected both growth and life span of the cancer cells at levels that demonstrated no cytotoxicity. No toxicity to normal cells was found at doses necessary to kill the human mammary cancer cells *in vitro*.[85]

Inhibition required at least 8.5g of GA/kg of consumed food in the diets of the laboratory animals challenged by carcinogens. Glucaric acid then represents 0.85% of the total diet. Because grapefruit has only 3.6g/kg (0.36%), it is not possible to reach this level by normal diet.[79] All of the studies listed above refer to highly challenged rats and mice. In most, the rats are treated with powerful doses of carcinogen and the tumors allowed to grow before the addition of GA to the diet. It may be that if foods like grapefruit and orange are consumed on a regular basis, adequate protection from the everyday exposure to background carcinogens may be attained. Another possibility is that GA acts in concert with other phytochemicals as it does with retinoids. Other interactions may yet be discovered.

CONCLUSION

It is generally accepted that consuming citrus fruit and drinking citrus juices are health promoting. Citrus is considered a nutritionally rich food containing several important vitamins. For some time this perception has been attributed to the presense of vitamins. Citrus phytochemicals may be another reason for the health promoting aspects. When a 240ml (8 fluid ounces) glass of orange juice is consumed, the following levels of phytochemicals are ingested: flavanone glycosides (70–190 mg), d-limonene (2–6 mg), limonoid glycosides (59–102 mg), limonin (02.-1 mg), methoxylated flavones (0.7–1.2 mg), and pectin (95–190 mg).[86] Glucaric acid levels in fresh fruit are 1.29 g/kg for a whole orange and 3.60 g/kg for a whole grapefruit.[79] Therefore, any person consuming citrus ingests a large amount of these phytochemicals. The anticancer and antiatherosclerotic activities of these compounds have been listed in the preceding parts of this publication. Perhaps the combination of citrus nutrients, flavonoids, limonoids, monoterpenes, dietary fiber, and glucaric acid create a food with greater health promoting ability than the sum of its parts.

REFERENCES

1. AFFANY, A., SALVAYRE, R., BLAZY, L. 1987. Comparison of the protective effect of various flavonoids against lipid peroxidation of erythrocyte membranes induced by cumene hydroperoxide. Fund. Clin. Pharm. 1: 451–457.
2. KUHNAU, J. 1976. The flavonoids, a class of semi-essential food components: Their role in human nutrition. World Rev. Nutr. Dig. 24: 117–119.
3. FISHER, K. D., SENTI, F. R., ALLISON, R.G., ANDERSON, S.A., CHINN, H.I., TALBOT, J.M. 1982. Evaluation of the Health Aspects of Hesperidin, Naringin, and Citrus Bioflavonoid Extracts as Food Ingredients. FDA Report No. FDA/BF-82/62, Federation of American Societies for Experimental Biology, Bethesda, MD, 34 pp.
4. RUSZNYAK, S., SZENT-GYORGYI, A. 1936. Vitamin P: flavonols as vitamins. Nature 27: 138.
5. VICKERY, H., NELSON, E., ALMQUIST, H., ELVEHJEM, C. 1950. Joint Committee on Nomenclature - Term "Vitamin P" recommendation to be discontinued. Science 112: 628.
6. KAUL, T.N., MIDDLETON, E. JR., OGRA, P.L. 1985. Antiviral effects of flavonoids on human viruses. J. Med. Virol. 15: 71–74.
7. ROBBINS, R.C., MARITIN, F.G., ROE, M.D. 1988. Ingestion of grapefruit lowers elevated hematocrits in human subjects. Int. J. Vit. Nutr. Res. 58: 414–417.
8. BOOTH, A.N., JONES, F.T., DEEDS, F. 1958. Metabolic fate of hesperidin, eriodictyol, homoeriodictyol, and diosmin. J. Biol. Chem. 230: 661–668.
9. BOOTH, A.N., JONES, F.T., DEEDS, F. 1958. Metabolic and glucosuria studies on naringin and phloridzin. J. Biol. Chem. 233: 280–282.
10. FUHR, U., KUMMERT, A.L. 1995. The fate of naringin in humans: A key to grapefruit juice-drug interactions? Clin. Pharm. Ther. 58: 365–368.
11. AMEER, B., WEINTRAUB, R.A., JOHNSON, J.V., YOST, R.A., ROUSEFF, R.L. 1996. Flavanone absorption after naringin, hesperidin, and citrus administration. Clin. Pharm.Ther. 60: 34–40.

12. FUHR, U., LKITTICH, K., STAIB, H. 1993. Inhibitory effect of grapefruit juice and its bitter principle, naringenin, on CYP1A2 dependent metabolism of caffeine in man. Br. J. Clin. Pharmacol. 35: 431–436.
13. BAILEY, D.G., ARNOLD, J.M.O., MUNOZ, C., SPENCE, J.D. 1993. Grapefruit juice-fleodipine interaction: Mechanism, predictability, and effect of naringin. Clin. Pharm. Ther. 53: 637–642.
14. BAILEY, D.G., ARNOLD, J.M.O., STRONG, H.A, MUNOZ, C, SPENCE, J.D. 1993. Effect of grapefruit juice and naringin on nisoldipine pharmocokinetics. Clin. Pharm. Ther. 54: 589–594.
15. MENON, L.G., KUTTAN, R., KUTTAN, G. 1995. Inhibition of lung metastasis in mice induced by B16F10 melanoma cells by polyphenolic compounds. Cancer Let. 95: 221–225.
16. SO, F., GUTHRIE, N., CHAMBERS, A.F., CARROLL, K.K. 1995. The effects of different combinations of flavonoids on the proliferation of MDA-MB-435 human breast cancer cells. Research Conference of the American Institue for Cancer research: Dietary Phytochemicals in Cancer Prevention and Treatment. Poster #13.
17. CARROLL, K.K., GUTHRIE, N., CHAMBERS, A.F. 1995 Effects of citrus flavonoids on proliferation of mda-mb-435 human breast cancer cells and on mammary tumorigenesis induced by DMBA in female Sprague-Dawley rats. Research Conference of the American Institue for Cancer Research: Dietary Phytochemicals in Cancer Prevention and Treatment. Poster #14.
18. TANAKA, T., MAKITA, H., OHNISHI, M., HIROSE, Y., WANG, A., MORI, H., SATOH, K., HARA, A., OGAWA, H. 1994. Chemoprevention of 4-nitroquinoline 1-oxide induced oral carcinogenesis by dietary curcumin and hesperidin: Comparison with the protective effect of B-carotene. Cancer Res. 54: 4653–4659.
19. MARTIN, M.J., MARHUENDA, E., PEREZ-GUERRERO, C., FRANCO, J.M. 1994. Antiulcer effect of naringin on gastric lesions induced by ethanol in rats. Pharmacology 49: 144–150.
20. ROBBINS, R.C. 1974. Action of flavonoids on blood cells: Trimodal action of flavonoids elucidates their inconsistent results. Int. J. Vil. Nutr. Res. 44: 203–216.
21. MIDDLETON, E. JR., DZREWIECKI, G. 1982. Effects of flavonoids and transitional metal cations on antigen induced histamine release from human basophils. Biochem. Pharmacol. 31: 1449–1453.
22. MIDDLETON, E. JR., FUJIKI, H., SAVLIWALA, M., DZREWIECKI, G. 1987. Tumor promotor-induced basophil histamine release: Effect of selected flavonoids. Biochem. Pharmacol. 36: 2048–2052.
23. WALL, M.E., WAN, M.C., MANIKUMAR, G., GRAHAM, P.A., TAYLOR, H., HUGHS, T.J., WALKER, J., MCGIVNEY, R.J. 1988. Plant antimutagenic agents: Flavanoids. J. Nat. Prod. 51: 1084–1091.
24. WATTENBERG, L.W. 1985. Chemoprevention of cancer. Cancer Res. 45: 1–8.
25. WATTENBERG, L.W. 1992. Chemoprevention of cancer by naturally occurring and synthetic compounds. In: Cancer Chemoprevention, (L.W. Wattenberg, M. Lipkin, C.W. Boone, G.J. Kelloff, Eds.), CRC Press, Boca Raton, FL. pp. 19–31.
26. BRACKE, M.E., VYNCKE, O.G., FOIDART, J.M., DEPESTEL, G., MAREEL, M. 1991. Effect of catechins and citrus flavonoids on invasion *in vitro*. Clin. Expl. Metastasis. 9: 13–25.
27. BRACKE, M.E., BRUYNEEL, E.A., VERMEULEN, S.J., VENNEKENS, K., VAN MARCK, V., MAREEL, M.M. 1994. Citrus flavonoid effect on tumor invasion and metastasis. Food Technology 47(11): 121–124.
28. TAKEICHI, M. 1991. Cadherin cell adhesion receptors as a morphogenetic regulator. Science 251: 1451–1455.
29. EDELMAN, G.M., CROSSIN, K.L. 1991. Cell adhesion molecules: Implications for a molecular histology. Annu. Rev. Immunol. 60: 155–190.
30. CHEN, W., OBRINCK, B. 1991. Cell-cell contacts mediated by E-cadaherin (uvomorulin) restrict invasive behavior of L-cells. J. Cell Biol. 114: 319–327.

31. MAREEL, M., BRACKE, M., VAN ROY, F. 1994. Invasion promoter versus invasion suppressor molecules: The pardigm of E-cadherin. Mol. Biol. Reports. 19: 45–67.
32. KANDASWAMI, C., PERKINS, E., SOLONUIK, D.S., DRZEWIECKI, G., MIDDLETON, E.J.R. 1991. Antiproliferative effects of citrus flavonoids on a human squamous cell carcinoma *in vitro*. Cancer Letters. 56: 147–152.
33. GUTHRIE, N., GAPOR, A., CHAMBERS, A.F., CARROLL, K.K. 1996. Effects of palm oil tocotrienols, flavonoids and tamoxifen on proliferation of MDA-MB-435 and MCF human breast cancer cells. Research Conference of the American Institue for Cancer Research: Dietary Fat and Cancer. Poster #8.
34. SUGIYAMA, S., UMEHARA, K., KUROYANAYI, M., UENO, A., TAKI, T. 1993. Studies on the differentiation inducers of myloid leukemic cells from citrus species. Chem. Pharm. Bull. 41: 714–719.
35. HIRANO, T., ABE, K., GOTOH, M., OKA, K. 1995. Citrus flavone tangeretin inhibits leukemic, HL-60 cell-growth partially through induction of apoptosis with less cytotoxicity on normal lymphocytes. Br. J. Cancer 72: 1380–1388.
36. LAM, L.K.T., ZHANG, J., HASEGAWA, S., SCHUT, H.A.J. 1993. Inhibition of chemically induced carcinogenesis by citrus limonoids. In: Food Phytochemicals for Cancer Prevention I: Fruits & Vegetables, (M.T. Haung, O. Osawa, C.T. Ho, R. Rosen, Eds.), ACS Symposium Series No. 546, American Chemical Society, Washington, DC, pp. 209–219.
37. FONG, C.H., HASEGAWA, S., HERMAN, Z., OU, P.J. 1989. Limonoid glucosides in commercial citrus juices. J. Food Sci. 54: 1505–1506.
38. HASEGAWA, S., OU, P., FONG, C.H., HERMAN, Z., COGGINS, C.W. Jr., ATKIN, D. 1991. Changes in the limonate a-ring lactone and limonin 17-ß-D-glucopyranoside content of navel oranges during fruit growth and maturation. J. Agric. Food Chem. 39: 262–265.
39. FONG, C.H., HASEGAWA, S., COGGINS, C.W. JR., ATKIN, D., MIYAKE, M. 1992. Contents of limonoids and limonin 17-ß-D-glucopyranoside in fruit tissue of valencia orange during fruit growth and maturation. J. Agric. Food Chem. 40: 1178–1181.
40. OZAKI, Y., AYANO, S., MIYAKE, M., MAEDA, H., IFUKU, Y., HASEGAWA, S. 1992. Limonoid glucosides in juices of *Satsuma mandarin* (Citrus unshiu Marc.). Biosci. Biotech. Biochem. 56: 836–837.
41. HAGA, M., WADA, K. Sept. 6, 1994. Central Nervous System Activatior and Taste Enhancing Additive. United States Patent #5,344,648.
42. LAM, L.K.T., LI, Y., HASEGAWA, S. 1989. Effects of citrus limonoids on glutathione S-transferase activity in mice. J. Agric. Food Chem. 37: 878–880.
43. LAM, L.K.T., HASEGAWA, S. 1989. Inhibition of benzo[a]pyrene-induced forestomach neoplasia in mice by citrus limonoids. Nutr. Cancer 12: 43–47.
44. MILLER, E.G., FANOUS, R., RIVERA-HIDALGO, F., GINNIE, W.H., HASEGAWA, S., LAM, L.K.T. 1989. The effect of citrus limonoids on hamster buccal pouch carcinogenesis. Carcinogenesis 10: 1535–1537.
45. HASEGAWA, S., MIYAKE, M., OZAKI, Y. 1993. Biochemistry of citrus limonoids and their anticarcinogenic activity. In Food Phytochemicals for Cancer Prevention I: Fruits & Vegetables, (M.T. Haung, O. Osawa, C.T. Ho, R. Rosen, Eds.), ACS Symposium Series No. 546, American Chemical Society, Washington, DC, pp. 198–208.
46. MILLER, E.G., GONZALES-SANDERS, A.P., COUVILLON, A.M., WRIGHT, J.M., HASEGAWA, S., LAM, L.K.T., SUNAHARA, G.I. 1993. Inhibition of oral carcinogenesis by green coffee beans and limonoid glucosides. In: Food Phytochemicals for Cancer Prevention I: Fruits & Vegetables, (M.T. Haung, O. Osawa, C.T. Ho, R. Rosen, Eds.), ACS Symposium Series No. 546, American Chemical Society, Washington, DC, pp. 220–229.
47. BURKITT, D.P. 1971. Epidemiology of cancer of the colon and rectum. Cancer 28: 3–13.
48. PAINTER, N.S., ALMEIDA, A.Z., COLEBORUNE, K.W. 1972. Unprocessed bran in treatment of diverticular disease of the colon. Br. Med. J. 2: 137–140.

49. BAKER, R.A. 1994. Potential dietary benefits of citrus pectin and fiber. Food Technol. 48(11): 133–139.
50. BAIG, M.M., CERDA, J.J. 1983. Citrus pectic polysaccharides-their *in vitro* interaction with low-density serum lipoproteins. In: Unconventional Sources of Dietary Fiber I. (I. Furda, Ed.), American Chemical Society, Washington, D.C., pp. 185–190.
51. KRITCHEVSKY, D. 1994. Dietary fiber and cancer. In: Nutrition and Disease Update: Cancer. (K.K. Carroll , D. Kritchevsky, Eds.), AOCS Press, Champaign, IL, pp. 1–23.
52. BAIG, M.M., CERDA, J.J. 1980. Studies on the role of citrus in health and disease. In: Citrus Nutrition and Quality, (S. Nagy, J.A. Attaway, Eds.), American Chemical Society, Washington, D.C., pp. 25–41.
53. KEYS, A., GRANDE, F., ANDERSON, J.T. 1961. Fiber and pectin in the diet and serum cholesterol concentrations in man. Proc. Soc. Exp. Biol. Med. 106: 555–558.
54. NATIONAL RESEARCH COUNCIL. 1989. Dietary fiber. In: Diet and Health. National Academy Press, Washington, D.C. pp. 291–310.
55. KAY, R.M., TRUSWELL, A.S. 1977. Effect of citrus pectin on blood lipids and fecal steroid excretion in man. An. J. Clin. Nutr. 30: 171–175.
56. LAFONT, H., LAIRON, D., VIGNE, S.L., CHANUSSOT, F., CHABERT, C., PORTUGAL, H., PAULI, A.M., CRETTE, C., HAUTON, J.C. 1985. Effect of wheat bran, pectin, cellulose on the secretion of bile lipids in rats. J. Nutr. 115: 849–855.
57. BACKEY, P.A., CERDA, J.J., BURGIN, C.W., ROBBINS, F.L. 1988. Grapefruit pectin inhibits hypercholesterolemia and atherosclerosis in minature swine. Clin. Cardiol. 11: 595–600.
58. CERDA, J.J., ROBBINS, F.L., BURGIN, C.W., BAUMGARTNER, T.G. 1988. The effects of grapefruit pectin on patients at risk for coronary heart disease without altering diet or lifestyle. Clin. Cardiol. 11: 589–594.
59. CERDA, J.J., ROBBINS, F.L., RICE, R.W., BURGIN, C.W. 1988. The effect of dietary grapefruit pectin and cholesterol on hepatic 7-hydroxy-cholesterol formation in the rat. Clin. Res. 36(3): 355A.
60. CERDA, J.J., ROBBINS, F.L., BURGIN, C.W., VATHUDA, S., SULLIVAN, M.P. 1991. Effectiveness of a grapefruit pectin product in lowering plasma cholesterol in guinea pigs. Clin. Res. 39: 80A.
61. BIAG, M.M., BURGIN, C.W., CERDA, J.J. 1983. Influence of dietary pectin and cholestyramine on the activity of hepatic 3-hydroxyl 3-methyl-glutaryl coenzyme A reductase in the rat. Fed. Proc. 42: 1787.
62. BAIG, M.M., BURGIN, C.W., CERDA, J.J. 1985. Hypocholesterolemic agents: A comparison of the relative effectiveness of cholestyramine and pectin in rats. Drug-Nut. Interact. 3: 109–113.
63. CERDA, J.J. 1993. Role of grapefurit pectin in health. In: America's Foods: Health Messages and Claims. (J.E. Tillotson, Ed.), CRC Press, Boca Raton, FL., pp. 203–208.
64. CERDA, J.J. 1996. The role of pectin in cholesterol regulation. In: Hypernutritious Foods. (J.W. Finley, A.A. Armstrong, S. Nagy, S. Robinson, Eds.), AgScience, Inc., Auburndale, FL, pp. 161–170.
65. NIH CONSENSUS CONFERENCE. 1985. Lowering blood cholesterol to prevent heart diesease. J. Am. Med. Assoc. 253: 2080–2086.
66. HARTMAN, P.E., SHANKEL, D.M. 1990. Antimutagens and anticarcinogens: A survey of putative interceptor molecules. Environ. Mol. Mutagenesis. 15: 145–182.
67. WATTENBERG, L.W. 1983. Inhibition of neoplasia by minor dietary constituents. Cancer Res. 43: (Suppl.) 2448s-2453s.
68. WATTENBERG, L.W. 1985 Chemoprevention of cancer. Cancer Res. 48: 1–8.
69. WATTENBERG, L.W., COCCIA, J. 1991. Inhibition of 4-(methylnitrosamino)-1(3-pyridyl)-1-butanone carcinogenesis in mice by d-limonene and citrus fruit oils. Carcinogenesis 12: 115–117.
70. ELSON, C.E., MALTZMAN, T.H., BOSTON, J.L., TABBER, M.A., GOULD, M.N. 1988. Anti-carcinogenic activity of d-limonene during initiation and promotion/progression stages of DMBA-induced mammary carcinogenesis. Carcinogenesis (Lond.). 9: 331–332.

71. MALTZMAN, T.H., HURT, L.M., ELSON, C.E., TANNER, M.A., GOULD, M.N. 1989. The prevention of nitrosomethylurea-induced mammary tumors by d-limonene and orange oil. Carcinogenesis (Lond.) 10: 781–783.
72. LAM, L.K.T., ZHENG, G-Q., ZHANG, J., KENNEY, P.M. 1996. Cancer chemoprevention by constituents of plant essential oils. In: Hypernutritious Foods. (J.W. Finley, D.A. Armstrong, S. Nagy, S.F. Robinson, Eds.), AgScience Inc., Auburndale, FL. pp. 123–139.
73. WATTENBERG, L.W., LIPKIN, M., BOONE, C.W., KELLOFF, G.J. 1992. Cancer Chemoprevention. CRC Press, Boca Raton, FL., pp. 22–30.
74. WATTENBERG, L.W., SPARNINS, V.L., BARANY, G. 1989. Inhibition of N-nitrosodiethylamine carcinogenesis by naturally occurring organosulfur compounds and monoterpenes. Cancer Res. 49: 2689–2692.
75. REGAN, J.W., BJELDANES, L.F. 1976. Metabolism of (+)-limonene in rats. J. Agric. Food Chem. 24: 377–380.
76. CROWELL, P.L., LIN, S., VEDEJS, E., GOULD, M.N. 1992. Identification of metabolites of the antitumor agent d-limonene capable of inhibiting protein isoprenylation and cell growth. Cancer Chemother. Pharmacol. 31: 205–212.
77. CROWELL, P.L., CHANG, R.R., REN, Z., ELSON, C.E., GOULD, M.N. 1991. Selective inhibition of isoprenylation of 21–26 k Da proteins by the anticarcinogen d-limonene and its metabolites. J. Biol. Chem. 266: 17679–17685.
78. MARSH, C.A. 1963. Metabolism of D-glucuronolactone in mammalian systems. II. Conversion of D-glucuronolactone into d-glucaric acid by tissue preparation. Biochem. J. 87: 82–90.
79. WALASZEK, Z., SZEMRAJ, J., HANAUSEK, M., ADAMS, A.K., SHERMAN, U. 1996. D-Glucaric acid content of various fruits and vegetables and cholesterol-lowering effects of dietary D-glucarate in the rat. Nut. Res. 16: 673–681.
80. ABOU-ISSA, H., MOESCHBERGER, M., EL-MASRY, W., TEJWANI, S., CURLEY, R.W. Jr., WEBB, T.E. 1995. Relative efficacy of glucarate on the initiation and promotion phases of rat mammary carcinogenesis. Anticancer Res. 15: 805–810.
81. HEERDT, A.S., YOUNG, C.W., BORGEN, P.I. 1995. Calcium glucarate as a chemopreventive agent in breast cancer. Israel J. Med. Sci. 31: 101–105.
82. ABOU-ISSA, H., DWIVEDI, C., CURLEY, R.W. Jr., KIRKPATRICK, R., KOOLEMANS-BEYNAN, A., ENGINEER, F.N., HUMPHRIES, K.A., EL-MASTRY, W., WEBB, T.E. 1993. Basis for the anti-tumor and chemopreventive activities of glucarate and the glucarate: retinoid combination. Anticancer Res. 13: 395–400.
83. WEBB, T.E., PHAM-NGUYEN, M.-H., DARBY, M., HAMME, A.T. II. 1994 Pharmacokinetics relevant to the anticarcinogenic and anti-tumor activities of glucarate and the synergistic combination of glucarate: retinoid in the rat. Biochem. Pharmacol. 47: 1655–1660.
84. WEBB, T.E., ABOU-ISSA, H., STROMBERG, P.C., CURLEY, R.C. Jr., NGUYEN, M.-H.P. 1993. Mechanism of growth inhibition of mammary carcinomas by glucarate and the glucarate: retinoid combination. Anticancer Res. 13: 2095–2100.
85. BHATNAGAR, R., ABOU-ISSA, H., CURLEY, R.W. Jr., KOOLEMANS-BEYNAN, A., MOESCHBERGER, M.L., WEBB, T.E. 1991. Growth suppression of human breast carcinoma cells in culture by N-(4-hydroxyphenyl)retinamide and its glucuronide and through synergism with glucarate. Biochem. Pharmacol. 41: 1471–1477.
86. WIDMER, W. W., MONTANARI, A. The potential for citrus phytochemicals in hypernutritious foods. In: Hypernutritious Foods; (J.W. Finley, D.J. Armstrong, S. Nagy, S.F. Robinson, Eds.), AgScience: Auburndale, Florida, 1996, pp. 75–90.

Chapter Three

CHEMOPREVENTIVE AGENTS IN FOODS

Bozidar Stavric

Food Research Division, Bureau of Chemical Safety
Health Protection Branch, Department of Health
Banting Research Centre, Tunney's Pasture, Post. Loc. 2203D, Ottawa, Ontario, K1A OL2, Canada

INTRODUCTION

In recent years, organizations which promote health, the scientific community, and nutritionists have directed attention toward achieving long-term health benefits through prevention of certain chronic diseases, like cancer or cardiovascular disorders. The current consensus is that this can be achieved by a combination of several basic factors - avoiding smoking, performing regular exercise, and consuming a proper diet. The protective role of naturally occurring food agents is the subject of this review.

Currently, it is being considered whether foods, in addition to their traditional nutritional values with traces of natural or man-made toxicants, contain certain non-nutritional components that may have long-term health-promoting attributes. Many of these "inert" or "semi-essential" components,[1] long neglected, have become the focus of epidemiological and experimental investigations. Research is being conducted to identify naturally occurring, health-promoting substances in the diet, to explain their mode of action, and to assess significance for health of the general population.

There is substantial evidence to show an association between eating patterns and cancer.[2–6] Much of the explanation for the relation between diet and cancer comes from the traditional dependence of humans on their food supply.[4] Chemoprevention is the use of a chemical substance as a means of preventing the development of a disease. Such agents may block or control undesirable effects of components in the diet or the environment. In addition, changes and damage at the cellular level, due to oxidative stress, common physiological events, or aging, can be controlled or reduced by dietary components with chemoprotective capacity. Among different types of chemopreventive compounds, the phytochemicals present in a variety of foods and beverages recently have received attention. It has been assumed that these food components, while capable of protecting or slowing the occurrence of certain chronic diseases, are by themselves not toxic.[7–14]

While there are different approaches for reducing the burden in suffering and expense due to chronic diseases, two strategies are prominent: (a) identification of the causes and (b) development of measures for prevention. There are many dietary factors involved in the appearance of chronic diseases - energy imbalance, imbalance of food/nutrient intake, and specific deficiencies of nutrients and "non-nutrients". The imbalance of "non-nutrient" food components, especially micronutrients, may play a role in the development of some chronic disorders.

The strongest evidence for the chemopreventive potential of micronutrients comes from the association of fruit and vegetable intake with reduced incidence of cancer.[2,6,14–16] Intervention studies with experimental animals clearly indicate the beneficial effects of such diets.[17] Numerous epidemiological studies also confirm an inverse association between consuming fruits and vegetables and the occurrence of cancer.[2,6,14,18–21] One explanation is that these diets are also low in fat and caloric intake. Such factors are thought to decrease cancer risk.

This paper discusses recent findings concerning phytochemicals and other food ingredients which are generally characterized as *chemopreventers, nutraceuticals* or *functional foods*. These can be grouped and discussed on the basis of: chemical entity; physiological activity; category of foods in which they appear; mechanism of action; or the disease that is prevented or organ affected. An attempt is made to highlight these different approaches and illustrate them with a few examples.

Chemopreventive agents are found in all categories of foods, with fruits and vegetables being the main source. Table 1 summarizes the major types of foods with confirmed chemopreventive components. One should remember that very often different chemopreventers act on different sites or simultaneously perform different actions. Table 2 illustrates the multifaceted mechanisms of action which act both extracellularly and intracellularly. Many chemopreventers act in concert, often by different mechanisms, resulting in synergistic effects.[7,14,17,19]

In this article, a prime focus is given to the beneficial chemopreventive activity of antioxidants since they are abundant in foods and are the most thoroughly investigated. In addition to certain enzymes which are present in tissues and cells and which are part of endogenous defence mechanisms, it appears that antioxidants represent a strong biological defense against free radicals and reactive oxygen species. They have attracted much interest among

Table 1. Categories of foods with the most prominent chemopreventive agents

Type of food	Chemopreventers
Fruits	Vitamins, flavonoids, polyphenolic acids, fiber, carotenes, monoterpenoids
Vegetables	Vitamins, flavonoids, plant phenolics, chlorophyll, fiber, aliphatic sulfides, carotenes, aromatic isothiocyanates, folic acid, dithiothiones, phytic acid, calcium
Cereals	Fiber, α-tocopherol, phytic acid, selenium
Meats, fish, eggs,	Conjugated isomers of linoleic acid, vitamins (A,E), poultry selenites
Fat/oil	Omega-3 fatty acids, vitamin E, tocotrienols
Milk	Fermentation products, calcium, free fatty acids
Nuts, beans, grains	Polyphenolics, fiber, vitamin E, phytic acid, coumarins folic acid, phytoestrogens, isoflavones, lignans
Spices	Coumarins, curcumin, sesaminol
Tea	Plant phenolics, epigallocatechin, catechins
Coffee	Polyphenolic acids, diterpene alcohol esters, melanoidins
Wine	Flavonoids
Water	Selenium

Adapted from reference # 166.

Table 2. Mechanisms of action of chemopreventers

Extracellular	Intracellular
• Reduce or inhibit the formation of mutagens and carcinogens during food preparation • Reduce the bioavailability of mutagens and carcinogens • Accelerate intestinal transit • Protect the intestinal mucosal barrier • Modify intestinal microbial flora • Inhibit the penetration of cells by mutagens and carcinogens	• Enhance the activities of enzymes involved in detoxification of mutagens and carcinogens • Inhibit the activities of enzymes involved in formation of mutagenic and carcinogenic metabolites • Scavenge reactive oxygen species • Inhibit metabolic activation • Promote immune system • Protect DNA from carcinogens • Inhibit the detrimental effects of procarcinogens on DNA

scientists and the general population alike.[5,7,22,23] I focus also on the interpretation of results and the ongoing controversy regarding the benefits of food supplementation with antioxidants versus the benefits of obtaining them in appropriate amounts from regular dietary sources.

ANTIOXIDANTS

Antioxidants interact with free radicals preventing oxidative damage to cells produced by regular cellular metabolism, environmental factors, and certain dietary components or aging. The role and characterization of food antioxidants has been recently reviewed.[24] Many, like vitamins C and E, β-carotene and other carotenoids, are abundant in fruits and vegetables. It has been postulated that supplementation of the regular diet will protect the human population from chronic diseases like cancer and cardiovascular problems.[5,14] However, recent results of case-control studies in subjects taking such supplements contradicts this and suggests that beneficial effects may not be applicable to all segments of the population.[20,25] The safety of taking supplementary antioxidants also is still uncertain.

Several vitamins play important roles in protecting the human body against cancer[25–28] (for reviews see also refs.[7,9,10,16]). In addition, there are other components in the diet, such as enzymes, minerals, polyphenols, spices and others with antioxidant properties.[29,30] The combined and complex activity of different antioxidants/chemopreventers reacting by different mechanisms may be important although *in vivo* this is difficult to investigate. Deshpande and associates reported an interrelationship of vitamin E, selenium, vitamin C, and glutathione-S-transferase (GST) in cell membrane protection.[29] Combinations of several chemopreventive agents that may have little benefit alone, particularly those acting by different mechanisms, have been shown in animal studies to result in significant antitumor effects.[7,31]

Vitamin A, β-Carotene, Carotenoids

Based on epidemiological data, vitamin A and its precursors, the carotenoids, have been postulated as important in prevention of human cancers. The data clearly indicate that low intake of fruits and vegetables is associated with increased cancer risk.[5,6,14] In addition, low levels of β-carotene in serum and plasma have been associated with development of lung cancer.[32] Observations from earlier studies that vitamin A intake was inversely associated with lung cancer have been extended to reduced risk of cancer at other sites, such as the colon, esophagus, bladder, rectum and cervix.[32] Further investigations indicated that β-carotene was the most effective carotenoid.[32–34] In fact, the chemopreventive potential of vitamin A has been attributed to β-carotene. Therefore, for simplification, in this review, the roles of vitamin A and β-carotene as chemopreventers are discussed together, although they act as distinctly different entities.

Epidemiologic studies have suggested that people who consume more fruits and vegetables containing β-carotene have lower risks of cancer and cardiovascular disease[6,33–36] (Table 3). One mechanism for the beneficial effect may be β-carotene action as an antioxidant and scavenger of free radicals.[32] Together with vitamin C, it may prevent carcinogenesis and atherogenesis by interfering passively with oxidative damage to DNA and lipoproteins.[27,37] Additional support for this hypothesis is the finding that people who develop tumors have lower blood levels of β-carotene, vitamins A and C, and selenium than controls.[38]

Table 3. Possible health effects of antioxidant nutrients and foods

	Cancer sites						
	Lung		Gastrointestinal				
Antioxidant nutrient	Non-smokers	Smokers	Upper	Lower (colon)	Cervical	Breast	CVD
β-carotene	+	–	+?	0	+?	+?	+?
Vitamin C	+?	–?	++	0	++	+?	+?
Vitamin E	0	?	?	+?		?	++
Vegetables & fruit	++	++	++	++	+	+?	++

++ strong evidence for a protective effect
+ may have a protective effect
0 does not seem to confer a protective effect
– may have a negative effect
? evidence conflicting or not complete
Used with permission from: National Institute of Nutrition, NIN REVIEW, No. 25, Winter, 1996. (Ref. 23)

The Iowa Women's Health Study investigated the relationship between intakes of antioxidant vitamins C and E and retinol with the risk of upper digestive cancers in a Western population.[39] More than 34,000 women were followed for seven years, during which time 26 developed stomach cancer and 33 cancers of the mouth, pharynges and esophagus. Although a non-significant inverse association was observed for cancers of the mouth, pharynges and esophagus for vitamin C, vitamin E and carotene, a significant inverse association was found for intake of carotene and stomach cancer.[39]

The beneficial effect of β-carotene on the cardiovascular system also was supported by several studies. In one study of 1,300 elderly subjects, β-carotene intake decreased cardiovascular mortality during 4.75 years of follow-up.[40] In another study of almost 40,000 men, the effects of dietary antioxidants vitamins A, C and E, and β-carotene in relation to the incidence of coronary disease were investigated. Relative risk was reduced in the group taking β-carotene and vitamin E supplements. The intake of vitamins A and C, however, was not significantly associated with the risk of coronary disease. Risk was strongest for current smokers compared to past smokers and those who never smoked.[41]

Three large, randomized, placebo-controlled trials in well-nourished populations, however, using β-carotene and/or α-tocopherol have recently reported interesting and contradictory results. 29,133 Finnish male smokers received β-carotene, vitamin E, both agents, or neither for an average of six years.[42] In a second study conducted in the USA, 18,000 men and women at high risk for lung cancer, due to their smoking habits or occupational exposure to hazards, received treatment with β-carotene and retinol for an average of less than four years.[43] A third study was conducted in the USA with 22,000 male physicians. They received β-carotene supplementation or placebo for 12 years.[44] All three studies found no benefits of antioxidant supplementation in terms of incidence of cancer or cardiovascular disease. Unexpectedly, the first two studies found somewhat higher rates of lung cancer and cardiovascular disease among subjects given β-carotene. The estimated excess risks were small, and it remains unexplained whether β-carotene indeed is harmful, and under what conditions. In any case, clinical trials with β-carotene have been discontinued, and supplementation by β-carotene in smokers is not recommended.

In a hospital-based, case-control study, no differences were found in dietary estimates of vitamin A intake or in plasma α-carotene and lycopene between 83 patients with breast cancer and 113 control subjects. However, subjects with breast cancer had lower plasma β-carotene concentrations than controls, suggesting an association.[45] In contrast, no clear association between dietary intake and serum levels of retinol and β-carotene in breast cancer patients was reported in two other case-control studies.[46,47]

In a study to investigate the relationship between vegetable consumption and lung cancer risk, Le Marchand et al. found that all vegetables showed a stronger inverse relationship with lung cancer risk than did β-carotene, suggesting that other constituents in vegetables, for example lutein, lycopene, and indoles, may have anticancer activity[48] (Table 3).

Some reports have indicated that retinol and/or carotenoids (especially β-carotene) increase the risk of prostate cancer.[49] However, others have shown that the intake of β-carotene containing foods has protective effects.[50] Here again is the possibility that the beneficial/adverse effect is not due to β-carotene but to the presence or absence of other constituents.

In the USA, more than 47,000 eligible subjects over a 6 year period were followed to explore the relationship between prostate cancer and consumption of vegetables, fruits, vitamin A, and various related carotenoids. Unexpectedly, out of 46 types of vegetables and fruits, only tomato sauce, tomatoes, and pizza from vegetable sources were significantly associated with lowering prostate cancer risk. Men who consumed a total of 10 servings/week of tomatoes, tomato sauce, and pizza had a lower risk of prostate cancer than men eating only 1.5 servings/week.[34] A number of carotenoids, (α- and β-carotene and lutein) were not associated with this reduced risk. However, lycopene, a carotenoid occurring in ripe tomatoes, was. This compound was concentrated in plasma and in the prostate gland. Lycopene cannot be converted to vitamin A, but is the most efficient scavenger of oxygen radicals among the common carotenoids.[34] Presently it is premature to state that carotenoids either increase or decrease the risk of prostate cancer.

Consumption of diets characterized by low amounts of fat and red meat and high amounts of vegetables, fruits and grains, substantially reduce colorectal cancer risk.[51] In a recent clinical trial with more than 400 patients, MacLennan et al. assessed the effects of the following treatment protocols on the incidence of adenomas: reducing dietary fat; supplementing the diet by wheat bran; fortifying the diet with β-carotene; or combining different combinations of these treatments.[52] There was no statistically significant prevention of new adenomas with any of the protocols.[52] It appears that β-carotene does not provide any substantial protective effect in colorectal cancer.[53]

Despite conflicting results, it appears that cancer patients exhibit a significant trend of lower serum levels of carotene, vitamin A and vitamin B with increasing stages of the disease.[54]

Vitamin C

Epidemiological and experimental evidence suggests that vitamin C has a protective effect against several cancers. These effects are attributed to the chemical properties of vitamin C and not to its biological function as a vitamin.[55] The main mechanism for the anticarcinogenic effect is believed due to its antioxidant properties. It acts effectively as a scavenger of nitrite in the stomach, thereby preventing formation of carcinogenic nitrosamine species.[56,57] Additional benefits are attributed to its enhancement of immune system function, participation in synthesis of hormones, and enhancement of the hepatic clearance of toxins via the cytochrome P-450 enzymatic system.[58,59] These beneficial functions may be viewed as individual steps, or as contributing factors in prevention of cancer, coronary artery disease and the aging process.[7] In experimental animals, chemically induced cancers at many sites have been inhibited by vitamin C, although in certain organs (e.g. urinary bladder) carcinogenicity was enhanced. Therefore, the role of ascorbic acid in chemically induced cancers is still unclear.[58,59]

The large body of literature on the role of ascorbic acid as a modulator of tumor development has not been precisely summarized. Cohen and Bhagavan critically reviewed the literature and found the "evidence for ascorbic acid as an inhibitor of carcinogenesis is stronger with regard to gastric cancer and weaker with regard to esophageal and colon/rectal cancer".[60] Block and co-workers reviewed approximately 200 studies that examined the relationship between fruit and vegetable intake and cancers of lung, breast, cervix, esophagus, oral cavity, stomach, bladder, pancreas and ovary.[21] A statistically significant protective effect of fruit and vegetable consumption was found in 128 of 156 dietary studies in which results were expressed in terms of relative risk.[21] Epidemiological data from several studies indicate that vitamin C levels are inversely associated with cholesterol levels and directly related to high density lipoprotein (HDL) levels. Evidence also suggests that vitamin C may improve cardiovascular health and reduce the risk of cardiovascular mortality.[29]

For the Canadian population (and similarly for the U.S. population), the recommended intake of vitamin C is 40–60 mg/day. Smokers are suggested to increase the intake by 50%, and additional intake also is recommended during pregnancy.[61] However, because a number of recent studies have shown that vitamin C as an oxidant may play a role in chemoprevention of cancer and may reduce cardiovascular mortality, an increase in the recommended doses should be considered.

Vitamin E

Of all dietary antioxidants, vitamin E appears to have special significance primarily because of its solubility in lipids. It is the major antioxidant soluble in the lipid phase of cells and protects fatty acids against peroxidation.[62] Since vitamin E reduces free radical formation, a decrease in carcinogenesis and other degenerative diseases may result.[63] Vitamin E inhibits platelet aggregation and thereby reduces the risk of coronary heart disease.[64] Supplementation with vitamin E reduces the risk of oral, lung, pharyngeal and prostate cancers.[25,65,66] A 34% reduction in prostate cancer was seen in a recent study conducted in Finland in men taking α-tocopherol supplements.[25] Although several epidemiological studies reported a beneficial effect of vitamin E against breast and some gastrointestinal cancers,[67,68] others indicate that supplementation for cancer prevention must await additional clinical studies.[29]

Two large studies in men and women indicated that large doses of vitamin E were associated with significant reduction (up to 40%) in coronary heart disease.[66,69,70] Two trials, including 34,000 postmenopausal women, found that subjects taking vitamin E supplements had lower mortality from coronary heart disease and reduced incidence of non-fatal myocardial infarction.[71,72] To clarify the uncertainty of the optimal level of vitamin E needed to decrease the susceptibility of low density lipoprotein (LDL) to oxidation, Jialal *et al.*[73] conducted an 8-week study with a limited number of human volunteers and found that the optimum dose was 400 IU/day.

An important study with a combination of dietary supplements (retinol, zinc, riboflavin, niacin, vitamin C, molybdenum, β-carotene, vitamin E and selenium) was performed with 29,584 residents of Linxian county in northern China.[74] After 5.25 years, a combination of antioxidants (β-carotene, vitamin E and selenium) reduced total mortality (9%), mainly due to a lower incidence (21%) of stomach cancer. Extrapolation of these results to other populations, however, must be done with caution. The diet in Linxian is limited in some essential phytochemicals, especially antioxidants. Therefore, what was intended as supplementation may have been, in fact, correction of deficiencies.[74]

Selenium

Considerable controversy has surrounded the role of selenium in carcinogenesis.[6] In animal models, higher doses were associated with the development of neoplastic lesions in the liver. However, in other studies, selenium proved to be an inhibitor of liver carcinogenesis.[75] The chemopreventive action is attributed to selenium antioxidant activity as a cofactor for glutathione-S-transferase (GST), an enzyme that protects against oxidative tissue damage.[76] In numerous epidemiological studies, a low level of dietary selenium, especially from forage content, is associated with increased cancer risk at many sites (lung, breast, bladder, esophagus, rectum).[77] Among the trace elements, the inhibitory effect of selenium on experimentally induced cancers is well documented.[78,79] However, some studies could not confirm an association between serum selenium levels and risk of cancer.[80,81]

Phytic Acid

Phytic acid is abundant (up to 5% by weight) in fruits, edible legumes, cereals, oil seeds, and nuts. In the past, this compound has also been described as an "antinutrient", owing to its ability to decrease the bioavailability of minerals, especially calcium and zinc.[82] Phytic acid, however, was found to decrease the incidence of colon cancer in laboratory animals,[83,84] and to be an effective inhibitor of human mammary cancer cell growth *in vitro*.[85] Additional data have shown a strong association between intake of phytic acid and reduction of human colon cancer.[86] A strong antioxidant activity of phytic acid on lipid peroxidation and degradation of ascorbic acid has been reported.[87,88] It appears that beneficial effects of phytic acid may outweigh potential risks.

POLYPHENOLS/FLAVONOIDS

Phenolic compounds, especially flavonoids, occur widely in plants. At least 5000 phenolics, including over 2000 different flavonoids have been identified.[13,89] Many properties are associated with these polyphenolic compounds. They contribute to the pigmentation of plant foods and defend injured plants against pathogens.[89] The beneficial health-related effects of many foods are associated with plant polyphenols.[90,91] They may correlate with the oxidative stability of food products. In addition, many possess important biological activity related to their inhibitory effects on mutagenesis and carcinogenesis.[91,92] Almost all possess several common properties such as: antioxidant activity; ability to scavenge active oxygen species and free radicals; inhibition of nitrosation; reduction of the bioavailability of carcinogens; chelation of metals and modulation of certain cellular enzyme activities.[89–92]

In addition to their major activity within cells, polyphenols may play a role as chemopreventers during the preparation of foods, e.g. as antioxidants to prevent the rancidity of lipids before or during digestion processes. They accelerate intestinal transit time, protect intestinal flora, can increase the uptake of certain beneficial components from the diet, and may reduce the bioavailability of certain food mutagens and carcinogens.[93,94] We investigated the effect of three phenolics-quercetin, chlorogenic acid, and ellagic acid (Fig. 1) on the bioavailability of benzo-*a*-pyrene (B(a)P) and 2-amino-3-methylimidazo(4,5-*f*)quinoline (IQ), using rats and mice in three different experimental protocols.[92,95] A brief summary is presented in Table 4. In all experiments, phenolics significantly reduced the uptake of B(a)P. This indicates that some phytochemicals can reduce the absorption and bioavailability of certain toxic components in the diet.

A number of flavonoids, in addition to their antitumor activities, possess anti-inflammatory and antiviral activities and increase capillary permeability.[96] Food polyphenols also have the ability to form insoluble complexes with proteins, essential minerals (e.g. iron), and carbohydrates, and therefore may interfere with the normal utilization of dietary components (for reviews see Refs. 9 and 89).

Among the polyphenols with biological effects on human health, flavonoids are the most abundant and important. However, several have been found to be mutagens *in vitro*. Of particular interest are quercetin and kaempferol, widespread in vegetables and fruits and frequently consumed (up to 1 g/person/day).[13] Quercetin (Fig. 1) has been described as a potent mutagen in *in vitro* experiments, but also as a probable anticarcinogen *in vivo*.[90,93,97,98] Several

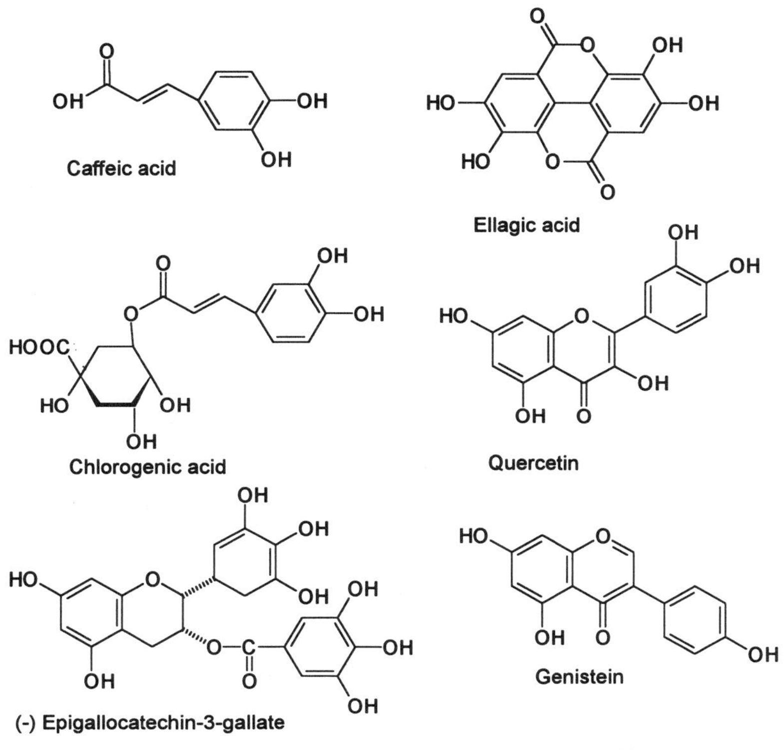

Figure 1. Selected food polyphenols with biological activities.

studies indicate that flavonoids in commonly consumed foods reduce the incidence of myocardial infarction and risk of death from coronary heart disease in elderly men.[99]

Many polyphenols occur as hydroxycinnamic acids, such as gallic, ellagic, chlorogenic, caffeic and ferulic acids. Chlorogenic acid is abundant in coffee beans (4% by weight), and occurs at higher concentration in instant coffee.[89] Gallates are abundant in tea.[100] These acids are also present in high amounts in prunes, blueberries, apples, pears and grapes.[90] They inhibit various types of tumors in several animal models.[7,101,102]

Caffeic, Ellagic, and Chlorogenic Acids (Fig. 1)

Caffeic acid is a naturally occurring compound found in a large number of plants and is ubiquitous in the food supply. In unprocessed food plants, caffeic

Table 4. The effect of phenolics (quercetin, chlorogenic acid and ellagic acid) on the bioavailability of benzo(a)pyrene, using rats and mice in 3 different experimental protocols

		Treatments				
Experimental protocols	Measurements	B(a)P only (control)	B(a)P + Q	B(a)P + CA	B(a)P + EA	Ref.
Bile cannulated *rats* were used to measure the percentage of radio-activity excreted in the bile	recovered radioactivity in the bile of rats.relative ratios	1.0	0.79	0.84	0.41	91
Measurements of the uptake of ^{14}c-b(a)p in *mice* in the presence of polyphenols	percentage of cpm/mg of liver tissue (mice)	100%	83%	71%	74%	90
The intrasanguineous host mediated assay with *mice* to evaluate *in vivo* mutagenicity of the test chemicals in presence of flavonoids	response ratio of the *in vivo* mutagenic response (liver, mice)	5.2	1.9	1.8	1.8	90

Q = quercetin; CA = chlorogenic acid; EA = ellagic acid
For chemical formulas see Fig. 1.

acid exists primarily as conjugates (esters and glucosides, e.g. chlorogenic acid).[6] Daily intake of caffeic acid from coffee in those who consume moderately high amount of coffees is estimated to be up to 9 mg/kg.[6] Chlorogenic acid also is a major constituent of coffee, and caffeic acid is readily formed from chlorogenic acid by hydrolysis. It is possible that these compounds contribute to the reported protective effects of coffee.[90,103]

Ellagic acid is a dimer of two molecules of gallic acid, while chlorogenic acid is a conjugate of quinic and caffeic acid. These are found in many foods, including apples, apricots, blueberries, blackberries, gooseberries, peaches, pears, raspberries, carrots and other vegetables.[89] For example, apple juice from MacIntosh apples contains a relatively high concentration (214 mg/L) of chlorogenic acid.[104] A cup of coffee may contain up to 325 mg of chlorogenic acid, with an average value (for USA consumers) of 190 mg/cup.[102]

In rats and hamsters, Tanaka and co-workers found that both, ellagic and chlorogenic acids had protective effects against chemically induced tumors of the liver, colon, and tongue.[105] Ellagic acid also inhibited lung tumorigenesis in mice,[106] and was antimutagenic to the food mutagen/carcinogen 2-amino-3-methylimidazo(4,5-*f*)quinoline (IQ), which was produced during heat treatment of meat.[107]

SELECTED FOODS

Beverages

Tea. Tea (*Camellia sinensis*) is one of the most popular beverages in the world.[100,102,108] A number of health benefits have been associated with drinking tea including prevention of atherosclerosis and reduction of serum cholesterol,[109–112] inhibition of platelet aggregation,[113] inhibition of nitrosamine formation,[114] and scavenging of lipid free radicals.[115] Lipid free radicals may be more harmful than oxygen free radicals in biological systems.[115] The antimutagenic activity of green tea polyphenols against polycyclic aromatic hydrocarbons, aflatoxin B_1 and some aromatic amines has also been reported.[116,117] Recent literature indicates that tea protects against tumorigenesis of several known carcinogens,[8,118–120] and inhibits mutagenesis due to heterocyclic aromatic amines (HAA).[121–123] HAAs are confirmed animal carcinogens which are formed during heat processing (cooking) of meat. Their role in human carcinogenicity, however, is not determined.[124,125]

A case-control population study in Shanghai, China suggested that drinking green tea protects against esophageal cancer.[126] A recent Japanese case-controlled study reported that consumption of tea provides a protective effect against lung cancer, especially in women. Daily consumption of partially fermented tea decreased the risk of squamous cell carcinoma by 50% for males and 92% for females.[127] The Chinese Academy of Medical Sciences' researchers, using both *in vitro* and *in vivo* tests, have confirmed that green tea epicatechin compounds inhibit some cigarette-smoke-induced harmful effects.[128]

With respect to heart disease, a Dutch epidemiological study found that dietary flavonoid intake, the majority of which came from tea, significantly reduced deaths from heart disease.[99] Several groups have investigated the mechanisms for these effects. Using a lipoprotein oxidation model that simulates the oxidation mechanism of atherogenesis, Vinson et al. found that flavonoids, especially tea flavonoids, are beneficial in the prevention of atherosclerosis and heart disease.[129]

Meanwhile, a recent epidemiological investigation (by prospective cohort study) of 58,279 men and 62,573 women in the Netherlands did not support the hypothesis that *black* tea protects against four major cancers (stomach, colorectal, lung and breast).[130] Although the study found that consumption of tea was inversely associated with stomach and lung cancers, this lower cancer risk was attributed to the fact that the subjects smoked less and ate more fruits and vegetables.[130] This study is another example illustrating how difficult it is to assess health benefits of food components and how controversial epidemiological studies often are. When the results were calculated taking into consideration consumption of tea and lung cancer, the risk in the highest category of tea consumption was about half that in those who did not drink tea. However, when the results of the same study were divided into three subgroups, current smokers, ex-smokers, and those who never smoked within each of these groups, the inverse association with tea consumption almost disappeared.[131]

Epigallocatechin gallate (EGCG) (Fig. 1) has been reported to be the principal component of green tea with anticarcinogenic or chemopreventive potential.[117,132] The anticarcinogenic activity may be due to its effect on tumor promotion,[133] DNA-adduct formation,[249] the scavenging activity of free radicals,[117] or the increase in antioxidant activities.[129,134] Black tea (consumed in N.America and Europe) is processed from *green* tea (consumed in Asiatic countries) *via* a fermentation process. This results in considerable changes in the appearance, aroma and chemistry of the tea. The fermentation significantly reduces the amount of EGCG in black tea.[108]

Disruption in communication between cells is considered an important step of tumor development. It appears that tea polyphenols, like catechin gallates, may protect cells by enhancing gap junction communication.[135] Additional mechanisms that could account for the protective effect of tea extracts could involve a direct interaction with the reactive genotoxic species of various promutagens,[136] inhibition of the cytochrome P-450-dependent bioactivation of promutagens,[116] and antioxidative and free radical scavenging.[137–139]

In our laboratory, the bacterial mutagenesis test was used to study the ability of water extracts of different teas to modify mutagenic activity of eight heterocyclic aromatic amines (HAAs). No difference in antimutagenic activity of non-fermented (green), semifermented (oolong), fermented (black) or decaffeinated teas were found among HAAs. Complete or substantial inhibition of the mutagenic activity with all tested HAAs was obtained with tea extracts equivalent to 50 mg of tea leaves/plate.[123] Therefore, there may be other beneficial components besides EGCG, which can reduce the mutagenicity of HAAs, and perhaps other mutagens.

Condensed tannins from tea may play a role in the prevention and treatment of clinical disorders attributed to the reaction of free radicals.[140] It has been reported that tea has a radioprotective capacity, for example in the treatment of sequelae (a morbid condition left as the result of a disease) of radiation therapy.[140,141]

Coffee. Research with coffee has been promoted primarily by concern for the health effects of caffeine, as well as suspicion of tumor induction.[102] However, in a number of epidemiological studies, coffee consumption was not correlated to an increase in cancer risk.[142,143] Nevertheless, in *in vitro* tests, several studies found that different types of mutagens are present in coffee.[102,144–147] Despite this, in mice fed coffee for 2 years, the overall tumor incidence was inversely correlated to coffee intake.[148] Coffee drinking also has been associated with reduced incidence of colon cancer in humans.[102,103,149]

The inhibitory effect of coffee extracts against some food mutagens has been reported by Yamaguchi and Iki.[150] We also have observed that certain fractions obtained from coffee brews inhibit the mutagenicity of a number of HAAs in the Ames *Salmonella* (TA98+S9) assay (B. Stavric et al., unpublished observation). A combination of coffee components, such as diterpene esters (cafestol and kahweol), may contribute to the observed anticarcinogenic potential.[151,152]

Wine. The phenomenon described as the "French Paradox"[153] has been associated with the apparent incompatibility of a high fat diet and regular consumption of red wine with a low incidence of coronary atherosclerosis.[154] There is evidence that the beneficial effects of red wine are at least in part contributed to by a combination of components other than alcohol.[155] Measuring the quality of the total diet might provide a new approach to the study of the French paradox.[156]

Wine, especially red wine, contains several polyphenols, such as quercetin, rutin and resveratrol, all of which have been found to possess antioxidant activity using an *in vitro* oxidation model for heart disease.[157] Such experiments have suggested that the antioxidants can bind to LDL and very low density lipoproteins (VLDL), and, therefore, are beneficial in preventing their oxidation. Consequently, there is a possibility that red wine can reduce atherosclerosis and morbidity and mortality from coronary artery disease.[154,158]

Vegetables and Grains

Garlic and Onion (Allium Vegetables). A significant reduction in gastric cancer risk in humans has been correlated with increasing consumption of *Allium* vegetables, namely garlic, onions and similar foods.[159,160] The beneficial effect of garlic on blood pressure and cholesterol was evaluated in recent meta-analyses (*meta-analysis* describes "the statistical analysis of a large collection of analyses results from individual studies for the purpose of integrating the findings").[161] It was found that garlic reduced the risk of cardiovascular disease.[162,163] In studies with rats, raw, instead of cooked, garlic and onion were found to be antithrombotic agents.[164]

The beneficial component with anticarcinogenic potential is thought to be diallyl sulfide.[165] Garlic contains other sulphur components (for reviews see Refs. 166,167). Enhancement of immune system function, scavenging of free radicals, direct cytotoxic effects on cancer cells and/or the effect on detoxification systems have all been considered as possible mechanisms for the chemopreventive activity of garlic.[168] Animal studies have been suggested as a model for use in developing chemopreventive agents.[169,170] However, in one study with rats, diallyl disulfide promoted hepatocarcinogenesis.[171] Foods contain a variety of components, and slight modification of any chemical structure may produce different effects in different organs. A recent editorial questioned the effectiveness of garlic in reducing cancer risk and blood cholesterol levels.[172]

In addition to organosulfur compounds, garlic (*Allium sativum*) and varieties of onion (*Allium cepa*), especially those with colored skins, have an exceptionally high flavonol content, (mainly quercetin in the form of aglycone).[173] These polyphenols may also contribute to the chemopreventive potential of these foods.

Cruciferous Vegetables. Consumption of broccoli, cabbage, or Brussels sprouts has been associated with a reduction in cancer incidence.[2,6] These vegetables contain a variety of chemicals (polyphenols, fiber, vitamin C, folic acid, carotenoids, dietary calcium and chlorophyllin), compounds known to have chemoprotective effects. Cruciferous vegetables, however, also contain large amounts of isothiocyanates and thiocyanates which may enhance or complement the beneficial effects. A number of isothiocyanates and thiocyanates possess antimutagenic activity which inhibit lung, liver and colon carcinogenesis in rats.[174,175] These agents seem to stimulate the production of protective enzymes like GST that detoxify carcinogens.[176]

Grains. In addition to their fiber, flavonoids, and phytic acid, grains contain a variety of phenolic compounds (caffeic, ferrulic, gallic and ellagic acids), which, as antioxidants, inhibit formation of carcinogenic metabolites.[177] Lignans in grains have been shown to possess a broad spectrum of biological activity, including antitumor, antimitotic and antiviral activity.[178–180] A combination of chemicals in grains, fruits and vegetables, may explain why vegetarians have reduced cancer risk.[6]

Soybeans/Phytoestrogens. The large scale consumption of soybean products in the Pacific basin has been correlated with the lower incidence of breast and prostate cancer observed there.[181–188] A large number of phenolic acids and other phenolic compounds are found in soybean products.[87] Of special interest are the isoflavones, daidzein and genistein (Fig.1). The content of isoflavones depends on cultivar, location and year. The concentration of isoflavone aglucones in food products depends on soaking conditions applied during preparation.[89]

In addition to the phytoestrogens, genistein and daidzein, soy products contain other components with anticarcinogenic potential, some of which are produced during fermentation. Fermented soy sauce is a form in which soybeans are frequently consumed in Japan.[189–191] Fermented soy products (*miso* and *natto*) contain higher amounts of genistein and its β-glucoside conjugate (genistin) than unfermented products such as soybean.[187] Genistein's antioxidant properties and antiproliferative effects (inhibits reproduction or multiplication of cells and morbid cysts) combined with its potential for inhibiting angiogenesis (formation of new blood capillaries) may be responsible for its anticarcinogenic effect.[192,193] The antiestrogenic activity suggests additional anticarcinogenic potential of a soy-based diet.[194] Genistein, when administered neonatally, manifested a protective effect against chemically induced mammary cancer in rats.[195] The high content of genistein in soybeans and its relatively high bioavailability favors soy products as promising candidates for foods with anticarcinogenic potential.

In laboratory animals, consumption of soy protein decreases serum cholesterol concentrations, but studies with humans are inconclusive.[196] Anderson *et al.* recently conducted a meta-analysis of 38 clinical studies in which dietary soy proteins were correlated with serum cholesterol and triglycerides.[197] They found that consumption of 17 to 25 g of soy protein/day significantly reduced LDL without lowering the levels of HDL. Serum triglyceride concentrations were also reduced. They concluded that "soy estrogens may be responsible for most of the hypocholesterolemic effects of soy protein".[197]

Flax seed and chaparral are also rich in phytoestrogens.[179,198] They are currently being marketed as dietary supplements with anticarcinogenic potential.[199] Recent research indicates that phytoestrogens produce a variety of adverse physiological effects in a number of experimental animals, including infertility in livestock.[188] Although evidence for beneficial effects is mounting, the potential for toxicity from high levels has not yet been adequately investigated.[198]

Protease Inhibitors and Saponins. A number of foods contain other bioactive components often characterized as antinutritional factors.[200] However, some, like protease inhibitors and saponins, also possess beneficial effects on human health. Many common foods (e.g. soy, kidney beans, grains, wheat, oats, rye) contain protease inhibitors. These enzyme inhibitors may impair protein utilization, and therefore can be considered as naturally ocuring food toxicants.[201] However, they may also inhibit tumor formation.[202] Protease inhibitors have been reported to be most effective as potential human cancer chemopreventive agents supressing radiation induced malignant transformation *in vitro*.[203]

Saponins are found in many vegetables and fruits (e.g. soy, alfalfa, spinach, broccoli, potatoes, apples).[201] They are capable of disrupting red blood cells and producing diarrhea and vomiting.[204] On the positive side, they may complex with serum cholesterol and bile acids,[184,205] thereby reducing blood serum cholesterol levels. *In vitro* and animal studies show that saponins reduce cell proliferation in the gut and decrease growth-rate and rate of DNA synthesis in some tumor cells.[184]

Palm and Fish Oils. Unprocessed palm oil has an antitumorigenic effect, apparently due to tocotrienols concentrated in the non-saponifiable matter.[206] This chemoprotective effect may be due to a combination of several components, e.g. vitamin E and other polyphenols.

Short-term trials in humans have shown that omega-3 fatty acids found in fish oil significantly reduce levels of plasma triglycerides and may increase levels of HDLs. No consistent effect on serum cholesterol levels, however, has been shown.[207] Although the omega-3 fatty acids appear to reduce risk of cardiovascular disease, their effects in reducing cancer risk are controversial.[208] However, the beneficial effect of the Mediterranean margarine in reducing mortality from cardiovascular disease has been attributed to its α-linolenic acid (α-LNA) content.[209] Eggs enriched in omega-3 as (α-LNA) when hens are fed flax,[210] are already available in Canadian supermarkets.

MISCELLANEOUS OTHER CHEMOPREVENTERS

Calcium and Vitamin D

Garland and Garland in 1980 postulated that sunlight reduces the likelihood of colon cancer.[211] They observed that people in northern latitudes had higher colon cancer rates than those living in areas exposed to more sun. Such an association might be attributed to the impact of sunlight on the formation of vitamin D, which subsequently increases the absorption of dietary calcium. The proposed mechanism for the chemoprotective effect is that ionized calcium inhibits the negative effects of fatty acids and bile acids by binding with them in the intestine to form insoluble calcium soaps.[212] Both types of acids appear in the intestine as a result of fatty diet intake. In a number of animal experiments, supplemental calcium and vitamin D inhibited chemically induced fat-promoted colon carcinogenesis.[213] Using a special *in vivo* bioassay (induction of foci of aberrant crypts, which indicate deviation from normality) in the colons of rats treated with known carcinogens, it was observed that while fat increased the number of aberrant crypts, a higher calcium intake, even with higher intake of fat, resulted in a lower number.[214] Experiments with human subjects at high risk for familial colon cancer demonstrated that the hyperproliferation of colon epithelial cells was reduced by supplementation of the regular diet with calcium.[215]

Folic Acid

Folic acid (or folacin) is an essential B-complex vitamin and together with vitamin B_{12} plays a role in promoting immune system function.[216] Folic acid is required for the regulation of normal gene expression and has been known for some time to have a chemoprotective role in prevention of birth defects like spina bifida, anencephaly, or cleft palate. Although a number of investigators have suggested a chemoprotective action against colon, rectum, and cervical cancers, others have not seen such a relationship.[6,217–219] Since folate deficiency has been associated with increase in chromosomal breaks, it may play a role in cancer prevention.[220] Recent research shows it can prevent the development of colorectal polyps in men. A folate effect may depend on sex-specific interactions with other nutritional or physiological factors.[221]

Diets rich in vegetables and whole grains provide enough daily folic acid. It is recommended that all women in their childbearing years consume 400 mg daily. Spina bifida and similar birth defects occur in the first two weeks of pregnancy.[222] There is a controversy regarding the fortification of cereals with folic acid.[223] Although it is non-toxic even in high doses,[208] high levels can "mask" vitamin B_{12} deficiency in patients with neuropathy and other neurological disorders.[224] Recently, the results of a fifteen-year follow up study on the effects of folic acid on coronary heart disease (CHD) mortality rate were reported. Low serum folate levels were associated with an increased risk of fatal CHD.[225]

Conjugated Linoleic Acid

Conjugated Linoleic Acid (CLA) is the only fatty acid with proven anti-carcinogenic effects in experimental animals. Various isomers of CLA, mainly present in bovine milk and meat from ruminant animals, were found to inhibit chemically induced carcinogenesis.[226,227] The mechanism appears to be *via* signal transduction pathways and effects on prostaglandin metabolism.[6]

Vanillin

Vanillin is used as a flavoring agent in many food products and possesses antimutagenic activity.[228] Vanillin has shown variable responses in combination with other food and environmental mutagens in mutagenesis tests, either enhancing or suppressing mutagenesis, depending upon the assay used.[19] Ohta and co-workers tested the effect of vanillin on responses caused by a number of different mutagens/carcinogens (nitroso derivatives, alkylating agents, HAAs). These tests produced a variety of responses: the mutagenicity of 17 mutagens was suppressed; 5 examples showed no effect; and in 9 assays mutagenicity was enhanced.[228] Therefore, despite the claim that vanillin is "an antimutagenic compound",[229] additional work is needed to explain the wide spectrum of effects. The mechanism of action of vanillin has not been explained, and its effectiveness in suppressing carcinogenicity in animals and man has not been adequately studied.[230]

Other Foods

Chemopreventive agents are found in all categories of foods (Table 1). The type and amount can vary considerably, not only by product, but also by subvarieties of product and/or the area of cultivation. This variability makes it easier for humans to obtain required mixtures of chemopreventive agents from different food sources, but at the same time makes it difficult for investigators to compare studies from different geographic regions.

Apart from the foods and agents described in this review (Table 1), there are a number of others which provide protective effects. The following is a partial list: probiotics (yogurt with living bifidobacteria); herbs and spices; chlorophylline; *d*-limonene in citrus peel oil; plant sterols; indoles and glucosinolates in cruciferous vegetables; glycyrrhetinic acid in licorice; turmeric and curcumin.[7,9,29,231]

INDUCTION OF DETOXIFYING ENZYMES

The deleterious effects of free radicals are normally controlled by a number of enzymatic systems built into natural protective cellular mechanisms. Some enzymes, like superoxide dismutase or catalase, are involved in catalyzing the reduction of oxygen containing free radicals, thereby preventing their accumulation within cells.[29] Others are involved in the metabolism of xenobiotics or toxic compounds mainly in the liver. Many dietary components influence enzymatic systems involved in the detoxification of food xenobiotics.

GST is involved in the detoxification of dietary mutagens/carcinogens.[29,232] Phenolic antioxidants and other food components activate GST, which in turn regulates synthesis of chemoprotective enzymes.[233] An increase in GST activity usually indicates an increase in detoxification of xenobiotics or carcinogens.[234] A number of food chemicals, especially components from cruciferous vegetables, stimulate the production of protective enzymes, such as GST. Chemopreventive effects of other food ingredients, such as turmeric, curcumin, *d*-limonene, dithiolthiones, sulphoraphane and others, are reported to be attributable to glutathione (GSH) (for review see Ref. 167).

INTERPRETING RESULTS FROM EPIDEMIOLOGICAL AND OTHER STUDIES – CONFOUNDING AND OTHER FACTORS

From epidemiological and observational studies, it is difficult to establish or evaluate the effects (beneficial or harmful) of a single compound in the diet. Large variations in types of foods, dietary habits, and life style make it difficult to pinpoint the particular component responsible. A myriad of substances could be responsible for reducing cancer risk. Assessing the quality of the total diet (e.g. "Diet Quality Index"), a relatively new approach in nutritional epidemiology, focuses more precisely on the correlation between diet and health.[235] In some studies, a low-score characterized by omission of food groups is associated with increased mortality from cancer and cardiovascular disease.[236] The complex and multidimensional nature of the human diet poses a challenge to investigators, especially when it comes to comparison between countries.[156]

Many components beside antioxidants contribute to the beneficial effects of fruits and vegetables, e.g. polyphenols and fibers, such as lycopene, lutein, indoles, and spices.[6,48] If the amount of a chemical in the diet is known, its bioavailability, which may vary under different circumstances, also needs to be known. Only long-term, large, randomized, well controlled trials with a single chemical, and which also control and record the quality of the total diet will provide answers to uncertainties.[237]

In studies using experimental animals, co-administration of test compounds with other chemicals or regular components of the diet introduces confounders. The same treatment also may affect different species or different organs in varying ways. In a limited number of cases, there are reports that naturally occurring compounds can inhibit tumors in one organ and, under different experimental conditions, promote carcinogenesis in others.

Plant polyphenols may be used as examples of the complexity of evaluating the real health benefits of experimental observations.[6,238,239] The case of quercetin, a potent mutagen *in vitro*, but with anticarcinogenic activity *in vivo*, was mentioned earlier.[98] Caffeic acid, depending on the experimental conditions, exhibits both carcinogenic and anticarcinogenic activity.[6] In addition to its reported antitumorigenic potential, in rats, particularly in combination with other components (e.g. sodium nitrite), caffeic acid induces forestomach carcinogenicity.[238] It is of interest to note that the International Agency for Research on Cancer, World Health Organization (IARC) concludes that caffeic acid is possibly carcinogenic to humans.[240]

Using a wide-spectrum organ carcinogenesis rat model, Hirose *et al.*[239] found that phytic acid was marginally effective in inhibiting hepatic and pancreatic carcinogenesis. However, it also increased the incidence of urinary bladder papillomas. The authors stressed that an ideal chemopreventer should not demonstrate adverse effects in any organs. Indol-3-carbinol (I3C) is still another example of duality in biological activity. I3C is found as a glucosinolate in cruciferous plants and has been shown to inhibit chemically induced neoplasms.[241,242] However, in some studies, I3C enhanced liver tumorigenesis in rainbow trout and colon carcinogenesis in rats.[7]

SUPPLEMENTS

The question of whether or not one should supplement regular diets with phytochemicals is still unresolved. Although a number of epidemiological studies indicate that dietary supplements are successful in reducing cardiovascular disease[243] and some forms of cancer,[39,244,245] many believe it is premature to make recommendations.[6,40,246,247] The unexpected increase in lung cancer observed in two clinical trials[43,44] for subjects taking supplements of β-carotene[15] indicate that supplementation on a large scale is not supported by scientific evidence, and therefore not recommended.[248] An ideal chemopreventer should not have any adverse effect. Although successes with supplements have been reported in human intervention studies, not all have been conducted as double blinds and results are difficult to compare.[249] Nevertheless, encouraging results have been obtained with individuals who are genetically predisposed to cancer or chronic disorders.[250,251]

In recent years, the effects of vitamins (particularly those with antioxidant activity) on human health and diseases have become the subject of numerous articles, monographs, reviews and books.[5,15–20,26,27,67,252–255] Special attention is given to interpretation of the roles of vitamins and to the adequacy or benefits of supplementation. This issue must be debated not only by researchers and medical personnel, but also among nutritionists, regulatory agencies, nurses, toxicologists, epidemiologists and companies who are supplying and promoting these supplements. Supplements can be used as replacements for nutrients missing from the regular diet, but they cannot replace a healthy diet. Many variables, including possible harmful effects, must be considered before supplementing. Often supplements are expensive, and many should be used only on the advice of a physician or qualified nutritionist.

CONCLUSION

Epidemiological evidence and experimental studies with animals indicate a correlation between the incidence of chronic diseases and some dietary factors. It is well accepted that some non-nutritional dietary components provide protection. It seems likely that relatively simple preventive measures may provide a primary means for significant reduction in mortality rates of many diseases in the Western world. In addition to smoking restrictions and control of alcohol use, other relatively easily applicable preventive measures could be achieved through dietary modification. For instance, colorectal cancer, the second leading cause of cancer death in the USA and many developed countries, is caused primarily by environmental and dietary factors (85–90% of all cases).[212] A diet high in dietary fibers, vegetables and micronutrients, and low in total calories and fat, could provide protection against this cancer.[6,29,35]

No single ingredient in the diet acts as a "magic bullet" and provides protection for a variety of health problems. To prevent different diseases, regular consumption of different types of chemopreventers will probably be required. A spectrum of such ingredients could be obtained by eating a variety of foods. It is possible that even minute amounts of different phytochemicals with antioxidant and other chemopreventive potentials may act in concert and produce synergistic effects explaining the repeatedly beneficial effects observed by consuming a diet containing fruits, vegetables and grains. Beneficial effects on health are more widespread when foods containing native chemopreventive nutrients are consumed, than when diets are supplemented with one or two particular antioxidants, minerals or functional food ingredients. The best advice for the consumer still appears to be to eat a wide variety of grains, fruits and vegetables. Foods containing ingredients with known beneficial components probably also contain other, still undiscovered beneficial components. These recommendations, expressed by a number of investigators and nutritionists, are also supported by the U.S. National Research Council[256] and by the Canadian Health Protection Branch.[61]

There are many unanswered questions about the functionality of phytochemicals with chemoprotective activity. Additional research is needed to confirm their protective efficacy and safety and to establish their "bioavailability, interactions with one another and other dietary factors, mechanisms of action, and methods of assessing their functional status".[251]

ACKNOWLEDGMENTS

The author thanks Dr. T.I. Matula for helpful discussions during the preparation of this manuscript, Dr. F.W. Scott for reading the manuscript and making suggestions for corrections and Dr. John T. Romeo for editorial help.

REFERENCES

1. KÜHNAU, J. 1976. The flavonoids. a class of semi- essential food components: their role in human nutrition. Wld. Rev. Nutr. Diet 24: 117–135.
2. National Research Council. 1982. Committee on Diet, Nutrition, and Cancer, Assembly of Life Sciences. Diet, Nutrition, and Cancer. National Academy Press, Washington, D.C., pp.1.1–1.16 and pp. 2.1–2.13.
3. DOLL, R., PETO, R. 1981. The causes of cancer: quantitative estimates of avoidable risks of cancer in the United States today. J. Natl. Cancer Inst. 66: 1191–1308.
4. POTTER, J.D. 1992. Epidemiology of diet and cancer: evidence of human maladaptation. In: Macronutrients, (M.S. Micozzi, T.E. Moon, eds.), Marcel Dekker Inc., New York, NY, pp. 55–84.
5. KRINSKY, N.I., SIES, H. (eds.) 1995. Antioxidant vitamins and β-carotene in disease prevention. Am. J. Clin. Nutr. 62: 1299S-1380S.
6. National Research Council. 1996. Carcinogens and Anticarcinogens in the Human Diet. National Academy Press, Washington, D.C., pp. 35–126.
7. TANAKA, T. 1994. Cancer chemoprevention by natural products (Review). Oncology Reports 1: 1139–1155.
8. KATIYAR, S.K., MUKHTAR, H. 1996. Tea in chemoprevention of cancer: epidemiologic and experimental studies. Intern. J. Oncol. 8: 221–138.
9. STAVRIC, B. 1994. Antimutagens and anticarcinogens in foods. Fd. Chem. Toxic. 32: 79–90.
10. STOKER, R., BOWRY V.W. 1996. Tocopherol-mediated peroxidation of lipoprotein lipids and its inhibition by co-antioxidants. In: Handbook of Antioxidants, (E. Cadenas and L.Packer, eds.), Marcel Dekker Inc., New York, pp.27–41.

11. BROWN L.A.S., JONES D.P. 1996. The biology of ascorbic acid. In: Handbook of Antioxidants, (E. Cadenas and L. Packer, eds.), Marcel Dekker Inc., New York, pp. 117–154.
12. SWANSON, J.E., PARKER, R.S. 1996. Biological effects of carotenoids in humans. In: Handbook of Antioxidants, (E. Cadenas and L. Packer, eds.), Marcel Dekker Inc., New York, pp. 337–367.
13. BORS, W., HELLER, W., MICHEL, C., STETTMAIER, K. 1996. Flavonoids and polyphenols: chemistry and biology. In: Handbook of Antioxidants, (E. Cadenas and L. Packer, eds.), Marcel Dekker Inc., New York, pp. 409–466.
14. MACKERRAS, D. 1995. Antioxidants and Health - Fruits and vegetables or supplements? Food Australia 47: issue 11 (Suppl.), pp. S.1-S.24.
15. SIES, H. KRINSKY, N.I. 1995. The present status of antioxidant vitamins and β-carotene. Am. J. Clin. Nutr. 62: 1299S-1300S.
16. PACKER, L. 1996. Antioxidant defenses in biological systems: An overview. In: Proceedings of the International Symposium on Natural Antioxidants. Molecular Mechanisms and Health Effects, (L. Packer, M.G. Traber, W. Xin, eds.) AOCS Press, Champaign, Illinois, pp. 9–23.
17. BIRT, D.F., BRESNICK, E. 1991. Chemoprevention by nonnutrient components of vegetables and fruits. In: Human Nutrition: A Comprehensive Treatise, (R. Alfin-Slater, D. Kritchewsky, eds.), Plenum, New York, NY, pp. 221–261.
18. POHAN, T.E., HOWE, G.R., FRIEDENRICH, C.M., JAIN, M., MILLER, A.B. 1993. Dietary fibre, vitamins A, C, and E, and risk of breast cancer: A cohort study. Cancer Causes and Control 4: 29–37.
19. FERGUSON, L.R. 1994. Antimutagens as cancer chemopreventive agents in the diet. Mutat. Res. 307: 395–410.
20. OMENN, G.S., GOODMAN, G.E., THORNQUIST, M.D., BALMES, J., CULLEN, M.R., GLASS, A., KEOGH, J.P., MEYSKENS, F.L., VALANIS, B., WILLIAMS, J.H., BARNHART, S., HAMMAR, S. 1996. Effects of a combination of beta-carotene and vitamin A on lung cancer and cardiovascular disease. New Engl. J. Med. 334: 1150–1155.
21. BLOCK, G., PATTERSON, B., SUBAR, A. 1992. Fruits, vegetables, and cancer prevention: a review of the epidemiological evidence. Nutr. Cancer. 18: 1–29.
22. STAVRIC, B., ALABASTER, O., BLUMBERG, J., STAMPFER, M.J. 1996. Do antioxidants really prevent disease? Patient Care/Canada 7: 29–49.
23. National Institute of Nutrition 1996. NIN Review No. 25, pp. 1–6.
24. HALLIWELL, B., AESCHBACH, R., LÖLIGER, J., ARUOMA, O.I. 1995. The characterization of antioxidants. Fd Chem. Toxic. 33: 601–617.
25. The Alpha-Tocopherol, Beta Carotene Cancer Prevention Study Group: The effect of vitamin E and beta carotene on the incidence of lung cancer and other cancers in male smokers. 1994. N. Engl. J. Med. 330: 1029–1035.
26. BERGER, M.R., BERGER, I., SCHMÄHL, D. 1991. Vitamins and cancer. In: Nutrition, Toxicology and Cancer, (I.R. Rowland, ed.), CRC Press, Inc., Boca Raton, FL, pp. 517–547.
27. BYERS, T., PERRY, G. 1992. Dietary carotenes, vitamin C, and vitamin E as protective antioxidants in human cancers. Annu. Rev. Nutr. 12: 139–159.
28. KORNHAUSER, A., LAMBERT, L.A., WAMER, W.G., WEI, R.R., LAVU, S., TIMMER, W.C. 1995. Antioxidants and cancer prevention. In: Nutrients in Cancer Prevention and Treatment, (K.N. Prasad, L. Santamaria, R.M. Williams, eds.), Humana Press, Totowa, NJ, pp. 83–100.
29. DESHPANDE, S.S., DESHPANDE, U.S., SALUNKHE, D.K. 1996. Nutritional and health aspects of food antioxidants. In: Food Antioxidants, (D.L. Madhavi, S.S. Deshpande, D.K. Salunkhe, eds.), Marcel Dekker Inc., New York, NY, pp. 361–469.
30. PRASAD, K.N. 1990. Nutrition and cancer. In: Nutrients and Cancer Prevention. (K.N. Prasad, F.L. Meyskens, Jr., eds.), Humana Press, Clifton, N.J., pp. xi-xvi.
31. NIGRO, N.D., BULL, A.W., WILSON, P.S., SOULLIER B.K., ALOUSI, M.A. 1982. Combined inhibitors of carcinogenesis: effect on azoxymethane-induced intestinal cancer in rats. J. Natl. Cancer Inst. 69: 103–107.

32. ZIEGLER, R.G. 1991. Vegetables, fruits and carotenoids and the risk of cancer. Am. J. Clin. Nutr. 53 (Suppl.): 251S-259S.
33. ZIEGLER, R.G. 1989. A review of epidemiologic evidence that carotenoids reduce the risk of cancer. J. Nutr. 119: 116–122.
34. GIOVANNUCCI, E., ASCHERIO, A., RIMM, E.B., STAMPFER, M.J., COLDITZ, G.A., WILLETT, W.C. 1995. Intake of carotenoids and retinol in relation to risk of prostate cancer. J. Natl. Cancer Inst. 87: 1767–1776.
35. PETO, R., DOLL, R., BUCKLEY, J.D., SPORN, M.B. 1981. Can dietary beta-carotene materially reduce human cancer rates? Nature 290: 201–208.
36. GAZIANO, J.M., MANSON, J.E., BURING, J.E., HENNEKENS, C.H. 1992. Dietary antioxidants and cardiovascular disease. Ann. N.Y. Acad. Sci. 669: 249–259.
37. JHA, P., FLATHER, M., LONN, E., FARKOUH, M., YUSUF, S. 1995. The antioxidant vitamins and cardiovascular disease: a critical review of epidemiological and clinical trial data. Ann. Intern. Med. 123: 860–872.
38. COMSTOCK, G.W., BUSH, T.L., HELZLSOUER, K. 1992. Serum retinol, beta-carotene, vitamin C and selenium as related to subsequent cancer sites. Am. J. Epidemiol. 135: 1068–1069.
39. ZHENG, W., SELLERS, T.A., DOYLE, T.J., KUSHI, L.H., POTTER, J.D., FOLSOM, A.R. 1995. Retinol, antioxidant vitamins, and cancers of the upper digestive tract in a prospective cohort study of postmenopausal women. Am. J. Epidemiol. 142: 955–960.
40. GAZIANO, J.M., MANSON, J.E., BRANCH, L.G., COLDITZ, G.A., WILLETT, W.C., BURING, J.E. 1995. A prospective study of consumption in fruits and vegetables and decreased cardiovascular mortality in the elderly. Ann. Epidemiol. 5: 255–260.
41. RIMM, E.B., ASCHERIO, A., WILLETT, W.C., COLDITZ, G.A., GIOVANNUCCI, E., STAMPFER, M.J. 1992. Dietary antioxidant intake and risk of coronary disease among men. Am. J. Epidemiol. 136: 985–986.
42. HEINONEN, O.P., ALBANES, D. 1994. The effect of vitamin E and beta carotene on the incidence of lung cancer and other cancers in prevention study group. N. Engl. J. Med. 330: 1029–1035.
43. OMENN, G.S., GOODMAN, G.E., THORNQUIST, M.D., BALMES, J., CULLEN, M.R., GLASS, A., KEOGH, J.P., MEYSKENS, F.L., VALANIS, B., WILLIAMS, J.H., BARNHART, S., HAMMAR, S. 1996. Effects of a combination of beta carotene and vitamin A on lung cancer and cardiovascular disease. N. Engl. J. Med. 334: 1150–1155.
44. HENNEKENS, C.H., BURING, J.E., MANSON, J.E., STAMPFER, M., ROSNER, B., COOK, N.R., BELANGER, C., LaMOTTE, F., GAZIANO, J.M., RIDKER, P.M., WILLETT, W., PETO, R. 1996. Lack of effect of long-term supplementation with beta carotene on the incidence of malignant neoplasms and cardiovascular disease. N. Engl. J. Med. 334: 1145–1149.
45. POTISCHMAN, N., McCULLOCH, C.E., BYERS, T., NEMOTO, T., STUBBE, N., MILCH, R., OARKER, R., RASMUSSEN, K.M., ROOT, M., GRAHAM, S., CAMPBELL, T.C. 1990. Breast cancer and dietary and plasma concentrations of carotenoids and vitamin A. Am. J. Clin. Nutr. 52: 909–915.
46. MARUBINI, E., DECARLI, A., COSTA, A., MAZZOLENI, C., ANDREOLI, C., BARBIERI, A., CAPITELLI, E., CARLUCCI, M., CAVALLO, F., MONFERRONI, N., PASTORINO, U., SALVINI S. 1988. The relationship of dietary intake and serum levels of retinol and beta-carotene with breast cancer. Cancer 61: 173–180.
47. LONDON, S.J., STEIN, E.A., HENDERSON, C., STAMPFER, M.J., WOOD, W.C., REMINE, S., DMOCHOWSKI, J.R., ROBERT, N.J., WILLETT, W.C. 1992. Carotenoids, retinol and vitamin E and the risk of proliferative benign breast disease and breast cancer. Cancer Causes & Control 3: 503–512.
48. Le MARCHAND, L., YOSHIZAWA, C.N., KOLONEL, L.N., HANKIN, J.H., GOODMAN, M.T. 1989. Vegetable consumption and lung cancer risk: A population-based case-control study in Hawaii. J. Natl. Cancer Inst. 81: 1158–1164.

49. OHNO, Y., YOSHIDA, O., OISHI, K., OKADA, K., YAMABE, H., SCHROEDER, F.H. 1988. Dietary β-carotene and cancer of the prostate: A case-control study in Kyoto, Japan. Cancer Res. 48: 1331–1336.
50. KOLONEL, L.N., NOMURA, A.M.Y. 1992. Dietary intervention trials on prostate cancer. In: Macronutrients, (M.S. Micozzi, T.E. Moon, eds), Marcel Dekker Inc., New York, pp.423–436.
51. POTTER, J.D. 1995. Risk factors for colon neoplasia - epidemiology and biology. Eur. J. Cancer 31A: 1033–1038.
52. MacLENNAN, R., MACRAE, F., BAIN, C., BATTTISTUTTA, D., CHAPIUS P., GRATTEN, H., LAMBERT, J., NEWLAND R.C., NGU, M., RUSSELL, A., WARD, M., WAHLQVIST, M.L. 1995. The Australian Polyp Prevention Project. Randomized trial of intake of fat fibre, and beta carotene to prevent colorectal adenomas. J. Natl. Cancer Inst. 87: 1760–1766.
53. SCHATZKIN, A., FREEDMAN, L.S., LANZA, E., TANGEA, J. 1995. Diet and colorectal cancer: still an open question. J. Natl. Cancer Inst. 87: 1733–1735.
54. RAMASWAMY, G., KRISHNAMOORTHY, L. 1996. Serum carotene, vitamin A, and vitamin C levels in breast cancer and cancer of the uterine cervix. Nutr. Cancer 25: 173–177.
55. HENSON, D.E., BLOCK, G., LEVINE, M. 1991. Ascorbic acid: biological functions and relation to cancer. J. Natl. Cancer Inst. 83: 547–550.
56. SCHORACH, C.J., SOBALA, G.M., SANDERSON, M., COLLIN, N., PRIMROSE, J.N. 1991. Gastric juice ascorbic acid: effects of disease and implications for gastric carcinogenesis. Am. J. Clin. Nutr. 53: 287S-293S.
57. TANNENBAUM, S.R., WISHNOK, J.S., LEAF, C.D. 1991. Inhibition of nitrosamine formation by ascorbic acid. Am. J. Clin. Nutr. 53: 247S-250S.
58. BLOCK, G. 1991. Vitamin C and cancer prevention: the epidemiologic evidence. Am. J. Clin. Nutr. 53: 270S-282S.
59. WEISBURGER, J.H., HARQUARDT, H., MOWER, H.F., HIROTA, N., MORI, H., WILLIAMS, G.M. 1980. Inhibition of carcinogenesis: Vitamin C and the prevention of gastric cancer. Prev. Med. 9: 352–361.
60. COHEN, M., BHAGAVAN, H.N. 1995. Ascorbic acid and gastrointestinal cancer. J. Am. Coll. Nutr. 14: 565–578.
61. Nutrition Recommendations 1990. The Report of the Scientific Committee, Health and Welfare Canada, Ottawa, 1990, pp.99–102.
62. KNEKT, P., AROMAA, A., MAATELA, J., AARAN, R.K., NIKKARI, T., HAKAMA, M., HAKULINEN, T., PETO, R., TEPPO, L. 1991. Vitamin E and cancer prevention. Am. J. Clin. Nutr. 53: 283S-286S.
63. PACKER, L. 1992. Interactions among antioxidants in health and disease: Vitamin E and its redox cycle. Proc. Soc. Exptl. Biol. Med. 200: 271–276.
64. BAYERS, T. 1993. Vitamin E supplements and coronary heart disease. Nutr. Rev. 51: 333–345.
65. GRIDLEY, G., MCLAUGHLIN, J.K., BLOCK, G., BLOT, W.J., GLUCH, M., FRAUMENI, J.F. Jr. 1992. Vitamin supplement use and reduced risk of oral and pharyngeal cancer. Am. J. Epidemiol. 135: 1083–1092.
66. LANDVIK, S.V., DIPLOCK, A.T., PACKER, L. 1996. Efficacy of vitamin E in human health and disease. In: Handbook of Antioxidants, (E.Cadenas, L.Packer, eds.), Marcel Deckker Inc., New York, pp. 63–87.
67. PACKER, L., LANDVIK S. 1989 Vitamin E: Introduction to biochemistry and health benefits. In: Vitamin E. Biochemistry and Health Implications (A.T. Diplock, L.J. Machlin, L. Packer, W.A. Pryor eds.), Vol 570. The New York Academy of Sciences, N.Y., pp. 1–6.
68. CORREA, P. 1995. The role of antioxidants in gastric carcinogenesis. Crit. Rev. Food Sci. Nutr. 35: 59–64.
69. RIMM, E.B., STAMPFER, M.J., ASCHERIO, A., GIOVANNUCCI, E., COLDITZ, G.A., WILLETT, W.C. 1993. Vitamin E consumption and the risk of coronary heart disease in men. N. Engl. J. Med. 328: 1450–1456.

70. STAMPFER, M.J., HENNEKENS, C.H., MANSON, J.A., COLDITZ, G.A., WILLETT, W.C. 1993. Vitamin E consumption and the risk of coronary disease in women. N. Engl. J. Med. 328: 1444–1449.
71. KUSHI, L.H., FOLSOM, A.R., PRINEAS, R.J., MINK, P.J., WU, Y., BOSTICK, R.M. 1996. Dietary antioxidant vitamins and death from coronary heart disease in postmenopausal women. New Engl. J. Med. 334: 1156–1162.
72. STEPHENS, N.G., PARSONS, A., SCHOFIELD, P.M., KELLY, F., CHEESEMAN, K., MITCHINSON, M.J. 1996. Randomized controlled trial of vitamin E in patients with coronary disease: Cambridge Heart Antioxidant Study (CHAOS). Lancet 347: 781–786.
73. JIALAL, I., FULLER, C.J., HUET, B.A. 1995. The effect of α-tocopherol supplementation on LDL oxidation. A dose-response study. Arterioscler. Thromb. Vasc. Biol. 15: 190–198.
74. BLOT, W.J., LI, J.Y., TAYLOR, P.R., GUO, W., DAWSEY, S., WANG, G.Q., YANG, C.S., ZHENG, S.F., GAIL, M., LI, G.Y., LIU, B.Q., TANGREA, J., SUN, Y.H., LIU, F., FRAUMENI, J.F. JR., ZHANG, Y.H., LI, B. 1993. Nutrition intervention trials in Linxian, China: Supplementation with specific vitamin/mineral combination, cancer incidence and disease-specific mortality in the general population. J. Natl. Cancer Inst. 85: 1483–1492.
75. EL-BAYOUMY, K. 1991. The role of selenium in cancer prevention. In: Cancer Principles and Practice of Oncology, 4th edition, (V. Davita, S. Helman, S.A. Rosenberg, eds.), Philadelphia, PA, pp. 1–15.
76. HOCMAN, G. 1988. Chemoprevention of cancer: selenium. Int. J. Biochem. 20: 123–132.
77. CLARK, L.C., CANTOR, K.P., ALLAWAY, W.H. 1991. Selenium in forage crops and cancer mortality in U.S. counties. Arch. Envir. Health 46: 37–42.
78. IP, C. 1986. The chemoprotective role of selenium in carcinogenesis. J. Am. Coll. Toxicol. 5: 7–20.
79. COMBS, G.F.J. 1989. Selenium. In: Nutrition and Cancer Prevention, (T.E. Moon, M.S. Micozzi, eds.), Marcel Dekker Inc., New York, NY, pp. 389–420.
80. NOMURA, A., HEIBRUN, L.K., MORRIS, J.S., STEMMERMANN, G.N. 1987. Serum selenium and risk of cancer, by specific sites: case-control analysis of prospective data. J. Natl. Cancer Inst. 79: 103–108.
81. VIRTAMO, J., VALKEILA, E., ALFTHAM, G., PUNSAR, S., HUTTUNEN, J.K. KARVONEN, M.J. 1987. Serum selenium and risk of cancer. A prospective follow-up of nine years. Cancer 60: 145–148.
82. CHERYAN, M. 1980. Phytic acid interactions in food systems. CRC Crit. Rev. Food Sci. Nutr. 13: 297–325.
83. SHAMSUDDIN, A.M., ULLAH, A. 1989. Inositol hexaphosphate inhibits large intestine cancer in F344 rats 5 months after induction by azoxymethane. Carcinogenesis 10: 625–628.
84. SHAMUSUDDIN, A.M., ULLAH, A., CHAKRAVARTHY, A.K. 1989. Inositol and inositol hexaphosphate suppress cell proliferation and tumor formation in CD-1 mice. Carcinogenesis 10: 1461–1463.
85. VUCENIK, I., CUMMINGS, P.J., YANG, G., SHAMSUDDIN, A.M. 1994. Inositol hexaphosphate inhibits growth of MCF-7 human mammary cancer cells and down-regulates *c-myc*, *c-jun* and *c-fos* expression. Proc. Am. Assoc. Cancer Res. 35: 624.
86. MESSINA, M., BARNES, S. 1991. The role of soy products in reducing risk of cancer. J. Natl. Cancer Inst. 83: 541–546.
87. EMPSON, K.L., LABUZA, T.P., GRAF, E. 1991. Phytic acid as a food antioxidant. J. Food Sci. 56: 560–563.
88. ZHOU, J.R., ERDMAN, J.W. 1995. Phytic acid in health and disease. Crit. Rev. Food Sci. Nutr. 35: 495–508.
89. SHAHIDI, F., NACZK, M. 1995. Food Phenolics: Sources, Chemistry, Effects, Applications. Technomic Publishing Co. Lancaster, pp. 171–198.

90. HUANG, M.-T., FERRARO, T. 1992. Phenolic compounds in food and cancer prevention. In: Phenolic Compounds in Food and Their Effects on Health. II. Antioxidants and Cancer Prevention. (M.-T. Huang, C.-T. Ho, C.Y. Lee, eds.), ACS Symposium Series 507, American Chemical Society, Washington, DC, pp. 8–34.
91. SHAHIDI, F., WANASUNDARA, P.K.J.P.D. 1992. Phenolic antioxidants. CRC Crit. Rev. Food Sci. Nutr. 32: 67–103.
92. STAVRIC, B., MATULA, T.I., KLASSEN, R., DOWNIE, R.H. 1990. Inhibitory effect of flavonoids against mutagenic and carcinogenic xenobiotics in foods. In: Flavonoids in Biology and Medicine III, (N.P. Das, ed.), National University of Singapore, Singapore, pp. 515–529.
93. DESCHNER, E.E. 1992. Dietary quercetin and rutin. In: Phenolic Compounds in Food and Their Effects on Health. II. Antioxidants and Cancer Prevention, (M.-T. Huang, C.-T. Ho, C.Y.Lee, eds.), ACS Symposium Series 507, American Chemical Society, Washington, D.C., pp. 265–268.
94. SHAHIDI, F., NACZK, M. 1995. Food Phenolics: Sources, Chemistry, Effects, Applications. Technomic Publishing Co. Lancaster, pp. 235–280.
95. STAVRIC, B., MATULA, T.I., KLASSEN, R., DOWNIE, R.H., WOOD, R.J. 1992. Effect of flavonoids on mutagenicity and bioavailability of xenobiotics in foods. In: Phenolic Compounds in Food and their Effects on Health II, (M.-T. Huang, C.-T. Ho and C.Y. Lee, eds.), ACS Symposium Series 507. American Chemical Society, Washington, DC, pp. 239–249.
96. HAVSTEEN, B. 1983. Flavonoids, a class of natural products of high pharmacological potency. Biochem. Pharmacol. 32: 1141–1148.
97. STAVRIC, B. 1984. Mutagenic food flavonoids. Fed. Proc. 43: 2454–2458.
98. STAVRIC, B. 1994. Quercetin in our diet: from potent mutagen to probable anticarcinogen. Clin. Biochem. 27: 245–248.
99. HERTOG, M.G.L., FESKENS, E.J.M., HOLLMAN, P.C.H., KATAN, M.B., KROMHOUT, D. 1993. Dietary antioxidant flavonoids and risk of coronary heart disease: The Zutphen elderly study. Lancet, 342: 1007–1011.
100. GRAHAM, H.N. 1992. Green tea composition, consumption, and polyphenol chemistry. Prevent. Med. 21: 334–350.
101. WATTENBERG, L.W., COCCIA, J.B., LAM, L.K.T. 1980. Inhibitory effects of phenolic compounds on benzo[a]pyrene-induced neoplasia. Cancer Res. 40: 2820–2823.
102. IARC 1991. Monographs on the Evaluation of Carcinogenic Risks to Humans. Coffee, tea, mate, methylxanthines and methylglyoxal. 51, IARC, Lyon, France, pp. 152–161.
103. ROSENBERG, L. 1990. Coffee and tea consumption in relation to risk of large bowel cancer: a review of epidemiologic studies. Cancer Lett. 52: 163–171.
104. LEE, H.S., WROLSTAD, R.E. 1988. Apple juice composition: sugar, nonvolatile acid and phenolic profiles. J. Assoc. Anal. Chem. 71: 789–794.
105. TANAKA, T., YOSHIMI, N., SUGIE, S., MORI, H. 1992. Protective effects against liver, colon, and tongue carcinogenesis by plant phenols. In: Phenolic Compounds in Food and Their Effects on Health. II. Antioxidants and Cancer Prevention. (M.-T. Huang, C.-T. Ho, C.Y. Lee, eds.), ACS Symposium Series 507, American Chemical Society, Washington, D.C. pp. 326–337.
106. BOUKHARTA, M., JALBERT, G., CASTONGUAY, A. 1992. Biodistribution of ellagic acid and dose-related inhibition of lung tumorigenesis in A/J mice. Nutr. Cancer 18: 181–189.
107. AYRTON, A.D., LEWIS, D.F.V., WALKER, R., IOANNIDES, C. 1992. Antimutagenicity of ellagic acid towards the food mutagen IQ. Fd Chem. Toxic. 30: 289–295.
108. YAMANISHI, T. (ed.) 1995. Special Issue on Tea, Food Rev. Internat. 11: No. 3.
109. KONO, S., SHINCHI, K., IKEDA, N., YANAI, F., IMANISHI, K. 1992. Green tea consumption and serum lipid profile: A cross-sectional study in Northern Kyushu, Japan. Prevent. Med. 21: 526–531.
110. STAMPFER, M.J., RIMM, E.B. 1993. A review of the epidemiology of dieting antioxidants and risk of coronary artery disease. Canad. J. Cardiol. 9 (Suppl B): 14B-18B.

111. STENSVOLD, I., TVERDAL, A., SOLVOLL, K., FOSS, O.P. 1992. Tea consumption, relationship to cholesterol, blood pressure, and coronary and total mortality. Prevent. Med. 21: 546–553.
112. IMAI, K., NAKACHI, K. 1995. Cross sectional study of effects of drinking green tea on cardiovascular and liver diseases. Br. Med. J. 310: 693–696.
113. SAGESAKA-MITANE, Y., MIWA, M., OKADA, S. 1990. Platelet aggregation inhibitors in hot water extract of green tea. Chem. Pharmaceut. Bull. 38: 790–793.
114. JAIN, A.K., SHIMOI, K., NAKAMURA, Y., KADA, T., HARA, Y. TOMITA, I. 1989. Crude tea extracts decrease the mutagenic activity of *N*-methyl-*N'*-nitro-*N*-nitrosoguanidine *in vitro* and in intragastric tract of rats. Mutat. Res. 210: 1–8.
115. XIN, W., SHI, H., YANG, F., ZHAO, B., HOU, J. 1996. A study on the effects of green tea polyphenols on lipid free radicals. In: Proceedings of the International Symposium on Natural Antioxidants. Molecular Mechanisms and Health Effects,(L. Packer, M.G. Traber, W. Xin, eds.), AOCS Press, Champaign, Illinois, pp.397–416.
116. BU-ABBAS, A., CLIFFORD, M.N., WALKER, R., IOANNIDES, C. 1994. Marked antimutagenic potential of aqueous green tea extracts: mechanism of action. Mutagenesis 9: 325–331.
117. WANG, Z.Y., CHENG, S.J., ZHOU, Z.C., ATHAR, M., KHAN, W.A., BICKERS, D.R., MUKHTAR, H. 1989. Antimutagenic activity of green tea polyphenols. Mutat. Res. 223: 273–285.
118. CONNEY, A.H., WANG, Z.-Y., HUANG, M.-T., HO, C.-T., YANG, C.-S. 1992. Inhibitory effect of green tea on tumorigenesis by chemicals and ultraviolet light. Prev. Med. 21: 361–369.
119. YANG, C.S., WANG, Z-Y. 1993. Tea and Cancer. J. Natl. Cancer Inst. 85: 1038–1049.
120. STICH, H.F. 1992. Teas and tea components as inhibitors of carcinogen formation in model systems and man. Prev. Med. 21: 377–384.
121. CONSTABLE, A., VARGA, N., RICHOZ, J., STADLER, R.H. 1996. Antimutagenicity and catechin content of soluble instant teas. Mutagenesis 11: 189–194.
122. YEN, G.-C., Chen, H.Y. 1996. Relationship between antimutagenic activity and major components of various teas. Mutagenesis 11: 37–41.
123. STAVRIC, B., MATULA, T.I., KLASSEN, R., DOWNIE, R.H. 1996. The effect of teas on the *in vitro* mutagenic potential of heterocyclic aromatic amines. Fd Chem. Toxic. 34: 515–523.
124. SUGIMURA, T., WAKAYABASHI, K. 1990. Mutagens and carcinogens in food. In: Mutagens and Carcinogens in the Diet, (M.W. Pariza, U.U. Aeshbacher, J. Felton, S. Sato, eds.), Wiley-Liss, New York, N.Y., pp. 1–18.
125. STAVRIC, B. 1994. Biological significance of trace levels of mutagenic heterocyclic aromatic amines in human diet: A critical review. Fd Chem. Toxicol. 32: 977–994.
126. GAO, Y.T., MCLAUGHLIN, J.K., BLOT, W.J., JI, B.N.T., DAI, Q., FRAUMENI, J.F.Jr. 1994. Reduced risk of esophageal cancer associated with green tea consumption. J. Natl. Cancer Inst. 86: 855–858.
127. OHNO, Y., WAKAI, K., GENKA, K., OHMINE, K., KAWAMURA, T., TAMAKOSHI, A., AOKI, R., SENDA, M., HAYASHI, Y., NAGAO, K., FUKUMA, S., AOKI, K. 1995. Tea consumption and lung cancer risk: A case-control study in Okinawa, Japan. Jpn. J. Cancer Res. 86: 1027–1034.
128. CHENG, S.J. 1996. Study on antimutagenicity and anticarcinogenicity of green tea epicatechins - A natural free radical scavenger. In: Proceedings of the International Symposium on Natural Antioxidants. Molecular Mechanisms and Health Effects, (L. Packer, M.G. Traber, W. Xin, eds.), AOCS Press, Champaign, IL, pp.392–396.
129. VINSON, J., DABBAGH, Y.A., SERRY, M.M., JANG, J. 1995. Plant flavonoids, especially tea flavonols, are powerful antioxidants using an *in vitro* oxidation model for heart disease. J. Agric. Food Chem. 43: 2800–2802.
130. GOLDBOHM, R.A., HERTOG, M.G.L., BRANTS, H.A.M., VAN POPPEL, G., VAN DEN BRANDT, P.A. 1996. Consumption of black tea and cancer risk: a prospective cohort study. J. Natl. Cancer Inst. 88: 93–100.

131. GOLDBOHM, R.A., VAN DEN BRANDT, P. 1996. Correspondence. J. Natl. Cancer Inst. 88: 768–769.
132. FUJIKI, H., YOSHIZAWA, S., HORIUCHI, T., SUGANUMA, M., YATSUNAMI, J., NISHIWAKI, S., OKABE, S., NISHIWAKI-MATSUSHIMA, R., OKUDA, T., SUGIMURA T. 1992. Anticarcinogenic effects of (-)epigallocatechin gallate. Prev. Med. 21: 503–509.
133. KLAUNIG, J.E. 1992. Chemopreventive effects of green tea components on hepatic carcinogenesis. Prev. Med. 21: 510–519.
134. KAHN, S.G., KATIYAR, S.K., AGARWAL, R., MUKHTAR, H. 1992. Enhancement of antioxidant and phase II enzymes by oral feeding of green tea polyphenols in drinking water to SKH-1 hairless mice: Possible role in cancer chemoprevention. Cancer Res. 52: 4050–4052.
135. RUCH, R.J., CHENG, S.-J., KLAUNIG, J.E. 1989. Prevention of cytotoxicity and inhibition of intercellular communication by antioxidant catechins isolated from Chinese green tea. Carcinogenesis 10: 1003–1008.
136. HAYATSU, H., INADA, N., KAKUTANI, T., ARIMOTO, S., NEGISHI, T., MORI, K., OKUDA, T., SAKATA, I. 1992. Suppression of genotoxicity of carcinogens by (-)epogallocatechin gallate. Prev. Med. 21: 370–376.
137. HO, C.T., CHEN, Q., SHI, H., ZHANG, K.Q., ROSEN, R.T. 1992. Antioxidative effects of polyphenol extract prepared from various Chinese teas. Prev. Med. 21: 520–525.
138. WEISBURGER, J.H. 1994. Practical approaches to chemoprevention of cancer. Drug Metabol. Rev. 26 (1 & 2): 253–260.
139. WEISBURGER, J.H. 1996. Tea antioxidants and health. In: Handbook of Antioxidants, (E. Cadenas, L. Packer, eds.), Marcel Dekker Inc., New York, NY, pp. 469–486.
140. UCHIDA, S., EDAMATSU, R., HIRAMATSU, M., MORI, A., NONAKA, G.-I., NISHIOKA, I., NIWA, M., OZAKI, M. 1987. Condensed tannins scavenge active oxygen free radicals. Med. Sci. Res. 15: 831–832.
141. AGARWAL, R., KATIYAR, S.K., KHAN, S.G., MUKHTAR, H. 1993. Protection against ultraviolet B radiation-induced effects in the skin of SKH-1 hairless mice by a polyphenolic fraction isolated from green tea. Photochem. Photobiol. 58: 695–700.
142. STEINHART, C.E., DOYLE, M.E., COCHRANE, B.A. (eds.) 1996. Naturally occurring toxicants and food constituents of toxicological interest. In: Food Safety 1995. Food Research Institute, University of Wisconsin, Madison, Wisconsin. Marcel Dekker Inc., New York, NY, pp. 326–376.
143. ABRAHAM, S.K. 1996. Anti-genotoxic effects in mice after the interaction between coffee and dietary constituents. Fd Chem. Toxicol. 34: 15–20.
144. STOLTZ, D.R., STAVRIC, B., KREWSKI, D., KLASSEN, BENDALL, R., JUNKINS, B. 1982. Mutagenicity screening of foods I. results with beverages. Environ. Mutagen. 4: 477–492.
145. KATO, T., TAKAHASHI, S., KIKUGAWA, K. 1989. Generation of heterocyclic amine-like mutagens during the roasting of coffee beans. Eisei Kakagu 35: 370–376.
146. STAVRIC, B., MATULA, T.I., KLASSEN, R., DOWNIE, R.H. 1992. Investigations for the presence of mutagenic heterocyclic aromatic amine-type compounds in coffee. Canadian Institute of Food Science and Technology, 35th Annual Conference, Ottawa, May 31-June 3. p.7.
147. JOHANSSON, M.A.E., KNIZE, M.G., JÄGERSTAD, M., FELTON, J.S. 1995. Characterization of mutagenic activity in instant hot beverage powders. Environ. Molec. Mutagen. 25: 154–161.
148. STALDER, R., BEXTER, A., WŰRZNER, H.P., LUGINBŰHL, H.A. 1990. A carcinogenicity study of instant coffee in Swiss mice. Fd Chem. Toxic. 28: 829–837.
149. JACOBSEN, B.K., KVALE, G., HEUCH, I. 1991. Coffee drinking and the risk of colon cancer - unfinished business. Epidemiology 2: 77–78.
150. YAMAGUCHI, T., IKI, M. 1986. Inhibitory effect of coffee extract against some mutagens. Agric. Biol. Chem. 50: 2983–2988.

151. WATTENBERG, L.W., LAM, L.K.T. 1984. Protective effects of coffee constituents on carcinogenesis in experimental animals. In: Banbury Report 17. Coffee and Health. Cold Spring Harbor Laboratory, Cold Spring Harbor, New York, pp. 137–145.
152. HUGGETT, A.C., SCHILTER, B. 1995. The chemoprotective effects of cafestol and kahweol: effects on xenobiotic metabolising enzymes. J. Cell. Biochem. 19A: 190.
153. FORD, G. 1993. The French Paradox & drinking for health. Wine Appreciation Guild, San Francisco, CA, pp. 3–25.
154. FRANKEL, E.N., KANNER, J., GERMAN, J.B., PARKS, E., KINSELLA, J.E. 1993. Inhibition of oxidation of human low-density lipoprotein by phenolic substances in red wine. Lancet 341: 454–457.
155. KINSELLA, J.E., KANNER, J., FRANKEL, E., GERMAN, B. 1992. Wine and health: The possible role of phenolics, flavonoids and other antioxidants. Proceedings, Potential health effects of components of plant foods and beverages in the diet. University of California, Davis, 14–15 August, pp. 107–121.
156. DREWNOWSKI, A., HENDERSON, S.A., SHORE, A.B., FISCHER, C., PREZIOSI, P., HERCBERG, S. 1996. Diet quality and dietary diversity in France: Implications for the French paradox. J. Am. Diet. Assoc. 96: 663–669.
157. VINSON, J.A., JANG, J., DABBAGH, Y.A., SERRY, M.M., CAI, S. 1995. Plant polyphenols exhibit lipoprotein-bound antioxidant activity using an *in vitro* oxidation model for heart disease. J. Agric. Food Chem. 43: 2798–2799.
158. FRANKEL, E.N., WATERHOUSE, A.L., KINSELLA, J.E. 1993. Inhibition of human LDL oxidation by resveratrol. Lancet 341: 1103–1104.
159. YOU, W.-C., BLOT, W.J., CHANG, Y.S., ERSHOW, A., YANG, Z.T., AN, Q., HENDERSON, B.E., FRAUMENI, J.F., JR. WANG, T.-G. 1989. Allium vegetables and reduced risk of stomach cancer. J. Natl. Cancer Inst. U.S.A. 89: 2399–2403.
160. LIU, J., LIN, R.I., MILNER, J.A. 1992. Inhibition of 7,12-dimethyl-benz[a]anthracene-induced mammary tumors and DNA adducts by garlic powder. Carcinogenesis 13: 1847–1851.
161. GLASS, G.V. 1976. Primary, secondary and meta-analysis of research. Education Research 5: 3–8.
162. WARSHAFSKY, S., KAMER, R., SIVAK, S. 1993. Effect of garlic on total serum cholesterol. A meta-analysis. Ann. Intern. Med. 119: 599–605.
163. SILAGY, C.A., NEIL, H.A.W. 1994. A meta-analysis of the effect of garlic on blood pressure. J. Hypertens. 12: 463–468.
164. BORDIA, T., MOHAMMED, N., THOMSON, M., ALI, M. 1996. An evaluation of garlic and onion as antithrombic agents. Prostaglandins, Leukotriens Essent. Fatty Acids 54: 183–186.
165. NAGABHUSHAN, M., LINE, D., POLVERINI, P.J., SOLT, D.B. 1992. Anticarcinogenic action of diallyl sulphide in hamster buccal pouch and forestomach. Canc. Lett. 66: 207–216.
166. BLOCK, E., E.M. CALVEY, C.W. GILLIES, J.Z. GILLIES AND P. UDEN. 1997. Peeling the onion: Organosulfur and selenium phytochemicals in genus *Allium* plants. In: Recent Advances in Phytochemistry, Vol. 31. Functionality of Food Phytochemicals (T. Johns and J.T. Romeo, eds.). Plenum, New York. pp. 1–30.
167. STAVRIC, B. 1993. Antimutagens and anticarcinogens in foods. Fd. Chem. Toxic. 32: 79–90.
168. DAUSCH, J.G., NIXON, D.W. 1990. Garlic: a review of its relationship to malignant disease. Prev. Med. 19: 346–361.
169. MENG, C.-L., SHYU, K.-W. 1990. Inhibition of experimental carcinogenesis by painting with garlic extract. Nutr. Cancer 14: 207–217.
170. WARGOVICH, M.J., WOODS, C., ENG, V.S., STEPHENS, L.C., GRAY, K. 1988. Chemoprevention of N-nitrosomethyl-benzylamine-induced esophageal cancer in rats by the naturally occurring thioether, diallyl sulphide. Cancer Res. 48: 6872–6875.
171. TAKAHASHI, S., HAKOI, K., YADA, H., HIROSE, M., ITO, N., FUKUSHIMA, S. 1992. Enhancing effects of diallyl sulphide on hepatocarcinogenesis and inhibitory actions of the

related diallyl disulphide on colon and renal carcinogenesis in rats. Carcinogenesis 13: 1513–1518.
172. Editorial. 1996. UC Berkeley Wellness Letter 12; Issue 9, 1–2.
173. HERMAN, K. 1976. Flavonols and flavones in food plants: a review. J. Fd Technol. 11: 433–448.
174. CHUNG, F.-L. 1992. Chemoprevention of lung carcinogenesis by aromatic isothiocyanates. In: Cancer Chemoprevention, (L. Wattenberg, M. Lipkin, C.W. Boone, G.J. Kelloff, eds.), CRC Press, Boca Raton, FL, pp. 227–245.
175. SUGIE, S., OKAMOTO, K., OKUMURA, A., TANAKA, T., MORI, H. 1994. Inhibitory effects of benzyl thiocyanate and benzyl isothiocyanate on methylazomethanol acetate-induced intestinal carcinogenesis in rats. Carcinogenesis 15: 1555–1560.
176. WATTENBERG, L.W., 1992. Inhibition of carcinogenesis by minor dietary constituents. Cancer Res. 52: 2085S-2091S.
177. NEWMARK, H.L. 1984. A hypothesis for dietary components as blocking agents of chemical carcinogenesis: Plant phenolics and pyrrole pigments. Nutr. Cancer 6: 58–70.
178. HARRIS, R.K., HAGGERTY, W.J. 1993. Assays for potentially anticarcinogenic phytochemicals in flaxseed. Cereal Foods World 38: 147–151.
179. JOHNSTON, P.V. 1955. Flaxseed oil and cancer: α-linolenic acid and carcinogenesis. In: Flaxseed in Human Nutrition, (S.C. Cunnane, L.U. Thompson, eds.), AOCS Press, Champaign, IL, 207–218.
180. THOMPSON, L.U. 1995. Flaxseed, lignans and cancer. In: Flaxseed in Human Nutrition, (S.C. Cunnane, L.U. Thompson, eds.), AOCS Press, Champaign, IL, pp. 219–236.
181. LEE, H.P., GOURLEY, L., DUFFY, S.W., ESTEVE, J., LEE, J., DAY, N.E. 1991. Dietary effects on breast-cancer risk in Singapore. Lancet 337: 1197–200.
182. WHELAN, S.L., PARKIN, D.M., MASUYER, E. (eds.) 1990. Patterns of Cancer in Five Continents. IARC Scientific Publications No. 102, International Agency for Research on Cancer, p. 30.
183. BOYD, N.F. 1993. Nutrition and breast cancer. J. Nat. Cancer Inst. 85: 6–7.
184. MESSINA, M., BARNES, S. 1991. The role of soy products in reducing risk of cancer. J. Natl. Cancer Inst. 83: 541–546.
185. MESSINA, M.J., PERSKY, V., SETCHELL, K.D.R., BARNES, S. 1994. Soy intake and cancer risk: A review of the *in vitro* and *in vivo* data. Nutr. Cancer 21: 113–131.
186. BARRETT, J. 1996. Phytoestrogens - friends or foes? Environ. Health Persp. 104: 478–482.
187. FUKUTAKE, M., TAKAHASHI, M., ISHIDA, K., KAWAMURA, H., SUGIMURA, T., WAKABAYASHI, K. 1996. Quantification of genistein and genistin in soybeans and soybean products. Fd Chem. Toxicol. 34: 457–461.
188. SHEEHAN, D.M. 1995. The case for expanded phytoestrogen research. Proc. Soc. Exptl. Biol. Med. 208: 3–5.
189. BARNES, S., GRUBBS, C., SETCHELL, K.D.R., CARLSON, J. 1990. Soybeans inhibit mammary tumours in models of breast cancer. In: Mutagens and Carcinogens in the Diet, (M.W. Pariza, H.U. Aeschbacher, J.S. Felton, S. Sato, eds.), Wiley-Liss, New York, NY, pp. 239–253.
190. NUNOMURA, N., SASAKI, M., ASAO, Y., YOKOTSUKA, T. 1976. Isolation and identification of 4-hydroxy-2(or5)-ethyl-5(or 2)-methyl-3(2H)-furanone, as a flavour component in shoyu (soy sauce). Agric. Biol. Chem. 40: 491–495.
191. HA, E.Y.W., MORR, C.V., SEO, A. 1992. Isoflavone aglucones and volatile organic compound in soybeans: Effect of soaking. J. Food Sci. 57: 414–417.
192. FOTSIS, T., PEPPER, M., ALDERCREUTZ, H., FLEISCHMANN, G., HASE, T., MONTESANO, R., SCHWEIGERER, L. 1993. Genistein, a dietary-derived inhibitor of *in vitro* angiogenesis. Proc. Natl. Acad. Sci. 90: 2690–2694.
193. WEI, H., BOWEN, R., CAI, Q., BARNES, S., WANG, Y. 1995. Antioxidant and antipromotional effects of the soybean isoflavone genistein. Proc. Soc. Exptl. Biol. Med. 208: 124–130.

194. SETCHELL, K.D.R., ADLERCRUETZ, H. 1988. Mammalian lignans and phytoestrogens: recent studies on their formation, metabolism and biological role in health and disease. In: Role of the Gut Flora in Toxicity and Cancer, (I.R. Rowland, ed.), Academic Press, New York, NY, pp. 315–345.
195. LAMARTINIERE, C.A., MOORE, J., HOLLAND, M., BARNES, S. 1995. Neonatal genistein chemoprevents mammary cancer. Proc. Soc. Exptl. Biol. Med. 208: 120–123.
196. KAROLL, K.K. 1982. Hypercholesterolemia and atherosclerosis: effects of dietary protein. Fed. Proc. 41: 2792–2796.
197. ANDERSON, J.W., JOHNSTONE, B.M., COOK-NEWELL, M.E. 1995. Meta-analysis of the effects of soy protein intake on serum lipids. N. Engl. J. Med. 333: 276–282.
198. OBERMEYER, W.R., MUSSER, S.M., BETZ, J.M., CASEY, R.E., POHLAND, A.E., PAGE, S.W. 1995. Chemical studies of phytoestrogens and related compounds in dietary supplements: flax and chaparral. Proc. Soc. Exptl. Biol. Med. 208: 6–12.
199. MacLEOD, G., AMES, J. 1988. Soy flavour and its improvement. CRC Crit. Rev. Food Sci. Nutr. 27: 219–400.
200. KOLODZIEJCZYK, P.P., FEDEC, P. 1995. Processing flaxseed for human consumption. In: Flaxseed in Human Nutrition. (S.C.Cunnane, L.U.Thompson, eds.), AOCS Press, Champaign, IL, pp. 261–280.
201. JONES, J.M. 1992. Food Safety, Eagan Press, St. Paul, Minnesota, pp. 69–105.
202. MERZ, B. 1983. Adding seeds to the diet may keep cancer at bay. J. Am. Med. Assoc. 149: 2746.
203. KENNEDY, A.R. 1990. Effects of protease inhibitors and vitamin E in the prevention of cancer. In: Nutrient and Cancer Prevention (K.N. Prasad, F.L. Meyskens Jr., eds.), Humana Press, Clifton, N.J., pp. 79–98.
204. OAKENFULL, D., SIDHU, G.S. 1989. Saponins. In: Toxicants of Plant Origin. Vol. 2 Phenolics. (P.R. Cheeke, ed.), CRC Press, Boca Raton, FL, pp. 97–142.
205. BIRK, Y., PERI, I. 1980. Saponins. In: Toxic constituents in Plant Foodstuffs, 2nd ed. (I.E. Leiner, ed.), Academic Press, N.Y., 161 pp.
206. NESARETNAM, K., KHOR, H.T., GANESON, J., CHONG, Y.H., SUNDRAM, K., GAPOR, A. 1992. The effect of vitamin E tocotrienols from palm oil on chemically-induced mammary carcinogenesis in female rats. Nutr. Res. 12: 63–75.
207. HOLUB, B.J. 1988. Dietary fish oils containing eicosapentaenoic acid and the prevention of atherosclerosis and thrombosis. Can. Med. Assoc. J. 139: 377–381.
208. PARIZA, M.W. 1988. Dietary fat and cancer risk: Evidence and research needs. Annu. Rev. Nutr. 8: 167–183.
209. CUNNANE, S.C., McINTOSH D. 1994. α-Linolenic acid and reduction in cardiac deaths. Lancet 344: 622.
210. FERRIER, L.K., CASTON, L.J., LEESON, S., SQUIRES, J., WEAVER, B.J., HOLUB, B.J. 1995. α-Linoleic acid- and docosahexaenoic acid-enriched eggs from hens fed flaxseed: influence on blood lipids and platelet phospholipid fatty acids in humans. Am. J. Clin. Nutr. 62: 81–86.
211. GARLAND, C.F., GARLAND F.C. 1980. Do sunlight and vitamin D reduce the likelihood of colon cancer? Int. J. Epidemiol. 9: 227–231.
212. VARGAS, P.A., ALBERTS, D.S. 1992. Primary prevention of colorectal cancer through dietary modification. Cancer 70: 1229–1235.
213. PENCE, B.C., BUDDINGH, F. 1988. Inhibition of dietary fat-promoted colon carcinogenesis in rats by supplemental calcium or vitamin D. Carcinogenesis 9: 187–190.
214. WEISBURGER, J.H., RIVENSON, A., HARD, G.C., ZANG, E., NAGAO, M., SUGIMURA, T. 1994.Role of fat and calcium in cancer causation by food mutagens, heterocyclic amines. Proc. Soc. Exptl. Biol. Med. 205: 347–352.
215. LIPKIN, M., NEWMARK, H. 1985. Effect of added dietary calcium on colonic epithelial cell proliferation in subjects at high risk for familial colorectal cancer. N. Engl. J. Med. 313: 1381–1384.

216. WAGNER, C. 1995. Biochemical role of folate in cellular metabolism. In: Folate in Health and Disease (L.B. Baily, ed.), Marcel Dekker Inc., New York, pp. 23–42.
217. FREUDENHEIM, J.L., GRAHAM, S., MARSHALL, J.R., HAUGHEY, B.P., CHOLEWINSKI, S., WILKINSON, E.D. 1991. Folate intake and carcinogenesis of the colon and rectum. Int. J. Epidemiol. 20: 368–374.
218. ZIEGLER, R.G., BRINTON, L.A., HAMMAN, R.F., LEHMAN, H.F., LEVINE, R.S., MALLIN, K., NORMAN, S.A., ROSENTHAL, J.F., TRUMBLE, A.C., HOOVER, R.N. 1990. Diet and the risk of invasive cervical cancer among white women in the United States. Am. J. Epidemiol. 132: 432–445.
219. MASON, J.B. 1994. Folate and colonic carcinogenesis. Searching for a mechanistic understanding. J. Nutr. Biochem. 5: 170–175.
220. AMES, B.N., GOLD, L.S., WILLETT, W.C. 1995. The causes and prevention of cancer. Proc. Natl. Acad. Sci. U.S.A. 92: 5258–5265.
221. BIRD, C.L., SWENDSEID, M.E., WITTE, J.S., SHIKANY, J.M., HUNT, I.F., FRANKL, H.D., LEE, E.R., LONGNECKER, M.P., HAILE, R.W. 1995. Red cell and plasma folate, folate consumption, and the risk of colorectal adenomatous polyps. Cancer Epidemiol. Biomarkers Prev. 4: 709–714.
222. BOWER, C., WALD, N.J. 1995. Vitamin B_{12} deficiency and the fortification of food with folic acid. Europ. J. Clin. Nutr. 49: 787–793.
223. BUTTERWORTH, C.E., TAMURA, T. 1989. Folic acid safety and toxicity: a brief review. Am. J. Clin. Nutr. 50: 353–358.
224. FDA, 1996. Food Additives Permitted for Direct Addition to Food for Human Consumption; Folic Acid (Folacin). Federal Register, March 5, 61: 8797–8807.
225. MORRISON, H.I., SCHAUBEL, D., DESMEULES, M., WIGLE D.T. 1996. Serum folate and risk of fatal coronary heart disease. J. Am. Med. Assoc. 275: 1893–1896.
226. PARIZA, M.W., HA, Y.L. 1990. Conjugated dienoic derivatives of linoleic acid: mechanism of anticarcinogenic effect. In: Mutagens and Carcinogens in the Diet, (M.W. Pariza, H.-U. Aeschbacher, S. Sato, eds.), Wiley-Liss, New York, NY, pp. 217–221.
227. IP, C., SINGH, M., THOMPSON, H.J., SCIEMECA, J.A. 1994. Conjugated linoleic acid suppresses mammary carcinogenesis and proliferative activity of the mammary gland in the rat. Cancer Res. 54: 1212–1215.
228. OHTA, T. 1993. Modification of genotoxicity by naturally occurring flavourings and their derivatives. Crit. Rev.Toxic. 23: 127–146.
229. OHTA, T., WATANABE, M., SHIRASU, Y., INOUE, T. 1988. Post-replication repair and recombination in *uvrA umuC* strain of *Escherichia coli* are enhanced by vanillin, an antimutagenic compound. Mutat. Res. 201: 107–113.
230. KOHLMEIER, L., SIMONSEN, N., MOTTUS, K. 1995. Dietary modifiers of carcinogenesis. Environ. Health Persp. 103: 177–184.
231. FISHER. C. 1992. Phenolic compounds in spices. In: Phenolic Compounds in Food and Their Effects on Health I, (C.-T. Ho, C.Y. Lee, M.-T. Huang, eds.), ACS Symposium Series No. 547. Amercan Chemical Society, Washington, DC, pp. 118–129.
232. GOUD, V.K., POLASA, K., KRISHNASWAMY, K. 1993. Effect of turmeric on xenobiotic metabolising enzymes. Plant Food Hum. Nutr. 44: 87–92.
233. TALALAY, P., De LONG, M.J., PROCHASKA, H.J. 1988. Identification of a common chemical signal regulating the induction of enzymes that protect against chemical carcinogenesis. Proc. Natl. Acad. Sci. U.S.A. 85: 8261–8265.
234. WATTENBERG, L.W. 1990. Inhibition of carcinogenesis by naturally-occurring and synthetic compounds. In: Antimutagenesis and Anticarcinogenesis Mechanisms II, (Y. Koroka, D.M. Shankel, M.D. Waters, eds.), Plenum Press, New York, pp. 155–166.
235. PATTERSON, R.E., HAINES, P.S., POPKIN, B.M. 1994. Diet Quality Index: capturing a multidimensional behavior. J. Am. Diet. Assoc. 94: 57–64.

236. KANT, A.K., SCHATZKIN, A., ZIEGLER, R.G. 1995. Dietary diversity and subsequent cause-specific mortality in the NHANES I Epidemiologic Follow-up Study. J. Am. Coll. Nutr. 14: 233–238.
237. BLOCK, G. 1995. Are clinical trials really the answer? Am. J. Clin. Nutr. 62 (Suppl.): 1517S-1520S.
238. ITO, N., HIROSE, M., SHIRAI, T. 1992. Carcinogenicity and modification of carcinogenic response by plant phenols. In: Phenolic Compounds in Food and Their Effects on Health. II. Antioxidants and Cancer Prevention, (M.-T. Huang, C.-T. Ho, C.Y. Lee, eds.), ACS Symposium Series 507, American Chemical Society, Washington, DC pp. 269–281.
239. HIROSE, M., OZAKI, K., TAKUBA, K., FUKUSHIMA, S., SHIRAI, T., ITO, N. 1991. Modifying effects of the naturally occurring antioxidants γ-oryzanol, phytic acid, tannic acid and n-tritriacontane-16,18-dione in a rat wide-spectrum organ carcinogenesis model. Carcinogenesis 12: 1917–1921.
240. IARC. 1993. Monographs on the Evaluation of Carcinogenic Risks to Humans. Some Naturally Occurring Substances: Food Items and Constituents, Heterocyclic Aromatic Amines and Mycotoxins. 56, IARC, Lyon, France, pp. 115–134.
241. WATTENBERG, L.W. 1985. Chemoprevention of cancer. Cancer Res. 45: 1–8.
242. TANAKA, T., MORI, Y., MORISHITA, Y., HARA, A., OHNO, T., KOJIMA, T., MORI, H. 1990. Inhibitory effect of sinigrin and indole-3-carbinol on diethylnitrosamine-induced hepatocarcinogenesis in male AC/N rats. Carcinogenesis 11: 1403–1406.
243. STAMPER, M.J., RIM, E.B. 1995. Epidemiologic evidence for vitamin E in prevention of cardiovascular disease. Am. J. Clin. Nutr. 62: 1365S-1369S.
244. BYERS, T., GUERRERO, N. 1995. Epidemiologic evidence for vitamin C and vitamin E in cancer prevention. Am. J. Clin. Nutr. 62: 1385S-1392S.
245. van POPPEL, G. GOLDBOHM, R.A. 1995. Epidemiologic evidence for beta-carotene and cancer prevention. Am. J. Clin. Nutr. 62: 1393S-1402S.
246. Editorial. 1993. Disease prevention effects of antioxidants explored at ILSI meetings in Washington and Stockholm. ILSI News 11: 2–6.
247. HUNTER, D.J., MANSON, J.A., COLDITZ, G.A.S., STAMPER, M.J., ROSNER, B., HENNEKENS, H.C., SPEIZER, F.E., WILLETT, W.C. 1993. A prospective study of the intake of vitamins C, E, and A and the risk of breast cancer. N. Engl. J. Med. 329: 234–240.
248. GREENBERG, E.R., SPORN, M.B. 1996. Antioxidant vitamins, cancer, and cardiovascular disease. N. Engl. J. Med. 334 1189–1190.
249. MORSE, M.A., STONER, G. 1993. Cancer chemoprevention: Principles and prospects. Carcinogenesis 14: 1737–1746.
250. KELLOFF, G.J., BOONE, C.W., MALONE, W.F., STEELE, V.E. 1992. Chemoprevention clinical trials. Mutat. Res. 267: 292–295.
251. BLUMBERG, J.B. 1995. Considerations of the scientific substantiation for antioxidant vitamins and β-carotene in disease prevention. Am. J. Clin. Nutr. 62: 1521S-1526S.
252. BASU, T.K., DICKERSON, J.W. 1996. Vitamins in Human Health and Disease. CAB International, Guildford, UK, pp. 125–190; 214–227; 252–266.
253. STAHL, W., SUNDQUIST A.R., SIES, H. 1994. Role of carotenoids in antioxidant defense. In: Vitamin A in Health and Disease. (R. Blomhoff, ed.), Marcel Dekker Inc., New York, pp. 275–288.
254. HUNTER, D.J., WILLETT, W. 1994. Vitamin A and cancer. In: Vitamin A in Health and Disease. (R. Blomhoff, ed.), Marcel Dekker Inc., New York, pp. 561–584.
255. SOMER, E., Health Media of America. 1995. The Essential Guide to Vitamins and Minerals. HarperCollins Publishers Inc., New York, pp. 17–84; 149–279.
256. Diet and Health, Implications for Reducing Chronic Disease Risk. 1989. Committee on Diet and Health, Food and Nutrition Board, and Commission on Life Sciences. National Academy Press, Washington, D.C., pp. 665–710.

Chapter Four

CONSTITUENTS OF WILD FOOD PLANTS

Angela Sotelo

Departamento de Farmacia
Facultad de Química
Universidad Nacional Autónoma de México
México, D. F., 04510 México

INTRODUCTION

Increasing food production for humans and domestic animals has long been a world-wide preoccupation of scientists. In countries with high population indices and economies which depend upon their raw materials, the problem is even more serious. This has stimulated investigators to learn more about exploitation of the natural resources of their own countries. There is an increasing need for new sources of protein and fat, both for the growing human population and for domestic animal feeding. In countries with poverty and high birth rates, plants are considered the major source of dietary protein.

Functionality of Food Phytochemicals
edited by Johns and Romeo, Plenum Press, New York, 1997

Food plants are defined as those crops which grow wild or are cultivated and gathered or harvested for food. Indigenous wild food plants are often called native or traditional plants. Wild plants can be considered under-exploited or neglected food sources. Many of them, however, locally provide complementary foods in several countries. Mainly in Africa, but also in other tropical countries, wild plants are used as food and play an important role in domestic animal feeding.[1] During the last 20 years, our research group has been interested in the study of Mexican wild flora from the nutritional point of view. The present paper is part of a comprehensive study of wild plants undertaken to: 1) establish the nutrient composition of wild seeds, mainly legumes; 2) detect toxins and antinutritional factors; and 3) select from the wild plants studied those proposed for human or animal use, by performing the necessary biological studies.[2–4] An important advantage of most wild or semi-wild plants is their tolerance to extreme natural conditions, such as poor soils, erratic rainfall, pests, and diseases.

MEXICAN PLANTS

Since it is located between a moderately northern climate and a southern tropical zone with an extensive subtropical area, Mexico possesses one of the most varied flora in the world. It has a diversity of climates and soils, with high jungles in the southeastern regions, northern desert areas in Chihuahua and Sonora, and high and cold regions in the volcanic chain.[5]

Legumes and cereals are some of the earliest food crops cultivated by different groups of ancient cultures. Legumes have been eaten by humans for over 8,000 years. In Mexico, beans (*Phaseolus vulgaris* and *P. lunatus*), together with corn, squash, and peppers, have been found in archeological zones proving that their consumption dates from 1500–4000 years B.C.[6] The *Leguminosae* family has a world-wide distribution and comprises 560–670 genera represented by 12000–17000 species.[7] Besides food, legumes provide important economic resources (*e.g.* drugs and dyes); their role, moreover, in soil nitrogen fixation, in symbiosis with *Rhizobium*, is of central importance for agriculture. Mexico has about 134 genera of legumes with 1707 species.[8] Most of these wild legumes are totally unexploited, leaving untapped a large potential source of protein.

Nutritive Value

The nutrients contained in wild plants are, in most cases, similar to those of the cultivated varieties belonging to the same species. The main difference is the higher content of antinutritional factors present in the wild representatives. The high protein content in edible legumes causes them to be labeled the "meat of the poor people". The chemical composition of wild legumes is comparable

Table 1. Proximate analysis of some Mexican wild seeds

Species	Moisture %	Dry basisg/100 g of sample				
		Crude protein	Fat	Ash	Fiber	CHO'S[1]
Wild legumes from Yucatan Peninsula[a]						
Acacia gaumeri	5.2	19.2	0.9	3.7	11.2	65.0
Acacia pennatula (WL)[2]	5.9	10.7	0.4	3.9	34.9	50.1
Albizia lebbeck	6.4	37.1	3.4	0.1	9.9	45.5
Caesalpinia vesicaria	8.0	14.6	4.3	3.3	11.6	66.3
Caesalpinia yucatanensis	10.7	22.5	15.6	4.6	7.4	49.9
Leucaena leucocephala	3.7	26.6	5.6	4.1	13.7	50.1
Lonchocarpus longystilus	3.5	24.5	31.3	3.6	8.7	31.8
Pithecellobium keyense	4.2	16.1	14.4	2.2	9.3	58.0
Pithecellobium saman	5.1	37.6	6.9	3.1	9.2	43.2
Senna atomaria	7.9	24.2	0.8	4.7	12.5	57.8
Wild lupins from Jalisco State [b]						
L. elegans	3.9	45.4	5.8	4.2	12.9	31.7
L. exaltatus	4.2	40.5	8.5	3.6	14.6	32.8
L. reflexus	3.1	37.3	7.9	3.6	16.6	34.6
L. rotundiflorus	3.4	42.8	5.5	4.0	15.1	32.6
L. simulans	2.9	40.7	6.3	3.6	14.4	35.0
L. splendens	10.1	37.0	8.9	3.3	12.7	38.1
L. spp	3.7	41.5	6.8	3.5	15.4	32.8

[1]CHO'S = Carbohydrate calculated by difference.
[2]WL = Whole legume seeds + pod.
[a]Ref 11
[b]previously unpublished data

to that of most of the edible ones. In some cases, *e.g. Gliricidia sepium* and some *Lupinus* species, they have a composition similar to soy bean, which is considered an exceptional legume having 38–40% protein and 20% fat.[9,10] Table 1 shows the nutrient composition of some Mexican wild legume seeds collected in Yucatan and also wild lupins from the state of Jalisco.[11] It is important to notice the high protein concentration in *Pithecellobium saman* and *Albizia lebbeck*. The protein content in the wild lupins is similar to that found in cultivated species (37–45% crude protein). As it is important to look for new sources of fat, wild seeds are potentially important in this regard. *Lonchocarpus longystilus* contains 31% fat, which makes this seed very important as a potential source. None of the studied lupin species, however, showed a significant fat content.

Phaseolus vulgaris

Of the nearly 40 species of *Phaseolus* beans found in Mexico and Central America, only 4 have been domesticated: *P. vulgaris, P. lunatus, P. acutifolius* and *P. coccineus*. There are multiple varieties of each, which have represented the most important protein sources in the diet of the people of Mesoamerica.[12] Today it is possible to find the wild ancestors of *P. vulgaris*. Additionally, the chemical variability between the wild and the cultivated species has been demonstrated.[13,14] Moreover, there are phenotipic differences, *e.g.* in shape and size of the seeds, germination rates, water permeability, growth and reproductive habits, resistance to disease, and plagues and parasites.

One indication of chemical variability in these species is the remarkable resistance of wild varieties to plague attack.[15] This protective property is presumably due to toxic natural products. Recently a study of Mexican wild and cultivated *Phaseolus vulgaris* was undertaken. Significant differences in regard to protein, fat, and ash and several antinutritional factors were found.[16] Wild beans had more crude protein, fiber, and ash than cultivated beans, which have more energy nutrients (fats and carbohydrates). All essential amino acids, however, were in higher concentration in cultivated beans than in the wild ones. In both wild and cultivated beans, the limiting amino acids were sulfur amino acids with cultivated beans having a higher content. The concentration of antinutritional factors was significantly higher in wild beans than cultivated varieties. Domestication, therefore, seems to have created a positive amino acid profile (better protein quality) and a lower antinutritional factor content.

Oil Seeds

There is an important deficit of fat production in Mexico and other countries. Therefore, wild plants are potentially important because of their high fat concentrations. After domestication, they can be useful as animal feeds and food for human consumption. Table 2 shows the proximate analysis (percentage of macronutrients) of some Mexican wild oily seeds. Fatty acid composition is shown in Table 3. In the group studied, nine belong to the Leguminosae family whose fatty acid composition showed the highest unsaturated fatty acid content. Physical and chemical characteristics are similar to those of edible oils.

Among the oily seeds it is important to mention *Theobroma*. Twenty-two species of this genus, one of them being the cacao bean, have been reported.[19] In Mexico, *T. bicolor* and *T. angustifolium* grow wild in the states of Chiapas and Tabasco and are used by natives for various purposes. The pulp of the fruit of *T. bicolor* is used to prepare a refreshing beverage, while the seeds are mixed with cacao beans as an adulterant to obtain a bitter chocolate. Less information exists on the traditional uses of *T. angustifolium*. The fat content of both wild *Theobroma* is lower than *T. cacao* (Table 2). However, the fat content varies with time of collection.[18,20,21]

Figure 1 illustrates the distribution of theobromine, caffeine, and theophylline, in three different plant parts of *T. cacao* and wild *Theobroma* species. In the two varieties of *T. cacao*, the seeds were the most important reservoir of alkaloids, while in *T bicolor*, the flowers and leaves have a higher concentration

Table 2. Proximate analysis of some wild oily seeds[1]

Sample	(g/100 g of sample)					
	Moisture	Protein	Fat	Fiber	Ash	CHO'S[2]
Bauhinia purpurea (Leguminosae)	2.1	29.9	14.8	3.8	3.8	44.5
Caesalpinia crista (Leguminosae)	3.9	23.5	23.5	4.7	5.1	39.3
Caesalpinia pulcherrima (Leguminosae)	4.1	21.0	24.1	7.2	3.8	39.8
Callophylum brasilensis (Guttiferae)	4.8	3.9	17.0	5.0	1.4	67.9
Castilla elastica (Moraceae)	9.9	12.8	30.5	18.0	2.9	30.9
Cymbopetalum penduliflorum (Leguminosae)	7.2	11.2	27.6	47.4	2.6	11.2
Entada polystachia (Leguminosae)	3.9	24.6	24.4	4.9	5.3	40.8
Erythrina americana (Leguminosae)	5.8	27.2	17.7	15.3	3.6	30.4
Erytrina breviflora (Leguminosae)	5.0	23.3	13.9	21.6	3.4	32.8
Gliricidia sepium (Leguminosae)	5.8	39.3	21.5	6.3	3.8	23.3
Guarea chichon (Meliaceae)	9.1	8.2	17.9	33.3	3.4	30.2
Guarea excelsa (Meliaceae)	8.7	9.5	14.5	19.8	3.4	44.1
Pithecellobium flexicaule (Leguminosae)	8.3	28.9	13.7	13.4	3.2	30.5
Theobroma angustifollium (Sterculiaceae)	3.2	10.5	46.0	21.7	4.5	14.1
Theobroma bicolor (Sterculiaceae)[3]	5.3	19.2	34.2	23.2	3.9	13.9
Theobroma cacao (Sterculiaceae)	3.9	13.8	51.0	8.4	3.2	19.7
Trichilia hirta (Meliaceae)	4.5	14.0	49.2	24.4	2.5	5.4

[1]Refs. 2, 9, 17, 18, 20
[2]CHO'S = Carbohydrates calculated by difference

than the seeds. In *T. angustifolium*, the flower contained most of the theophylline.[21]

Leaves and Flowers

In Mexican traditional culinary practices, leaves and flowers from wild plants are eaten. Many of them have toxic compounds, but the people have learned how to eliminate or diminish their effects, mainly by discarding the cooking broth. Table 4 shows the nutritive value of some edible leaves and flowers. When the nutrient content is presented on a dry weight basis, the protein content of the leaves is similar to or higher than that of legume seeds, and no diferences in the chemical composition, compared with other edible cultivated leaves, are found. In most cases, the "true protein" (nitrogen from protein) was 42 - 88% of the total nitrogen. Limiting sulfur amino acids were found in the protein of all leaves and flowers studied. There is a high concentration of Na, K, Ca, and Mg in leaves of wild plants, but the minerals are generally in lower concentration in flowers.[22,23]

Table 3. Fatty acid composition of some Mexican wild oily seeds[1]

Sample	Fatty acid composition %								
	C_{16}	C_{18}	C_{20}	C_{22}	$C_{18:1\Delta}^{9}$	$C_{18:2\Delta}^{9,12}$	$C_{18:3\Delta}^{9,12,15}$	Not identified	Unsaturated %
Caesalpinea crista	8.06	3.08			17.38	71.48			88.86
Castilla elastica	39.77	1.42			16.78	41.69	0.34		58.81
Callophylium brasilensis	36.24	8.32			48.58	6.07	0.79		55.44
Gliricidia sepium	19.76	18.61	3.37		17.12	40.91			58.03
Guarea chichon	18.04	10.43			36.73	28.93	5.86		71.53
Guarea excelsa	26.13	6.79	1.43		42.11	12.85	6.97	3.72	65.47
Pithecellobium flexicaule	15.34	9.39	1.38	1.03	32.60	39.27		0.39	72.96
Theobroma angustifolium	5.00	25.00	11.29	3.91	49.67	5.13			54.80
Theobroma bicolor	6.11	50.44	1.26		39.38	2.81			42.19
Trichilia irta	25.72	9.54			43.13	18.79	1.94	0.63	64.49

[1]Ref. 18

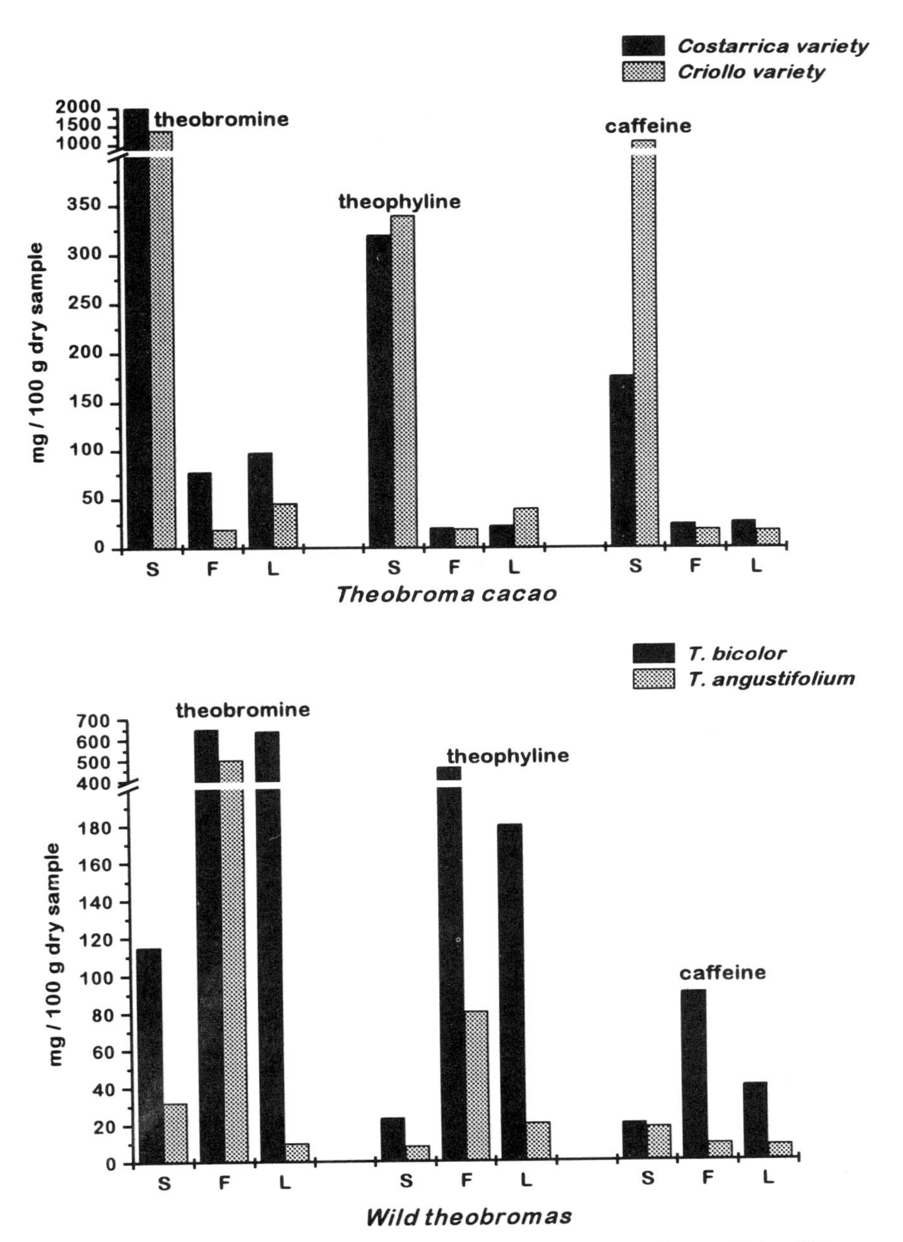

Figure 1. Alkaloid concentration in seeds (S), flowers (F) and leaves (L) of two varieties of *T. cacao* and two wild species.

Table 4. Proximate analysis and true protein of edible leaves and flowers of wild and cultivated plants[1]

Scientific name	Local name	Moisture %	g/100 g dried sample					
			Crude protein	Crude fat	Fiber	Ash	Carbohydrates	True protein[2]
Leaves								
Wild plants								
Crotalaria pumila	Chipilin	83.6	36.2	1.3	11.7	9.4	41.3	23.8
Dondia diffusa	Romeritos	89.4	28.1	1.2	8.2	32.9	29.6	14.2
Onidoscolus aconitifolius	Chaya	76.3	33.8	3.4	10.7	12.0	40.0	23.9
Piper auritum	Hoja Santa	80.4	27.0	1.9	15.2	15.6	40.2	23.8
Portulaca oleracea	Verdolaga	90.5	28.9	2.3	10.6	29.0	29.2	12.7
Solanum americana	Hierba Mora	85.8	35.0	1.1	13.6	17.5	32.8	17.1
Taraxacum officinale	Diente de León	86.7	21.0	3.0	16.6	20.1	39.3	12.1
Cultivated plants								
Beta vulgaris	Salt-wort	88.8	27.8	1.0	12.4	20.9	37.9	12.6
Medicago sativa	Lucern	79.5	28.5	1.7	18.2	13.3	32.2	12.0
Nostrurtium officinales	Water cress	90.0	33.1	1.5	13.8	12.0	40.5	25.2
Spinacea oleracea	Spinach	86.5	35.7	1.6	9.2	20.8	32.8	23.2
Flowers								
Agave atrovirens	Flor de Maguey	86.8	14.1	3.0	13.8	7.0	61.9	6.4
Chamaedorea tepejilote	Pacaya, Tepejilote	86.8	29.2	1.3	10.4	15.3	43.7	15.9
Cucurbita pepo	Flor de Calabaza	91.7	26.5	4.1	9.7	14.1	45.5	14.7
Erythrina americana	Colorín	91.2	23.2	1.8	21.6	8.5	55.0	13.7
Yucca elephantipes	Yuca, Izote	87.3	23.0	1.9	12.7	9.9	52.5	11.4

[1]Previously Unpublished data
[2]Lucas et al 1988 (Ref. 79)

ANTINUTRITIONAL FACTORS AND TOXINS

There is no clear division between natural antinutritional factors (ANF) and toxins. ANFs are those natural compounds that interfere with nutrient availability. These substances produce nutrient inbalance which in the long run promotes a particular pathology. Toxins are compounds that have a toxic effect that cannot be reduced by nutrient supplementation.[24] Protease and amylase inhibitors, tannins, saponins, lectins, and phytate are considered ANFs. Alkaloids, cyanogenic glycosides, toxic amino acids are considered toxic compounds. Some workers include alkaloids and cyanogenic glycosides in the ANF's group and others consider lectins as toxins since some of them can kill experimental animals.[25,26]

Many antinutritional factors and toxins are found in edible and cultivated plants but people have often learned how to eliminate them. In wild plants, they are usually found in higher concentrations. Sometimes they are heat resistant and difficult to eliminate. Some antinutritional factors and toxins distributed in the plant kingdom and their effects are shown in Table 5.

Cyanogenic Glycosides

Cyanogenesis, known for more than a century, is the production of the respiratory poison hydrogen cyanide (HCN) by biological organisms. Since its

Table 5. Natural antinutritional factors and toxins in edible plants[a]

ANF	Distribution	Physiological effect
Amylase inhibitors	Legumes and cereals	Interference with starch digestion
Lectins	Most legumes	Depressed growth; death
Protease inhibitors	Most legumes	Depressed growth; pancreatic hypertrophy, hyperplasia
Amino acid analogues		
β-N-methylamino-L-alanine	*Cycas circinalis*	Neurotoxin
β-N-oxalyl-α, β propionic acid	*Lathyrus sativus*	Lathyrism
Glycosides		
Alkaloids	Lupins, potatoes	Depressed growth
Cyanogen	Lima bean	Respiratory failure
Cycasin	Cycas circinalis	Carcinogen
Glucosinolates	Cruciferous seeds	Gointrogenic and hepatotoxic effect
Gossypol	Cotton seeds	Infertility in male
Oligosaccharides	Most legumes	Flatulence
Phytates	Legumes and cereals	Interference with mineral availability
Saponins	Most legumes	Affects intestinal permeability
Tannins	Widley distributed	Interference with protein digestibility
Vicine/convicine	*Vicia faba*	Hemolytic anemia

[a] I. Liener. Refs. 25 and 58

discovery in plants by Liebig and Wohler in 1837, over 2500 cyanogenic plants have been described. Families with major cyanogenic activity are the Rosaceae, Leguminosae, Gramineae, Araceae, Compositae, Euphorbiaceae, and Passifloraceae.[27,28] The ability to produce HCN is exhibited by at least 1000 species of plants representing approximately 90 families and 250 genera. Cyanogenic plants have been responsible for many cases of acute and chronic cyanide poisoning in animals and humans.[29] HCN is an effective inhibitor of many metalloenzymes. Cytochrome oxidase, the terminal oxidase of aerobic respiration, is the primary site of action, provoking death due to oxygen starvation at the cellular level.

In normal conditions, although the hydrolytic enzyme which liberates HCN from glycosides is found in the same vegetable material, hydrolysis is prevented by subcellular and/or tissue compartmentation. The enzyme acts only when the tissue is disrupted and the cyanogenic glycosides are rapidly catabolized by two distinct enzymes, namely β-glycosidase and α-hydroxynitrile lyase. Since large-scale HCN release occurs only upon disruption of the plant tissue, it has been suggested that cyanogenic glycosides constitute defense compounds against herbivore predation.[30] Cyanogenic glycosides are derived from amino acids tyrosine, phenylalanine, valine, isoleucine, and leucine.

Plant breeders have been successful in breeding low cyanide varieties. Conn[28] has reported that there is an interesting correlation between the amount of cyanogenic glycosides in the seed and the color and size of the bean. Similar

Table 6. Toxins and antinutritional factors, of wild and cultivated lima beans (*Phaseolus lunatus*)[a]

Sample	Hydrocyanic acid (mg HCN/100 mg sample)	Trypsin inhibitors TUI[b]/mg	Hemagglutinins[c]		
			Cow	Human "A"	Rabbit
Wild beans					
IBES-CHOH-524-B	69.11	47.50	—	9	—
IBES-CHOH-527	24.57	82.05	—	10	—
IBES-CHOH-572	46.90	86.25	—	10	—
Del monte beans (raw)	158.00	94.50	—	9	—
Del monte beans (cooked)	0.54	3.82	—	—	—
Cultivated beans					
IBES-503-B	1.06	56.40	—	9	—
Munición morado 562	2.03	82.50	—	9	—
Xolis-IBES-500 (raw)	1.18	57.50	—	4	—
Xolis-IBES-500 (cooked)	—	—	—	—	—

[a]Ref 31
[b]Trypsin units inhibited
[c]The highest dilution that produces agglutination of RBC of the three different species.

results have been found by De la Vega and Sotelo[31] with wild, semi wild, and cultivated lima beans (*P. lunatus*). A high concentration of cyanogenic glycosides present in wild beans elevates total nitrogen levels. For this reason, it is important to measure protein when cyanogenic glycosides are detected. Table 6 shows that cyanogenic glycosides are present in many wild and cultivated lima bean varieties.

Protease Inhibitors

Protease inhibitors inhibit the proteolytic activity of enzymes. They are found in many tissues of plants, animals, and microorganisms. The most studied protein inhibitors from plants are those that act against pancreatic proteases: trypsin, chemotrypsin, elastase, and carboxyipeptidases. There are scores of protease inhibitors of plant origin dispersed among different botanical families. Many are found not only in Leguminosae,[25,32,33] but also in potato tubers and cereals.[34–36] The first report was published in 1938, and the first isolation of a plant protease inhibitor from soy bean was performed by Kunitz in 1945.[37]

The protease inhibitors present in food plants and their nutritional and toxicological significance have been reviewed by Le Guen.[38] There are marked differences in the degree to which pancreatic proteases of different animals are inhibited by various legume protease inhibitors.[39–41] The growth rate of rats, food digestibility, and metabolic utilization are affected by the ingestion of protease inhibitors.[38] Most of the *in vivo* activity has been studied by using soy bean extract or purified proteins. Orally ingested inhibitors inactivate trypsin and chemotrypsin in the intestine and block protein digestion. The inhibition of proteases in the small intestine stimulates, by feed back control mechanisms, pancreatic enzyme secretion. As pancreatic enzymes are rich in sulfur amino acids, this stimulation creates a loss of methionine and cysteine for body tissue synthesis. During food processing, most protease inhibitors can easily be destroyed by wet heating.[32]

The physiological role of protease inhibitors in plants is unclear, but there are indications of some functions. Some investigations have suggested they play a role in controlling endogenous proteases which are involved in plant defense against insects.[42] Others suggest they have a storage role, mainly as a sulfur depot, because they are rich in sulfur amino acids.[43]

There are many reports of the disappearance of protease inhibitors during germination.[44,45] Recently, our group measured trypsin inhibitor concentration of two varieties of *Phaseolus. vulgaris* throughout the development of the seed: dry seed, germination phase, seedling, different parts of the developing plant (leaves, root and flowers), string bean, mature seed (physiological ripe), and final dry seed. Figure 2 shows the concentrations during different stages of development of "canario" and "negro jamapa" beans. There is a decay of trypsin inhibitors in the germination phase. They nearly disappear in the seedling and

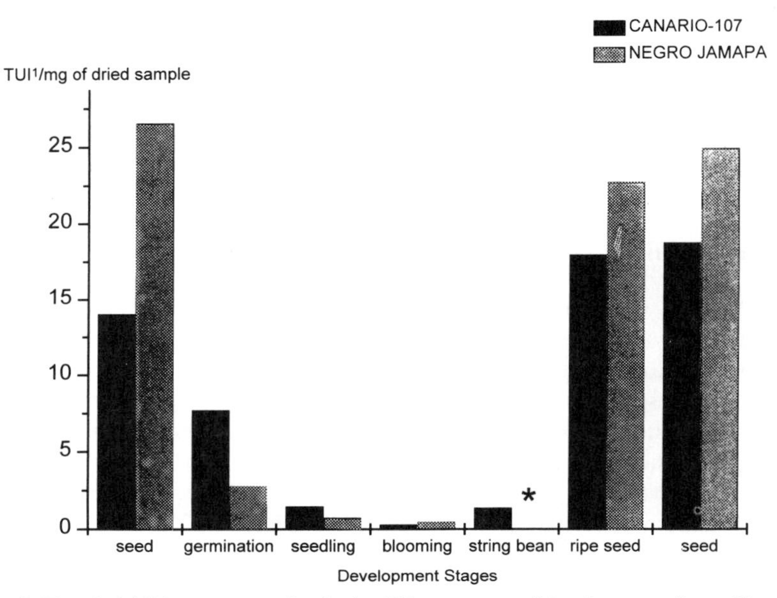

Figure 2. Trypsin inhibitor concentration in the different stages of development of two *Phaseolus vulgaris* bean varieties. TUI[1] = Trypsin units inhibited (Ref. 33). * = Not detected.

blooming phases, and then begin to increase in the "canario"string bean (they are absent in the "negro-jamapa" variety in this phase). Finally, in the mature seed, they appear again in high concentration in both varieties. One interpretation of these results is that the protease inhibitors may be controling the activity of endogenous proteases involved in the germination process. (unpublished data).

Table 7 shows trypsin inhibitors present in several non-cultivated legumes and potatoes. Interestingly, empty pods of *Pithecellobium keyense* show a high trypsin inhibitor activity, superior to that found in seeds. All the wild Mexican lupins have a very low content of trypsin inhibitors. Alkaloids appear to be major toxic compounds in these legumes. In contrast, wild potatoes have trypsin inhibitor concentrations similar to those of wild legumes (Table 7). Trypsin inhibitors may possibly be a defense against insect larvae and microorganisms present in the soil.

Lectins

Goldstein defined lectins as sugar-binding proteins or glycoproteins of non-immune origin which agglutinate cells and/or precipitate glycoconjugates.[46] Most lectins have been studied from higher plants, but they have also been found in animals and microorganisms.[47] Stillmark, in 1888, was the first to describe the agglutination of animal erythrocytes by ricin, a protein isolated from the seed of *Ricinus communis*.[48] Hellin, in 1891, and Elfstrand, in 1898, discovered abrin in

Table 7. Antinutritional factors and toxins present in some wild legume and potatoes

Sample	Trypsin inhibitorsTUI[1]/mg dry sample	Lectins hamster RBC[2]	Alkaloids qualitative test	Saponins (hemolysis)[2]
Wild legumes from Yucatan Peninsula (Sotelo et al. 1993). Seed and empty pod[2]				
Acacia gaumeri				
Seed	23.07	0	–	5
Pod	1.38	0	–	0
Acacia pennatula (seed + pod)	8.12	10	–	0
Albizia lebbeck				
Seed	25.83	0	–	5
Pod	0.63	3	–	0
Caesalpinia vesicaria				
Seed	6.33	6	–	0
Pod	1.5	6	–	0
Caesalpinia yucatanensis				
Seed	60.13	3	–	0
Pod	10.53	3	–	0
Leucaena leucocephala				
Seed	6.80	4	–	0
Pod	2.50	3	–	0
Lonchocarpus longystilus				
Seed	0.33	0	–	0
Pod	0.42	0	–	0
Pithecellobium keyense				
Seed	15.14	10	–	10
Pod	169.31	17	–	0
Pithecellobium saman				
Seed	10.14	0	+	0
Pod	1.54	12	+	5
Senna atomaria				
Seed	0.33	3	–	0
Pod	5.91	4	–	0
Wild lupinus from Jalisco State Ruiz and Sotelo (unpublished data)				
Lupinus elegans	1.09	5	+	*
Lupinus exaltatus	0.87	4	+	*
Lupinus reflexus	0.91	6	+	*
Lupinus rotundiflorus	0.78	6	+	*
Lupinus simulans	1.06	6	+	*
Lupinus splendens	1.19	6	+	*
Lupinus spp	1.66	4	+	*
Wild potatoes from the states of San Luis Potosí and Zacatecas in dry basis (unpublished data)				
Solanum cardiophyllum	11.6	0	+	
Solanum ehrenbergii	23.7	3	+	*
Solanum polytrichon	5.4	0	+	*
Solanum staloniferum	22.5	1	+	*

[1]Trypsin units inhibited (Kakade et al. 1974)
[2]Titer = Maximum dilution where agglutination or hemolysis was observed.
*Not determined

Abrus precatorius and crotin in *Croton tiglium* respectively.[49] Lectins, from the latin *legere,* means to choose or to pick out, and is a name given for these agglutinins. The terms phytohemagglutinin (PHA) and phytoagglutinin are also used for plant lectins.

Lectins have been detected in a large number of angiosperms. Of 147 families, 79 have been shown to contain PHA. In the Leguminosae, 56% of 198 genera and 40% of 1597 species and varieties tested have proven to be PHA positive.[50] The genus *Phaseolus* is of considerable importance as a source of PHA' s. They are found in the seeds of *P. vulgaris, P. coccineus*, and *P. acutifolius,* which has toxic activity.[51,52] Lectins are found generally in higher concentration in seeds where they constitute up to 20% of the protein content.[53]

Lectin presence is detected by agglutination of mammalian erythrocytes. Under suitable conditions they agglutinate at concentrations as low as 0.1 to 1mg per ml. The surface structures to which lectins bind are carbohydrate moieties of glycoprotein or glycolipids that protrude from the cell.[54] Some lectins may bind to cells without causing agglutination. In such cases, binding is demonstrated by using lectins labeled with radioactive isotopes or with compounds visible under the microscope, such as fluorescein, ferritin, peroxidase and hemocyanin.[54]

The most extensive biological activity studies of lectins come from cell agglutination and mitogenic stimulation.[54] Because of their carbohydrate-binding properties, lectins are useful for the separation and characterization of glycoproteins and glycopeptides, and for observing changes that occur on cell surfaces during physiological and pathological processes ranging from cell differentiation to cancer. They also stimulate lymphocytes, allowing one to assess the immune state of patients, and can be used for chromosome analysis in human cytogenesis.[55] Lectin distribution during the life cycle of a plant has been much studied.[55] According to some investigators,[56] lectins disappear after seed germination or decrease to low levels during the first days of germination. We have studied lectin content of 2 varieties of *Phaseolus vulgaris* ("canario" and "negro jamapa" beans) throughout the life cycle (Fig. 3). Lectins did not decrease during germination, but they disappeared in the seedling and blooming stages. They reappeared in the string bean or green pods in one variety and reached the highest concentration in the dry seeds. These findings support the results of Etzler,[57] and suggest a possible fungistatic or bacteriostatic role.

As previously mentioned, *Phaseolus vulgaris* provides a significant source of protein in Latin America and other areas of the world. The toxicity of lectins in raw or inadequately cooked beans has been proven by many authors such as Liener and Jaffe,[58,59] and more recently by Pusztai and his group who have studied lectin-gut interactions and lectin effects on the utilization of food.[60–63] Carbohydrates on gut epithelial cell membranes, depending on their sugar specificity, react with lectins in different functional "compartments" which interferes with the absorption of nutrients.

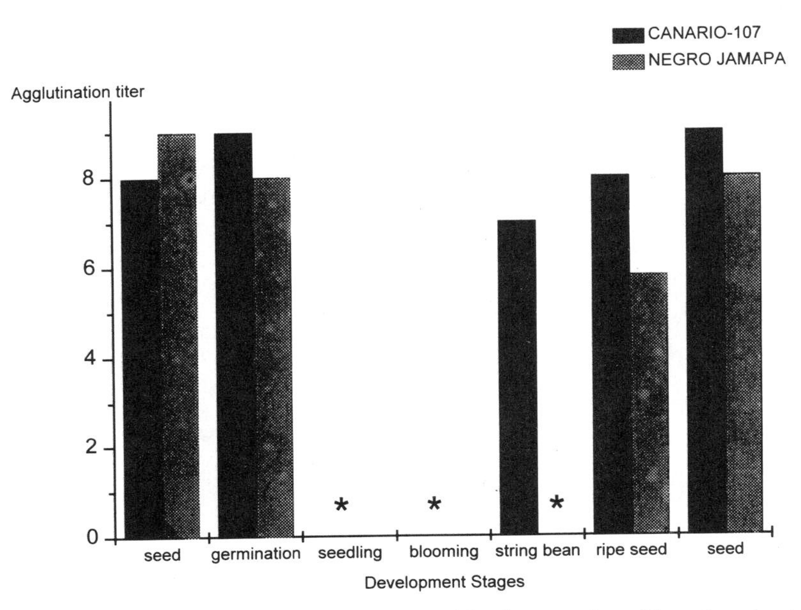

Figure 3. Lectin concentration in the different stages of development of two *Phaseolus vulgaris* bean varieties. * = Not detected.

Figure 4 compares the toxic effects of raw Escumite bean (*Phaseolus acutifolius*) extract and chickpea extracts on rat intestinal epithelial cells (*in vitro* and *in vivo* experiments). In all cases, differences are significant, but intragastric administration demonstrated the clearer toxic effect. Both the viability of cells and their number diminished. The ultrastructural changes produced by the damage of microvilli and organelles of the rat intestinal cells interfere with nutrient absorption.[51,52] These findings support the results of Pusztai. Table 7 shows the lectin content of some Mexican wild plants. It is interesting that, in *Pithecellobium saman* and *P. Keyense*, the lectins are in higher concentration in pods than in the seeds.

Alkaloids

Alkaloids are heat stable compounds present in some vegetable species used as food or feed. In those plants containing alkaloids, the concentration is usually very low. They are somewhat water soluble and can be diminished or eliminated by boiling and discarding the broth. Alkaloids are found in potatoes, but in most cultivated species mainly in the peel; thus, their presence is not significant.

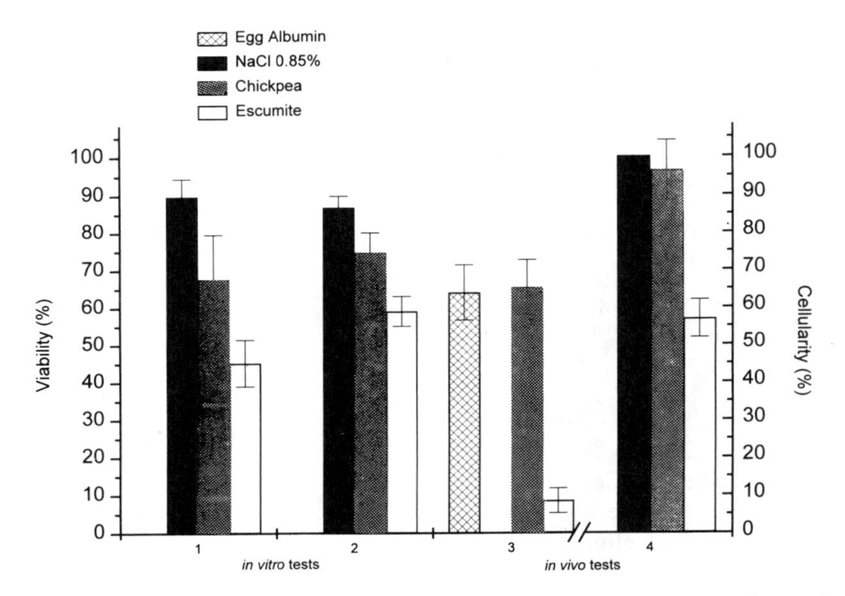

Figure 4. Effect of legume extracts on the viability and cellularity of the epithelial cells: 1. using a smear of epithelium; 2. using free cells; 3 and 4. (*in vivo* tests) legume extracts were given to the rats by intragastric tube. The animals were killed and the viability and percentage of intact cells (cellularity) were determined ($p < 0.001$). (Ref. 51).

Lupin is an ancient legume known in many countries. It has been used as human food and animal feed for more than 4000 years. More than 500 lupin species have been described. Only about 12 species occur in Europe and Africa, but 200–300 species inhabit North and South America. Lupins rely substantially on quinolizidine alkaloids (QA) for chemical defense against herbivores.[64] The genus *Lupinus* typically contains 36–52% protein and 5–20% fat, a composition similar to that of soy bean. Lupin species often substitute for soy bean in human food since they grow in areas where soy beans cannot because of their tolerance of poor soil quality and adaptability to temperate climate.[65] The main cultivated species are the white *L. albus,* the yellow *L. luteus, L. angustifolius*, and *L. mutabilis*.

Lupin alkaloids are based on a bicyclic quinolizidine ring. The most common QA, in lupins are lupinine (bicyclic), cytisine (tricyclic) and lupanine, sparteine and anagyrine (tetracyclic). The latter is considered important because it has a teratogenic effect in cattle.[66] The pharmacological effects of quinolizidine alkaloids have not been completely determined, but it is known that these alkaloids affect the nervous system and, in humans, can produce nausea, visual disturbances, ataxia, progresive weakness, and respiratory arrest. As with many plant toxins, they are metabolized in animals by undergoing biotransformation in the liver, usually mediated by the cytochrome P-450 system.[67,68]

Although about 90 species of wild lupins in Mexico have been reported, very few have been chemically studied. Table 7 shows alkaloid presence in seven wild species, natives of the state of Jalisco. These same plants have minimum amounts of trypsin inhibitors, but contain lectins in significant amounts. Due to their high protein and fat composition, lupins are the alkaloid containing legumes most often consumed, and there are many low alkaloid species (sweet lupins) with less than 0.002% alkaloids. Lupins are deficient in sulfur amino acids. A 0.15% methionine supplementation significantly increases the efficiency of protein utilization from lupins.[69]

Erythrina is a genus of the Fabaceae consisting of more than 150 species with a wide distribution in tropical regions. As a result of studies by Krukoff and Barneby,[70] this genus of beautiful trees, shrubs, and a few herbs is probably better known in most aspects than any other of comparable size and distribution. In Mexico, the genus is widespread. About 25 species are distributed throughout the country and also in Central America and the Caribbean area.[71] Nearly all have alkaloids, mainly in the seeds, but with lesser amounts in flowers. The boiling of flowers with broth elimination makes them edible. Many studies on *Erythrina* have been conducted resulting in the structural characterization of the alkaloids.[68,72–74] Sotelo and co-workers[75] have determined the nutritive value of the detoxified seed flour of both *E. Americana* and *E. breviflora*. Whole seeds of both species showed high protein and high fat content (27, 18 and 23, 14%, respectively). The protein and fiber contents were remarkably increased in the defatted and detoxified flours (42, 26 and 38, 31%, respectively). The nutritional evaluation by biological tests with rats indicated that both detoxified flours had good protein quality, similar to the casein diet used as a pattern.[76] Table 8 shows the antinutritional and toxic factors present in the whole and detoxified flour. Trypsin inhibitors remained in the detoxified flour since the temperature used for alkaloid extraction (60º C) was not high enough to destroy them, but lectins were inactivated at that temperature. Alkaloids were detected in the lipid fraction in both *Erythrinas*. Considering the high protein and fat concentration of *Erythrina* seeds, the adequate profile of amino acids, and the biological protein quality evaluation of detoxified flour, it would seem important to develop an inexpensive method of detoxifying the seeds. Future work might focus on controlling the biosynthesis of the alkaloids during development so that the seeds become harmless and thus able to be used for animal feeding.

REDUCTION OF ANTINUTRITIONAL FACTORS

Progress has been made in breeding programs oriented to reduce antinutritional factors and toxin content. Some have been eliminated or reduced by plant breeding *e.g.* tannins in peas, lentils, and faba beans; *e.g.* alkaloids in lupins; *e.g.* trypsin inhibitors in soybean. This strategy has not always been

Table 8. Antinutritional and toxic factors content in the whole, detoxified flour and lipid fraction of the *Erythrina americana* and *E. breviflora* seeds[1]

Sample	Trypsin inhibitors TUI[2]/mg of dry sample	Lectins[3] Hamster RBC	Cyanogenic glycosides mg of HCN/100 g of sample	Alkaloids
E. americana				
Whole flour	68.00[a]	5	0	+
Detoxified flour	49.24[b]	0	0	–
Lipid fraction	–	–	–	+
E. breviflora				
Whole flour	63.9[a]	0	0	+
Detoxified flour	62.9[a]	0	0	–
Lipid fraction	–	–	–	+

[1]Ref 74
[2]Trypsin units inhibited (Kakade et al. 1974). Means in column without common letters differ significantly ($P<0.05$).
[3]Titer: the highest dilution at which agglutination is found (Jaffe et al. 1972)

widely accepted. The objectives of plant breeders are to improve yield and disease resistance, but these characteristics often are adversely affected when ANFs and toxins are eliminated since they protect the plant against predation by insects, bacteria and fungi.[25] The elimination could increase the need for application of pesticides and create further pollution of the environment.[75] Reduction may benefit some users, but not others; *e.g.* seeds whose tannins have been removed often do not have enough taste; *e.g.* in ruminants, tannins have positive effects in protecting proteins from rapid degradation in the rumen, but in monogastric animals, the reduction of protein digestibility is not desirable. It might be more expedient to eliminate undesirable ANFs and toxins by technological or biotechnological techniques in the seeds such as dehulling, heating, and soaking, as proposed by many.[76,77] It is necessary to keep the effects and needs of ANFs and toxins under constant review.

CONCLUSIONS

The wild flora can be considered a great reservoir of protein and fat with enormous potential for use in human and animal feeding. Many Mexican wild plants already studied can be considered convenient for ruminant and monogastric animals. The seeds of some wild legumes are similar to soybean in their nutrient composition with the advantage that the wild species grow in poor soils and have great adaptability to dryness and temperate climates. Their disadvantage is mainly the presence of heat stable ANFs and toxins. Future use as food or feed requires the participation of plant breeders and agriculturists in the domestication of these wild plants. Many could play an important role in the feeding of poor farmers in local areas or during conventional food shortages.

ACKNOWLEDGMENTS

This paper is dedicated to Professor Francisco Giral, who introduced me to the fascinating world of the Mexican wild flora. This research was supported by grants from the Consejo Nacional de Ciencia y Tecnología (CONACyT), the Organization of American States (OAS) and the Dirección General de Asuntos del Personal Académico (DGAPA) de la Universidad Nacional Autónoma de Mexico. I express gratitude to my co-workers S. Flores, M.T. González Garza, B. Lucas, I. Montalvo and V. Sousa and students G. Alvarez., M. Arteaga, F. Blanc, M. Carrillo, E. Contreras, A. de la Vega, I. Frías, L. Garza, L. Gil, R. Nava, O. Rodríguez, L. Romero, H. Sousa, and A. Uvalle.

REFERENCES

1. Food and Agriculture Organization of the United Nations (FAO) 1988. Traditional food plants. Food and Nutrition Paper 42. Rome.
2. GIRAL, F., SOTELO, A., LUCAS, B., DE LA VEGA, A. 1978. Chemical composition and toxic factors content in fifteen leguminous seeds. Quart J. Crude Drug Res. 16:143–149.
3. SOTELO, A. 1981. Leguminosas silvestres, reserva de proteinas para alimentación del futuro. Inf. Científica Tecnol. 3:28–34.
4. SOTELO, A. 1996. The nutritive value of wild Mexican Legumes with a potential for nutritional use. In: Food and Feed from Legumes and Oilseeds (E. Nwokolo and J. Smartt eds.), Chapman & Hall, pp. 33–48.
5. MIRANDA, F., HERNANDEZ, X.E. 1985. The types of vegetation of Mexico and their classification I. Xolocotzia. Rev. Agricola. México, pp. 41–45.
6. AYKROYD, W.R., DOUGHTY, J. 1982. Legumes in human nutrition. FAO. Food and Nutrition Paper 20. Rome.
7. HEYWOOD, V.H. 1971. The leguminosae. A systematic purview. In: Chemotaxonomy of the leguminosae (J.B. Harborne, D. Boulter, B.L. Turner, eds.), Academic Press, London pp. 1–23.
8. SOUSA, M., DELGADO, A. 1991. Mexican Leguminosae: Phytogeography endemism and origin. In: Biological Diversity of Mexico: Origin and Distribution. (T. Ramamoorthy, R. Bye, A. Lot and J. Fa., eds.), Oxford University Press. New York pp. 459–511.
9. SOTELO, A., LUCAS, B., BLANC, F., GIRAL, F. 1986. Chemical composition of seeds of *Gliricidia sepium*. Nutr. Rep. Int. 34:315–321.
10. TODOROV, N.A., PAVLOV, D.C., KOSTOV, K.D. 1996. Lupin. In Food and Feed from Legumes and Oil seeds. (E. Nwokolo and J. Smartt, eds.), Chapman & Hall. London pp. 113–123.
11. SOTELO, A., CONTRERAS, E., FLORES, S. 1995. Nutritional value and content of antinutritional compounds and toxics in ten wild legumes of Yucatan Peninsula. Plant Food Hum. Nutr. 47:115–123.
12. DELGADO, S.A., MANZANILLA, L. 1990. El frijol prehispánico. Información Científica y Tecnológica 12:52–56.

13. DELGADO, S.A., BONET, A., GEPTS P. 1988. The wild relative of *Phaseolus vulgaris* in Middle America. In: Genetic Resources of *Phaseolus* Beans. (P. Gepts ed.), Kluwer Academic Publishers, Dordrecht, pp. 163–184.
14. KAPLAN, L. AND KAPLAN L.N. 1988. *Phaseolus* in archeology. In: Genetic Resources of *Phaseolus* Beans. (P. Gepts ed.), Kluwer Academic Publishers, Dordrecht, pp. 185–214.
15. BRÜCHER, H. 1988. The wild ancestor of *Phaseolus vulgaris* in South America. In: Genetic Resources of *Phaseolus* Beans. (P. Gepts ed.), Kluwer Academic Publishers, Dordrecht, pp. 195–214.
16. SOTELO, A., SOUSA, H., SÁNCHEZ, M. 1995. Comparative study of the chemical composition of wild and cultivated beans (*Phaseolus vulgaris*). Plant Foods Hum. Nutr. 47: 93–100.
17. SOTELO, A., LUCAS, B., UVALLE, A., GIRAL, F. 1980. Chemical composition and toxic factors content in sixteen leguminous seeds (II). Quart J. Crude Drug Res. 18:9–16.
18. SOTELO, A., LUCAS, B., GARZA, L., GIRAL, F. 1990. Characteristics and fatty acid content of the fat of seeds of nine wild Mexican plants. J. Agric. Food Chem. 38:1503–1505.
19. CUATRECASAS, J. 1964. Cocoa and its allies. A taxonomic revision of the genus *Theobroma*; Contribution National Herbarium U.S. National Museum 35; Washington, D.C., Part 6, pp. 1–1235.
20. DE LA VEGA, A., SOTELO A. 1985. Estudio comparativo de la composición química del cacao (*Theobroma cacao*) y de dos *Theobromas* silvestres. Rev. Soc. Quim Mex. 29:325–328.
21. SOTELO, A., ALVAREZ, R.G. 1991. Chemical composition of wild *Theobroma* species and their comparison to the cacao bean. J. Agric. Food Chem. 39:1940–1943.
22. CARRILLO, M., ROMERO, L. 1991. Hojas comestibles de plantas silvestres y cultivadas. Thesis. Facultad de Química, Universidad Nacional Autónoma de México. 88 pp.
23. GIL, L. 1992. Valor nutritivo de algunas flores comestibles. Thesis. Facultad de Química, Universidad Nacional Autónoma de México. 76 pp.
24. MITJAVILA, S. 1990. Sustancias nocivas en los alimentos. In: Toxicología y Seguridad de los Alimentos (Original edition Toxicologie et Sécurité des Aliments.) (R. Derache, ed.) Omega, Barcelona. pp 109–112.
25. LIENER, I.E. 1989. Antinutritional factors in legume seeds: state of the art. In: Recent Advances of Research in Antinutritional Factors in Legume Seeds, (J. Huisman, T.F.B. Van der Poel, I.E. Liener, eds.), Pudoc, Wageningen, pp. 6–14.
26. GUENGUEN, J., VAN OSRT, M.G., QUILLIEN, L., HESSING, M. 1993. The composition, biochemical characteristics and analysis of proteinaceous antinutritional factors in legume seeds. A review. In: Recent Advances of Research in Antinutritional Factors in Legume Seeds (A.F.B. van der Poel, J. Husman, H.S. Saini, eds.), Wageningen Pers., Wageningen. pp. 9–30.
27. CONN, E.E. 1969. Cyanogenic glycosides. Ann. Rev. Plant Physiol. 31:433–451.
28. CONN, E.E. 1980. Ann. Rev. Plant Phys. A comprehensive treatise (E.E. Conn, ed.), Academic Press, New York, pp. 479–500.
29. POULTON, J.E. 1983. Handbook of Natural Toxins. (R.F. Keeler, A.T. Tu, eds.), Marcel Dekker Inc. New York, pp. 118–157.
30. JONES, D.A. 1981. Cyanide in Biology (B. Vennsland, E.E. Conn, C.J. Knowles, J. Westley, F. Wissing, eds.), Academic Press. New York, pp. 495–508.
31. DE LA VEGA, A., SOTELO, A. 1986. The nutritional quality and toxin content of wild and cultivated lima beans (*Phaseolus lunatus*). Qual. Plant. Plant. Foods Hum. Nutr. 36:75–83.
32. LIENER, I.E., KAKADE, M.L. 1980. Protease inhibitors. In: Toxic Constituents of Plant Foodstuffs, (I.E. Liener, ed.), Academic Press, New York, pp. 7–71.
33. KAKADE, M.L., RACKIS, J.J., Mc GHEE, J.E., PUSHI, G. 1974. Determination of trypsin inhibitor activity of soy products. A collaborative analysis of on improved procedure. Cereal Chem. 51:376–382.
34. RICHARDSON, M., BARKER, R.D.J., Mc MILLAN, R.T. COSSING, L. 1977. Identification of the reactive (inhibitory) sites of chymotryptic inhibitors I from potatoes. Phytochemistry 16:837–839.

35. BOISEN, S. 1983. Protease inhibitors in cereals. Acta Agric. Scand. 33:369–381.
36. SAINI, H.S. 1989. Thermal stability of protease inhibitors in some cereals and legumes. Food Chem. 32:59–67.
37. KUNITZ, M. 1945. Crystalization of a trypsin inhibitor from soy beans. Science 101:668–669.
38. LE GUEN, M.P., BIRK Y. 1993. Protein protease inhibitors from legume seeds: Nutritional effects, mode of action and structure-function relationship. In: Recent Advances of Research in Antinutritional Factors in Legume Seeds. Proceedings of the Second International Workshop on Antinutritional Factors (ANFs) in Legume Seeds, (A.F.B. van der Poel, J. Huisman, H.S. Saini, eds.), Wageningen Pers, Wageningen, pp. 157–171.
39. WEDER, J.K.P. 1986. Inhibitors of human proteinase by grain legumes. Adv. Exp. Med. Biol. 199:239–279.
40. BELITZ, H.D. WEDER, J.K.P. 1990. Protein inhibitors of hydrolases in plant foodstuffs. Food Rev. Int. 6:151–211.
41. WEDER, J.K.P., KAHLEYSS, R. 1993. Differences in mode of action of lentil proteinase inhibitors against human and bovine chemotrypsin and trypsin. In: Recent Advances of Research in Antinutritional Factors in Legume Seeds. (T.F.B. Van der Poel, J. Huisman, H.S. Saini, eds.), Wageningen Pers., Wageningen, pp. 201–204.
42. RICHARDSON, M. 1977. The proteinase inhibitors of plants and microorganisms. Phytochemistry 16:159–169.
43. PUSZTAI, A. 1972. Metabolism of trypsin-inhibitors proteins in the germinating seeds of kidney bean (*Phaseolus vulgaris*). Planta 107:121–129.
44. SHAIN, Y., MANER, A.M. 1968. Activation of enzymes during germination. A trypsin like enzyme in lettuce. Phytochemistry 7:1491–1998.
45. GODBOLE, S.A., KRISHNA, T.G., BHATA, C.R. 1994. Changes in protease inhibitory activity from pigeon pea (*Cajanus cajan, L., Mill sp.*) during seed development and germination. J. Sci Food. Agric. 66:497–501.
46. GOLDSTEIN, I.E., HUGHES, R.C., MONSIGNY, M., OSAWA, T., SHARON, N. 1980. What should be called a lectin? Nature (London), 285:66.
47. ANDERSEN, O., LAURSEN, S.B., SVEHAG, S.E., HOLMSKOV, U., THIEL, M. 1991. Mammalian lectins in defence mechanisms against microorganism. In: Lectin Reviews 1, (D.C. Kilpatrick, E. Van Driessche and T.C. Bog-Hansen eds.), Sigma Press. St. Louis. pp. 41–42.
48. FRANZ, H. 1988. The ricin story. In: Advances in Lectin Reserch 1. (H. Frans ed.), VEB Verlag Volk und Gesundheit. Berlin. pp. 10–25.
49. JAFFE, W.G. 1969. Hemagglutinins. In: Toxic Constituents of Plant Foodstuffs, (I.E. Liener ed.), Academic Press, London, pp. 69–94.
50. TOMS, G.C. 1971. Phytohaemagglutinins. In: Chemotaxonomy of the Leguminosae. (J.B. Harborne, D. Boutler, B.L. Turner, eds.), Academic Press, New York, pp.367–462.
51. SOTELO, A., ARTEAGA, M.E., FRIAS, M.I., GONZÁLEZ-GARZA, M.T. 1980. Cytotoxic effect of two legumes in epithelial cells of the small intestine. Qual. Plant. Plant Foods Hum. Nutr. 30:79–85.
52. SOTELO, A., GONZÁLEZ-LICEA, A., GONZÁLEZ-GARZA, M.T., VELASCO, E., FERIA-VELASCO, A. 1983. Ultrastructural changes of epithelial intestinal cells induced by the ingestion of raw *Phaseolus acutifolius*. Nutr. Rep. Int. 27:329–337.
53. VAN DRIESSCHE, E. 1988. Structure and function of leguminosae lectins. In: Advances in Lectin Research 1, (H. Franz ed.), VEB Verlag Volk und Gesundheit, Berlin. pp. 73–134.
54. STOJANOVIC, D., FERNANDEZ, M., CASOLE, I., TRUJILLO, D., CASTES, M., 1994. Characterization and mitogenicity of a lectin from *Erythrina velutina* seeds. Phytochemistry 37:1069–1074.
55. SHARON, N., LIS, H. 1990. Legume lectins-a large family of homologous proteins. The FASEB J. 4:3198–3208.
56. PUSZTAI, A. 1991. Plant Lectins. Cambridge University Press. Cambridge. pp. 39–56.

57. ETZLER, M.E. 1986. Distribution and function of plant lectins. In: The Lectins, Properties, Functions and Application in Biology and Medicine (I.E. Liener, N. Sharon, I.J. Goldstein, eds.), Academic Press, Inc. Orlando pp.371–435.
58. LIENER, I.E. 1986. Nutritional significance of lectins in the diet. In: The Lectins. (I.E. Liener, N. Sharon, I.J. Goldstein, eds.), Academic Press, Orlando pp. 527–552.
59. JAFFE, W.C. 1949. Toxicity of raw Kidney beans. Experientia 5:81.
60. PUSZTAI A. 1986. The biological effects of lectins in the diet of animal an man. In: Lectins, Biology, Biochemistry, Clinical Biochemical. (T.C. Bog-Hansen and E. van Ddriessche eds.), Walter de Gruyter, Berlin and New York, pp. 317–327.
61. PUSZTAI, A., GRANT, G., EWEN, S.W.B., BROWN, D.S., BARDOCZ, S. 1990. *Phaseolus vulgaris* Lectin (PHA) induces growth of rat small intestine. Proc. Nutr. Soc. 49:145A.
62. BANWELL, J.G., HOWARD, R., KABIR, I., ADRIAN, T.E., DIAMOND, R.H., ABRAMOWSKY, C. 1993. Small intestine growth caused by feeding red kidney bean phytohemagglutinin lectin to rats. Gastroenterology, 104:1669–1677.
63. GRANT, G., VAN DRIESSCHE E., 1993. Legume lectins:physicochemical and nutritional properties. In: Recent Advances of Research in Antinutritional Factors in Legume Seeds. Proceeding of the Second International Workshop on "Antinutritional factors (ANFs) in legume seeds". (A.F.B. van der Poel, J. Huisman, H.S. Saini, eds.), Wageningen Pers, Wageningen, pp. 219–233.
64. WINK, M., MEISNER, C., WITTE, L. 1995. Patterns of quinolizidine alkaloids in 56 species of the genus lupinus. Phytochemistry 38:139–153.
65. MOHAMED, A.A., RAYAS-DUARTE, P. 1995. Composition of *Lupinus albus*. Cereal Chem. 72:643–647.
66. KELLER, R.F., GRAN, R. 1980. The total alkaloid and anagyrine contents of some bitter and sweet selections of lupin species used as food. J. Environ. Path. Toxical. 3:333–340.
67. CHEEKE, P.R., KELLY, J.D. 1989. Metabolism, toxicity and nutritional implications of quinolizidine (lupin) alkaloids. In: Recent Advances of Research in Antinutritional Factors in Legume Seeds. Proceedings of the First International Workshop on Antinutritional Factors in Legume Seeds. (J. Huisman, T.F.B. van der Poel, I.E. Liener eds.), Pudoc Wageningen, pp. 189–201.
68. CHEEKE, P.R., SHULL, L.R. 1995. Natural toxicants in feeds and poisonous plants. AVI Publ. Co., Wesport, C.T. USA, pp. 127–130.
69. SGARBIERI, V.C., GALEAZZI, M.A.N. 1985. Some physico-chemical and nutritional properties of a sweet lupin (*Lupinus albus* var. Multolupa) protein. J. Agric. Food Chem. 26:1438–1442.
70. KRUKOFF, B.A., BARNEBY, R.C. 1974. Conspectus of species relationships in *Erythrina*. Lloydia 37:332–459.
71. NEILL, D.A. 1988. Experimental studies on species relationships in *Erythrina*. (*Leguminosae papilionideae*) Ann. Mo. Bot. Gar. 75:886–969.
72. SOTO, H.M., JACKSON, A.H. 1994. *Erythrina* alkaloids. Isolation and characterization of alkaloids from seeds of *Erythrina* species. Planta Medica 60:175–177.
73. GARCIA MATEOS, R., LUCAS, B., ZENDEJAS, M., SOTO, M., MARTINEZ, M., SOTELO, A. 1996. Variation of total nitrogen, non-protein nitrogen content an types of alkaloids at different stages of development in *Erythrina americana* seeds. J. Agric. Food Chem. 44:2987–2991.
74. SOTELO, A., SOTO, M., LUCAS, B., GIRAL, F. 1993. Comparative studies of the alkaloidal composition of two mexican *Erythrina* species and nutritive value of the detoxified seeds. J. Agric. Food Chem. 41:2340–2343.
75. BOND, D.A., DUC, G. 1993. Plant breeding as a means of reducing antinutritional factors in grain legumes. In: Recent Advances of Research in Antinutritional Factors in Legume Seeds. Proceedings of the Second International Workshop on Antinutritional Factors (ANFs) in

Legume seeds. (A.F.B. van der Poel, J. Huisman, H.S. Saini, eds.), Wageningen Pers, Wageningen, pp.379–396.

76. MELCION, J.P., VAN DER POEL, A.F.B. 1993. Process technology and antinutritional factors: principles, adequacy and process optimization In: Recent Advances of Research in Antinutritional Factors in Legume Seeds. Proceedings of the Second International Workshop on Antinutritional Factors (ANFs) in Legume Seeds. (A.F.B. van der Poel, J. Huisman, H.S. Saini, eds.), Wageningen Pers, Wageningen, pp. 419–434.
77. CLASSEN, H. L., BALNAVE, D., BEDFORD, M.R. 1993. Reduction of legume antinutritional factors using biotechnological techniques. In: Recent Advances of Research in Antinutritional Factors in Legume Seeds. Proceedings of the Second International Workshop on Antinutritional Factors (ANFs) in Legume Seeds. (A.F.B. van der Poel, J. Huisman, H.S. Saini, eds.), Wageningen Pers, Wageningen, pp. 501–516.
78. PELLET, P.L., YOUNG, V.R. 1980. Nutritional evaluation of protein foods; The United Nations University World Hunger Programme. Food and Nutrition Bulletin Supplement 4; The United Nations University: Tokyo.

Chapter Five

BIOLOGY AND BIOCHEMISTRY OF UNDERGROUND PLANT STORAGE ORGANS

Hector E. Flores[1,2,3] and Tere Flores[2]

[1] Plant Pathology Department
[2] Intercollege Graduate Program in Plant Physiology
[3] Science, Technology, and Society Program
The Pennsylvania State University
University Park, Pennsylvania 16802

INTRODUCTION

Higher plants have evolved various ways of accumulating large amounts of assimilates, including both primary and secondary metabolites (micro- and macromolecules), in a single organ/location. Specialized cells and tissues such as trichomes, nectaries, and resin canals synthesize compounds which may protect the plants against pathogens and pests. A major mode of assimilate accumulation is the formation of seeds and fruits, which in most cases are the end result of sexual reproduction, although seeds can also originate through apomixis and fruits through treatment with growth regulators. A substantial amount of information is now available on the developmental and biochemical aspects of such organs,[1] including the accumulation of starch,[2] proteins,[3] and a wide diversity of phytochemicals.[4] In contrast, much less is known about the phytochemistry, biochemistry, and development of underground storage organs. This is surprising, considering that storage roots and tubers such as cassava, sweet potato, and potato constitute important staples for people in developing as well as developed countries in both tropical and temperate areas. This review summarizes some selected aspects of the phytochemistry, biochemistry, and ethnobotany of underground plant storage organs with emphasis on roots and tubers.

MEDICINAL PLANT STORAGE ORGANS

This review focuses on edible storage organs, but it is also important to mention that many such structures have been used for millennia in the treatment of various ailments. Some of the classic examples include taproots shaped into forms resembling humans, such as the mandrake (*Mandragora officinalis*) of

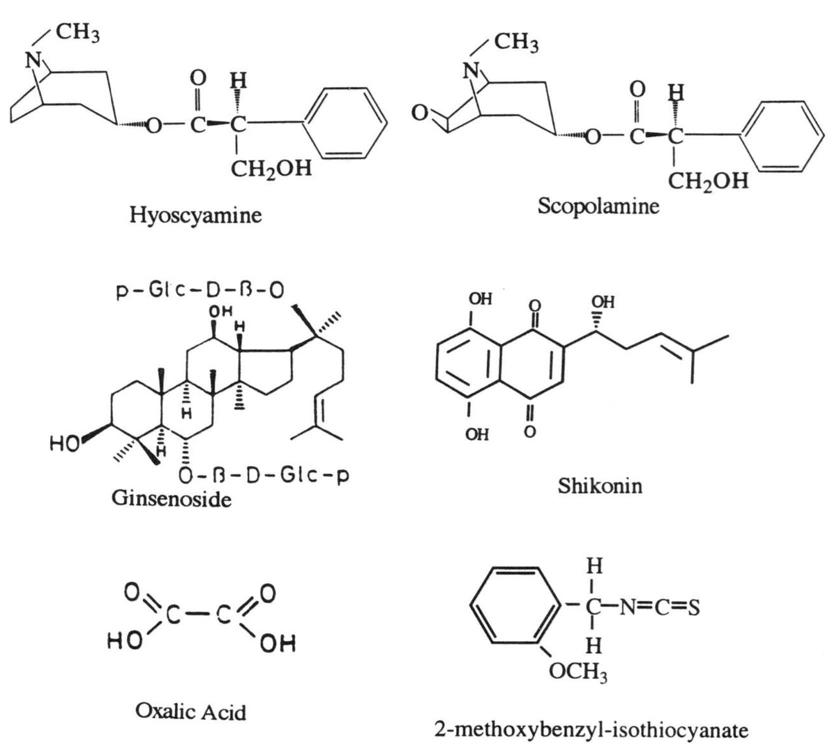

Figure 1. Representative structures of compounds from tubers.

central Europe and the ginseng (*Panax ginseng*) of China. In both cases, the magical properties were attributed to the male or female shapes of these roots, but their physiological effects have a defined biochemical basis. In the case of mandrake, the active substances are mainly two tropane alkaloids, hyoscyamine and scopolamine (Fig. 1). In ginseng, the adaptogenic effects are due to a rather complex set of triterpenoid saponins (ginsenosides) (Fig. 1). The scarce information available indicates that the highest levels of the corresponding bioactive compounds are correlated with the onset of secondary growth and subsequent radial expansion. Because most of these medicinal taproots are "matured" in the soil over a period of years before commercialization, one may also presume the existence of seasonal mechanisms of phytochemical accumulation. For example, the root of ko-shikon (*Lithospermum erythrorhizon*), a source of antibiotic naphthoquinones also used as dyes (Fig. 1), is collected only after three to five years of growth. For the most part, however, mechanisms of compound accumulation and pattern development remain unknown. The reader is referred to a recent review on the biochemistry of root specific metabolism for further examples.[5]

UNDERGROUND CROPS

In addition to medicinal uses, people have domesticated numerous storage organs for edible purposes over many millennia. This is especially prevalent in tropical regions (cassava and sweet potato) where agriculture tends to be based on vegetatively propagated crops,[6] but it also happens in every single major center of crop domestication. For example, carrots (*Daucus carota*) and beets (*Beta vulgaris*) are of Mediterranean origin and exemplify how domestication has affected the nutritional properties of these roots. The modern varieties of carrots have been selected for high content of beta-carotenes, whereas "primitive" varieties were either purple (reflecting high accumulation of anthocyanins rather than carotenes) or white to cream colored.[7] The common salad beet has been domesticated for high accumulation of betacyanins and related tyrosine derivatives. Many heirloom varieties, however, show lower contents and much more distinct concentric patterns of pigmentation. In contrast, the sugar beet, a variety of the same species, was selected for high concentrations of sucrose, which in the modern cultivars now surpasses 20% on a dry weight basis. This selection occurred over a period of less than 150 years, starting with varieties which contained less than 2% sucrose, and it constitutes one of the most striking examples of domestication involving a single primary metabolite. It is also a prime example of the enormous potential for metabolic flexibility present in tap roots and other storage organs, the biochemical and molecular basis for which remain unknown.

STORAGE PROTEINS

Storage proteins from underground storage organs are less well known than those from seeds. The best characterized tuber storage proteins are sporamin and patatin, from sweet potato and potato, respectively. Storage proteins provide a store of nitrogen, sulphur, and carbon to plants.[3] They are usually synthesized in reproductive organs, but can also be found in vegetative structures used by plants for propagation and dispersal. The most thoroughly studied are those found in seeds of grasses and legumes. The germination or sprouting of such tissues is accompanied by proteolysis in which structural and metabolic proteins are digested.[8] True storage proteins can be characterized by their high concentrations, their presence in discrete deposits (protein bodies), and their usual lack of any biological activity. However, in a growing number of cases, storage proteins have been shown to have biological activity, as in the case of patatin and sporamin.[9] Most of the storage proteins are synthesized only in one organ, although they may be synthesized in other tissues under exceptional circumstances.[3] A high level of polymorphism is characteristic of seed storage proteins and may arise either from the presence of multigene families or from post-translational processing. The importance of these mechanisms varies between species and protein types.[10] In addition to polymorphism within a single genotype, there is variation between the polypeptide compositions of storage protein fractions from different genotypes within a single species.[3] The most important groups, in terms of their distribution in plants, are prolamins, globulins (7S and 11S), and the 2S albumins. Prolamine, albumin, and globulin storage proteins are synthesized on the rough endoplasmic reticulum (ER) with signal peptides of about 20 residues which direct the nascent polypeptides into the ER lumen. The signal peptide is then cleaved, providing fully mature protein.[3]

Prolamins are the major storage proteins in cereals, except in rice and oats. These proteins are characterized by their extraction and solubility in alcohol/water mixtures, and by their insolubility in either water or dilute salt solutions. Prolamins have high contents of proline and amide nitrogen, and low levels of acidic or basic amino acids;[11] they are also low in lysine and deficient in threonine (in wheat and barley), and tryptophan (in maize). Cereal prolamines vary widely in amino acid composition and sequence.[3] The sulphur-rich prolamins account, in most cases, for 80–90% of the total content for this protein group.[3] In contrast, the oat (*Avena sativa)* prolamins are monomeric proteins with inter-chain disulfide bonds and account for about 10% of total proteins.[12] As in oats, the prolamins of rice (*Oryza sativa*) are minor components, accounting for about 5% of total proteins.[3]

The 2S albumins are characterized by their extraction with and solubility in water. They are present in species as diverse as sunflower (Compositae), mustard (Cruciferae), brazil nut (Lecythidaceae), and lupin (Leguminosae). The 2S albumins from the Cruciferae have been studied most widely, particularly from oilseed rape (*Brassica napus*), and consist of two polypeptide chains with a molecular weight of 9000 (86 residues) and 4000 (29 residues).[13] The castor bean albumin was the first 2S albumin to be completely sequenced at the protein level,[3] and consists of two subunits with 34 and 61 residues. The 2S albumins account for about 30% of the total seed protein in Brazil nut with 13 mol % cysteine and 17 mol % methionine. The 2S albumins appear to be compact globular proteins, and, as most other seed storage proteins, they are transported from the lumen of the endoplasmic reticulum to the vacuole, which is probably the site for post-translational processing.[14] The major role of the 2S albumins is as storage proteins, however, a few appear to have activity as inhibitors of hydrolytic enzymes. The 2S albumins of *Brassica napus* inhibited bovine trypsin, and the 2S albumins of radish were able to inhibit the growth of pathogenic fungi *in vitro*. Radish and oilseed rape albumins synergistically enhance the antifungal activity of wheat and barley thionins by permeabilizing the hyphal plasma lemma.[3]

The globulins are characterized by extraction with and solubility in dilute saline solutions. They are divided into two groups: legumins, with a sedimentation coefficient of 11; and vicilins, with a sedimentation coefficient of 7.[3] The legumin-type 11S globulins are the most widely distributed of all the storage protein groups, being major components in seeds of dicotyledonous plants. The 7S globulins are less widespread, but have been reported in species as diverse as cereal embryos and cotton seed.[15] The 11S globulins are assembled in the endoplasmic reticulum into trimers and only attain their mature hexameric structure in the vacuole. The 7S globulins are assembled into their mature trimeric structure in the endoplasmic reticulum.[16] In contrast to the seed storage proteins, much less is known about their analogs in underground storage organs. The best characterized systems are potato and sweet potato.

ANDEAN ROOT AND TUBER CROPS

The Andean region is recognized as one of the most important centers of crop origin and diversity in the world.[17,18] The Andean agricultural complex includes about 200 plant species domesticated in Pre-Columbian times, offering a highly balanced and diversified diet of roots, tubers, grains, legumes, vegetables, fruits, nuts, etc. The geographical trademark of this agricultural system is the presence of the second highest mountain chain in the world. Agriculture at the base of the narrow Andean valleys is based, as in most other systems, on the classic complement of grains and legumes, in this case corn (*Zea mays*) and common beans (*Phaseolus* spp.) supplemented with high altitude domesticates (*Chenopodium*, *Amaranthus*, *Lupinus*). However, there are two features of the Andean agroecosystem which have no parallel in any region in the world. First, about 25 species of root and tuber crops were domesticated in the Andes, forming the largest known concentration of these crop types.[19,20] The genetic diversity represented by these storage organ plants is incredible. Thousands of varieties have been uniquely adapted to a wide range of agroecological niches, resulting in the maximized use of the vertical space.[21] Thus, it is not surprising that the potato, the major contribution of the Andean region to world agriculture, is a tuber crop.

The second, and perhaps most remarkable, aspect, is that crop diversity evolved in one of the most inhospitable areas for agriculture. Pre-Columbian people made extremely efficient use of what would be considered marginal lands by today's agricultural standards.[22] The steep slopes of the Andes are prone to erosion, subject to extreme fluctuations in rainfall and temperature, and contain poor soils. Crops grown in this environment must cope with long periods of drought, frost damage, and high UV irradiation. Pre-Columbian agricultural technology was probably unsurpassed in its ability to grow food crops adapted to extreme conditions. It is estimated that the agricultural economy of the Inca empire sustained a population of 10–12 million people,[19] and Andean roots and tubers (ARTs) were central to their diet. Again, this is not surprising because underground storage organs are a very efficient strategy evolved by plants for survival in challenging environments. Roots and tuber crops also show the highest yields in calories produced per area of cultivation, so their adaptation and diversification in the Andes also made ecological and economic sense.

ARTs are common staples for an estimated 20 million people in the highlands, but their area of influence in South America probably comprises at least 100 million people in Ecuador, Colombia, Peru, Bolivia, Argentina, and Chile. The major ARTs are listed in Table 1. Their area of cultivation extends from Merida (Venezuela) to the northern Andes of Chile and Argentina (3 - 18 °S) at an altitude range of 1,500 to 4,500 m (ahipa, *Pachyrhyzus ahipa*; achira, *Canna edulis*; yacon, *Polymnia sonchifolia*; and oca, *Oxalix tuberosa* can also grow at sea level). Maca (*Lepidium meyenii*) cultivation is restricted to the central Andes of Peru; this and bitter potatoes (several *Solanum* species) are the only crops which can survive over 3,900 m. Mauka (*Mirabilis expansa*) is reduced to isolated pockets in Ecuador, Peru, and Bolivia. Several root crops: arracacha, ahipa, yacon, and the corm-producing achira are grown in the warmer valleys of the Eastern Cordilleras at 1,000 to 3,000 m. Most importantly, these crops are not grown in isolation from each other but as part of a system of mixed crops and crop rotations which continues to evolve and persist in the Andean highlands.[23]

The ARTs have a fascinating variety of features and uses. Ulluco (*Ullucus tuberosum*) for example, is a delicate tasting and nutritious tuber second only in importance to potatoes in the highlands.[24] It is also one of the few indigenous crops that has increased its range over the last century. It is reported to ease childbirth. The mashua (*Tropaeolum tuberosum*), closely related to the garden nasturtium, produces a high protein tuber preferred over others because it is easy

Table 1. Features of the major Andean roots and tuber crops

Family	Species	Common name	Altitude range (m)	Chemical constituents
Roots				
Apiaceae	*Acarracia xanthorriza*	Arracacha	600-300[9]	High in calcium and Vitamin A.[10]
Cruciferae	*Lepidium meyenii*	Maca	3500-4500[9,15]	High in iron, iodine, alkaloids, tanins and saponins.[10,15]
Asteraceae	*Polymnia sonchifolia*	Yacon	900-2700[9,50]	Fructan, inulin.[10,48]
Convolvulaceae	*Ipomoea batata*	Sweet potato	900-1200[35]	Sporamin (may play a role in defense against pathogens).[3,18]
Nyctaginaceae	*Mirabilis expansa*	Mauka	2200-3500[9]	RIPs (may have synergistic antifungal activity in combination with chitinases).[50,57]
Tubers				
Solanaceae	*Solanum tuberosum*	Potatoes	900-2700[19]	Patatin (lipid acyl hydrolase and acyl transferase function).[32,34]
Oxalidaceae	*Oxalis tuberosa*	Oca	0-4000[9]	Oxalic acid (may play a role in defense against pathogens).[44]
Basellaceae	*Ullucus tuberosum*	Ulluco	2000-3500[9,28]	Vitamin C, saponins.[9,42,48]
Tropaeolaceae	*Tropaeolum tuberosum*	Mashua	2400-4300[14,15]	Glucosinolates.[15,47,48]

to grow and can be stored in the ground.[24] Four tuber crops (potato, ulluco, oca, mashua) account for 70% of the diet of Andean highlanders.[19] The root crops show an array of interesting properties (Table 1). Arracacha, for example, belongs to the carrot family and is high in Vitamin A. It is also a good source of highly digestible starch and is preferred by infants and the elderly.[25] The yacon, a species in the sunflower family, is high in inulin, a fructose-based non-digestible carbohydrate used by diabetics and with potential as a diet food. The maca, a member of the radish family, is the only crop in the world capable of growing at 4,500 m.[22] The dried roots are rich in starch and proteins and can be stored for years. Maca is believed to enhance fertility in humans and animals. The ARTs represent a vast and mostly untapped depository of variation in the properties and content of starch, amino acids, nutritional factors, and natural pesticides. They are adapted to harsh environments, including poor soils, frost, drought, etc. Several have potential for growth outside their area of origin. For example, oca is now grown in Australia and New Zealand, and large plantations of arracacha have been established in Brazil.[18] Several others have recently been introduced to Mexico, Central America, U.S., Europe, and Australia (Table 1). In the sections below, we briefly describe some of these ARTs, with emphasis on their ethnobotany and agronomic potential. A short discussion on sweet potato (*Ipomoea batatas*) is included for comparison purposes. The emphasis of the biochemical aspects is on storage proteins. For a more detailed discussion on the carbohydrates found in these ARTs, the reader is referred to two recent compendia.[18,19]

Tuber Crops

Potato (*Solanum tuberosum* L.) (Solanaceae) is a well-known crop not only in the Andes, but worldwide. It has become one of the 20 major food crops of the world. There are at least five other species of cultivated potatoes and more than 100 species of wild potatoes.[9] Among these are yellow potatoes and blue potatoes, with unique characteristics that capture a very small portion of present markets.[27] Collectively, these are adapted to a wide array of climates and contain a wealth of diversity and disease resistance. The farmers have more than 200 unique names for the different sizes, colors, and textures of potatoes found in the Andes.[28] They can be less watery than common potatoes, and some of them possess considerable resistance to various diseases, insects, and nematodes.[29] The potato is not only high in caloric content, but it also contains high-quality protein and substantial amounts of essential amino acids, minerals, and trace elements. Amino acid analyses have shown high amounts of lysine and low amounts of the sulphur containing amino acids, methionine and cysteine.[9] The three major tuber storage proteins are the 40 kDa glycoprotein, patatin, the 22 kDa complex protein group, and the proteinase inhibitors.[30] Patatin accounts for 40% of the total soluble protein and corresponds to the major storage protein.[3]

It consists of about 360 amino acid residues and is synthesized with N-terminal signal peptides of 23 residues. It is encoded by two sub-families of mRNA and genes (class I and class II patatins) that are 98% homologous,[31] but show differential expression. Class I patatin is present in tubers. Class II is present in roots and in low levels in tubers.[9] Patatin gene expression is induced by sucrose,[32] and is expressed in stems and petioles if tubers and axillary buds are removed.[33] Apart from its storage role, patatin has biological activity as a lipid acyl hydrolase and acyl transferase.[34]

Sweet Potato (*Ipomoea batatas* L.) (Convolvulaceae) is a root crop domesticated in the tropical areas of South America. The root is consumed as a staple in many poor countries throughout the tropics, subtropics, and at least half of the temperate zones.[35] One important quality factor of sweet potato is the amino acid composition. It contains an excess of all essential amino acids except tryptophan and total sulfur amino acids.[36] The presence of trypsin inhibitors in sweet potato may contribute to decreased protein quality. It has been suggested that the presence of these polypeptides is partially responsible for the disease enteritis necrosis in man and animals.[36] Protein content in sweet potato roots varies widely, ranging from 2% to 10% on a dry weight basis.[37] The major storage protein is a globulin, initially termed ipomoein,[38] but now known as sporamin. It has a molecular weight of 20 kDa and accounts for 60–80% total soluble protein.[36] Sporamin is not detected, or is present only at very low levels, in non-tuber parts of sweet potato plants grown under normal field conditions.[39] However, it may be synthesized in vegetative tissues under special conditions, such as when plantlets are micro propagated *in vitro*,[40] and in excised leaves and petioles. The mature protein is encoded by two subfamilies (A: 22 cDNAs, and B: 27 cDNAs).[39] The amino acid sequences of sporamin A and sporamin B have homology with the Knuitz-type serine protease inhibitors of legumes,[18] and are also related to wound-inducible inhibitors, such as the *Win 3* gene product of poplar and the cathepsin D inhibitor of potato tubers.[3] Sporamin has not been found to possess protease inhibitory activity. However, these homologies and the fact that its synthesis in leaf petiole cuttings is induced by polygalacturonic acid[41] indicate that it may play a role in defense against pests and pathogens.

Oca (*Oxalis tuberosa* Molina) (Oxalidaceae) is an important annual tuber crop from the Andes. This plant is also found in Mexico and has become popular in New Zealand. Oca is found from Venezuela to Argentina growing between 2,800–4000 m elevation, although in New Zealand it grows at near sea level.[18] The oca tuber is resistant to low temperatures, but frost kills the foliage.[42] Oca is an annual plant with clover like leaves, usually 20–30 cm high with stems that vary in color from yellow and green to purplish red. Oca tubers have a firm, white flesh and shiny skins in colors ranging from white to red (Fig. 2). Oca yields about 7–10 tons per hectare under traditional Andean husbandry.[23] The major limitation for this crop is the attack of a tuber-boring beetle related to the potato

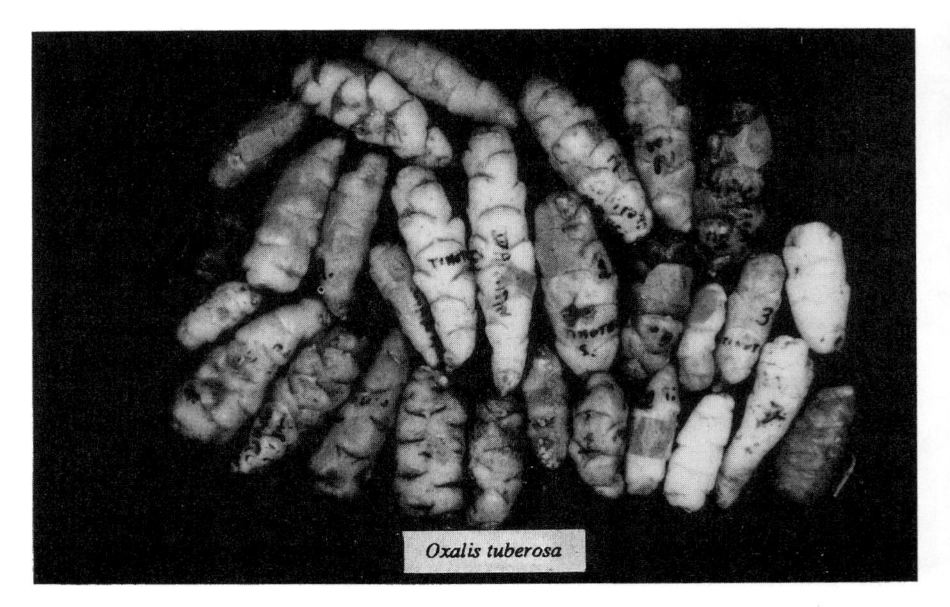

Figure 2. Different varieties of oca tubers produced in a single plot in Cuzco, Peru. Oca tubers have distinctive wrinkles and coloration patterns, and some varieties are rich in oxalic acid.

beetle.[18] Although various fungi occur on the oca plant, they seem unimportant in the field. However, molds can cause major losses.[43] The Andean farmers recognize about a dozen cultivars and more than 50 distinguishable types.[18] Oca is used in many ways: boiled, baked, fried, mixed fresh with salads, or pickled in vinegar. Most, however, are added to stews and soups. In the Andes, the tubers are placed in the sun for a few days during which they become sweet, and the amount of glucose can nearly double. Bitter varieties are almost always converted into dry products called *ccaya*, in which the bitterness disappears leaving bland-tasting products that can be stored without refrigeration.[19] The bitter tubers contain amounts of oxalic acid (Fig. 1) varying up to 500 ppm, while sweet types have only traces (79 ppm). There is no information available about the function of oxalic acid in the plant, however, it is possible that it acts as a defense against pathogens, since the sweet varieties are more susceptible to the attack of nematodes and virus than the bitter varieties (personal observations). Oca tubers contain 70–80 % moisture, 11–22 % carbohydrate, and about 1% each of fat and fiber. It is being examined as a potential commercial source of flour, starch, and alcohol.[44] Amino acid analysis indicates that glutamic acid/glutamine are high; however, leucine, isoleucine, and lysine levels are low. Protein levels vary among different varieties; on average tubers contain 5% on a dry-weight basis. A general screen of oca tuber protein from different varieties consistently showed a protein

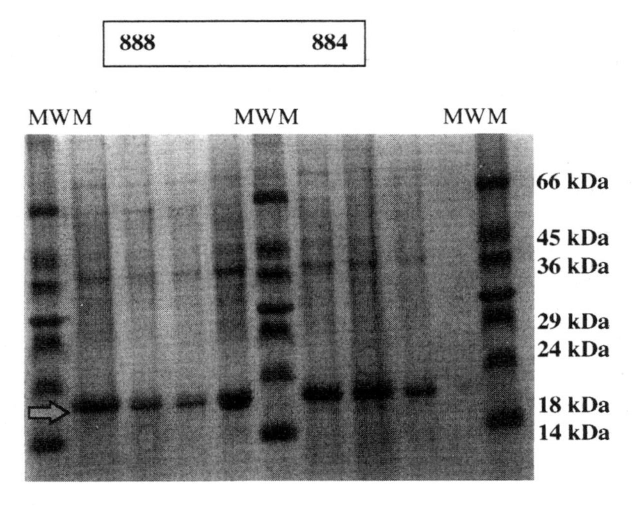

Figure 3. SDS-PAGE of total crude extract from oca tuber proteins. Molecular weight markers (MWM). Amounts of 10, 15 and 20 μl of protein (1 μl/μg) for each clone (888 and 884) were loaded in the gel.

band with a molecular weight of 18 kDa and a pI of 4.8 (Fig. 3). This 18 kDa protein was absent from the leaves, roots, and stems, suggesting that it is specific for oca tubers. It accounts for approximately 30 to 40% of the total soluble protein in tubers. The sequence of its amino acids is currently underway (Flores, unpublished results). Preliminary studies on the biological activity of oca tuber proteins showed that under *in vitro* conditions oca proteins inhibited the hyphal growth of a *Rhizoctonia solani* strain (Flores and Michaels, unpublished observations).

Mashua (*Tropaeolum tuberosum* Ruiz & Pavon*)* (Tropaeolaceae) is an edible tuber of the Andes mountains. Wild plants as well as numerous cultivated clones are known and two species have been distinguished.[45] Mashua is a frost tolerant crop cultivated in small plots in Argentina, Colombia, Ecuador, Peru, and Bolivia at altitudes up to 4000 m. The plant tolerates light frost and is unaffected by temperatures as low as 4^0 C. Mashua is a perennial herbaceous, semiprostrate climber occasionally reaching about 2 m in height. Flowers are solitary, axial, and orange to scarlet in color.[23] The tubers vary in color from white to yellow. Occasionally, the skin is purplish or red. Often tubers are mottled or striped with red or purple, especially below the eyes. The flesh is usually yellow, and the sharp flavor makes the tuber unsuitable for eating raw, so it is usually boiled with meat into stews. In addition to the tubers, the tender leaves are eaten as a boiled green vegetable. More than 100 varieties have been recognized.[46] In spite of the fact that mashua is apparently heavily infected with plant viruses, most of which are undescribed, the plant is extremely resistant to diseases and insects. It contains nematocidal, bactericidal, and insecticidal compounds (glucosinolates), thus making it a valuable plant to intercrop with other species.[47] Clinical tests of its pharmacological constituents and activity have been conducted and lend credence to its use in Andean folk medicine.[48] Test of the nematocidal and insecticidal properties of the isothiocyantes present in cultivars also support its use in protecting other cultivars from pathogen attack.[24] In spite of its productivity, pest resistance, and popularity, mashua is not widely commercialized either in its native land or elsewhere.[10] Mashua roots release the following glucosinolates: benzyl-(1) and 2-methoxybenzyl-isothiocyanate (Fig. 1) as major constituents, and 2-methyl-isothiocyanate in lesser amounts.[47] The relationship between aromatic isothiocyanates and human reproduction is not clear. However, a negative effect of mashua on male reproductive processes in rats supports the belief of Andean people that the plant has an antiaphrodisiac effect on males.[47] In addition to its food value, mashua is regarded as having positive fertility effects on females. Studies of the amino acid profile indicated that aspartic acid/asparagine, valine, and histidine are quite high; however, leucine is limited in some varieties. Mashua is relatively nutritious for a tuber crop; about 16% of the dry tuber is protein. Research carried out by King and Gershoff[49] has shown that there is a high amount of variation both in the quantity and quality of protein within varieties (14–16%). Protein studies from one variety of mashua showed the presence of 13 major proteins with molecular weights varying from 14 to 35 kDa (unpublished data).

Ulluco (*Ullucus tuberosus* Loz.) (Basellaceae) is an important mid- to high-altitude crop from Venezuela to Chile. However, it has also been grown at sea level in Canada, England, and Finland.[18] The plant grows well in cool, moist conditions and is frost resistant. Wild forms of ulluco are prostrate, while cultivated forms vary from prostrate or semi-climbing vines to dense, compact, bush-like mounds up to 50 cm tall.[28] The plant forms tubers on long stolons both below and above the ground. The tuber can be 2–15 cm long, and the most common type is conical and lemon yellow, but pigmentation can vary from white to pink, orange, red, or magenta, variegated, etc.[50] Based on tuber appearance, at least 50–70 distinct clones have been recognized.[19] Productivity sometimes reaches 5 to 9 tons per hectare. Ulluco tubers have a wide range of culinary uses. Because of their high water content, they are most often boiled, shredded, grated, mashed, pickled or mixed with hot sauces, or used to thicken soups and stews.[51] The tubers are sometimes freeze-dried, have a much stronger taste than the fresh ones, and can be stored for several years.[15] Amino acid analysis studies indicate that tryptophan is limiting in varieties. The titers of dicarboxylic amino acids and histidine are quite high in ulluco. Fresh tubers are about 85 % moisture, 14 % starches and sugars, and 1–2 % protein.[18] They are high in vitamin C, and contain a gum, but no fat and almost no obvious fiber.[52] Most diseases that are known to attack ulluco seem to be specific to this plant. The bitterness is probably due to saponins,[48] which may play a role in defense against pathogens.

Root Crops

Maca (*Lepidium meyenii* Walp.) (Brassicaceae) is a root crop found at altitudes up to 4,500 m. Maca is distinct from other Andean crops in both its limited distribution and method of propagation.[25] This species has been declared in danger of extinction,[53] grows in very poor soils, and requires intense sunlight and below freeing temperatures.[18] The leaves grow as a rosette close to the ground and as the older ones die they are replaced by new leaves at the center. The edible part is derived from the tuberous hypocotyl and may reach 8 cm in diameter.[9] The flesh is pearly white and has a marbled appearance. It consists of two fairly well-defined parts, a creamy colored outer section rich in sugars, and an inner firmer section which is particularly rich in starches.[54] There are four traditionally recognized types, all based on the color of the roots which can be cream-yellow, yellow banded with a purple waist, purple, and black. Maca is usually planted in small plots, often surrounded by stone fences that protect plants from desiccating wind. The roots are harvested 6–7 months after planting, yielding on average less than 3 tons per hectare. Maca roots are similar to the related radish and are yellow, purple, or yellow with purple bands. They are rich in sugars, starches, protein, and essential minerals, particularly iron and iodine.[50] No part of the plant is wasted, and often the leaves are eaten. The fresh roots are baked or roasted in ashes, and dry roots are mainly boiled in milk or water. It also makes a popular sweet, fragrant, fermented drink (*maca chicha*) that is often mixed with hard liquor. The dry roots store very well, and the flavor remains strong for two years and often for much longer. The nutritional value of maca is poorly studied but it is known that the dried roots are approximately 13–16% protein;[24] nutritional value stays high during storage, in contrast with other ARTs such as potato and oca. Seven year old roots still have a high caloric content as well as 9–10 % protein.[52] Maca has a high content of calcium (258 mg/g) and iron (15.4 mg/g) and is rich in carbohydrates, alkaloids, tanins, and starch.[19] Maca is further valued because it reputedly enhances fertility in both humans and livestock.[18] Preliminary tests support its effect on increased reproduction in domesticated Andean rodents such as guinea pigs fed a maca diet.[55] This may be related to the production of glucosinolates, specifically the release of aromatic isothiocyanates.[48]

Mauka *(Mirabilis expansa* Ruiz & Pavon) (Nyctaginaceae) is a root crop grown at 2200 to 3500 m and currently restricted to the north central Andes of Peru.[18] There are no well defined varieties, but several genotypes differing in root flesh color (white, salmon, yellow) have been recognized. Mauka is a low, compact plant, not exceeding 1 m in height. The stems are cylindrical with opposite, ovoid leaves. Bolivian types have purple flowers and Ecuadorian types are white.[50] The edible parts are the root and lower part of the stem which are boiled or fried and served as vegetables. The roots, which may reach several feet in length, are high in moisture, rich in starch, and contain little fiber.[9] The stems and the roots are high in carbohydrates (87% on a dry weight basis) and contain up to 7% protein. Mauka is rich in calcium, phosphorus, and potassium.[50]

We have recently found that mauka storage roots contain high levels of a 28 kD protein with biological activity.[58] Based on Western blot analysis, the protein, which accounts for over 50% of the total soluble polypeptides, has been identified as a type I (monomeric) ribosome inactivating protein (RIP) (Vivanco and Flores, unpublished results). These are widely distributed among higher plants and act by deglycosylating the 28S ribosomal RNA, thus preventing the binding of the elongation factor 2[56] and terminating protein synthesis. Potential uses of RIPs include application as antivirals, antifungals, immunotoxins,[56] and inhibitors of HIV replication.[57] In addition, the RIPs from *Mirabilis* spp. may have interesting potential agricultural applications. The effects of root extracts of *Mirabilis jalapa*, an ornamental species closely related to mauka, were tested against infection by two potato viruses (PVX and PVY) and one viroid (PSTVd). RIP-containing *Mirabilis* root extracts sprayed 24 hours before virus or viroid inoculation inhibited viral replication almost 100%. The antiviral activity was confirmed against mechanically transmitted viruses, but was not observed against aphid-transmitted viruses.[58]

An antiviral protein (MAP) with RIP properties has been purified from roots of *Mirabilis jalapa* and its molecular weight was confirmed about 28 kD, as described by Kubo *et al.*[59] Purified MAP showed the same antiviral effect as the crude extracts. We used polyclonal antibodies against MAP to determine that it was present at high concentrations in the roots of the white-flowered variety of mauka. The presence of MAP in species of the family Nyctaginaceae and other families was screened. A protein of similar size to MAP was detected by Western blots in leaf extracts of mauka (*Mirabilis expansa*) and in *Mirabilis multiflora* (Fig. 4), a species native to the southwest U.S. which has been used as food and medicine by native American Indians. Short term use could include preventive application of mauka extracts to control virus infection. In the long term, transformation of plants with the gene encoding for this RIP should be feasible.

Our studies with mauka have interesting similarities to others carried out in our laboratory with a Cucurbitaceae species, *Trichosanthes kirilowii*. The storage roots of this and related species of Chinese origin have been used to treat diabetes and to induce abortion. The abortifacient activity is due to a type I RIP named trichosanthin, which also shows inhibiting activity against HIV replication in cell cultures. This has led to recent clinical trials and to the identification of potentially therapeutic proteins in a related genus, *Momordica*.[5] Our studies

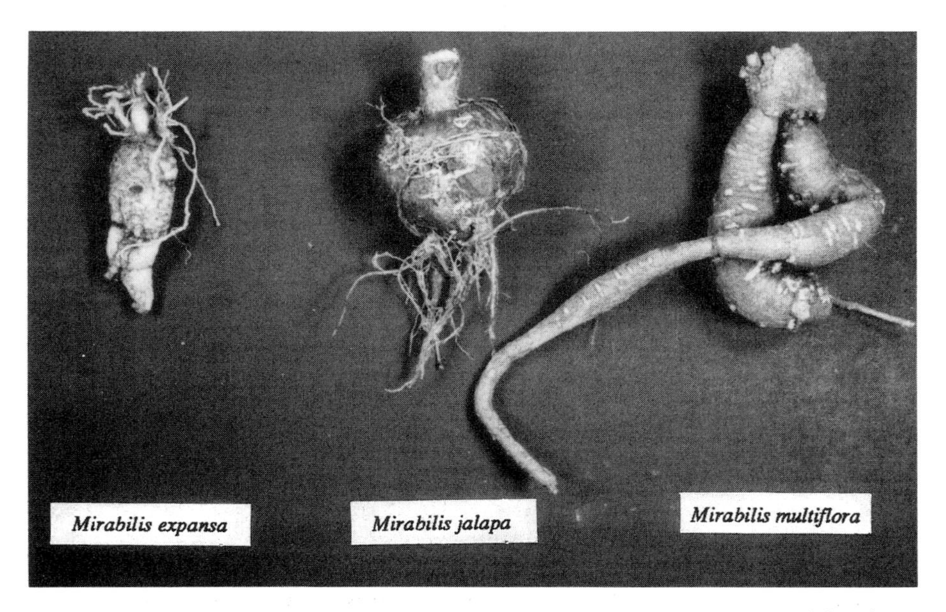

Figure 4. Storage roots of three different species of *Mirabilis*: *Mirabilis multiflora*, *Mirabilis jalapa* and *Mirabilis expansa*, from greenhouse-grown plants.

also indicate that the accumulation of trichosanthin in roots is tightly correlated with the initiation of storage root formation.[60,61] Thus, in fully developed storage roots of *Trichosanthes* species, RIPs account for 50–80% of the total soluble protein, as is also the case for *Mirabilis*. The existence of these correlations in widely divergent species suggests a fundamental biological role for these root-expressed RIPs, perhaps a dual function as a storage protein and defense agent. *In vitro* assays have shown that RIPs from barley roots can have synergistic, antifungal activity in combination with chitinases.[5] It is also intriguing, both in the case of *Trichosanthes* and *Mirabilis*, that the use of the same part of the plant for similar purposes has been independently discovered in the traditional pharmacopoeias of both Oriental and New World cultures. For example, the roots of a New World Cucurbitaceae, *Gurania* spp., have been used in Central America to induce abortions, suggesting the presence of novel RIPs in this genus. How indigenous cultures are able to accommodate these strong biological effects with a food use for the storage root is deserving of close study.

Arracacha (*Acarracia xanthorrhiza*. Bancroft) (Umbelliferae) is a root crop found at altitudes from 3,200 to 600 m. Arracacha is a perennial herb only known in South America and a few parts of Central America and the Caribbean.[24] Stems and leaves usually attain a height of about 1 m; flowers are purple or yellow, small and formed in flat clusters on stalks.[18] Although the plant is usually fertile, arracacha is generally harvested before completing a seed cycle. Roots are cylindrical with numerous lateral roots and the flesh ranges in color from white to yellow or purple, with a creamy white exterior.[51] Arracacha grows in sandy soils with pH of 5 or 6, needs an even distribution of rainfall and a temperature range of 14 to 21 °C. Lower temperatures delay maturity.[18] In the Andes, three different varieties have been distinguished based on the color of the root: white, yellow, and purple. Certain cultivars differ in flavor, texture, and length of time to maturity.[51] Arracacha is attacked by spider mites and is susceptible to nematodes in some regions. The roots have a relatively short storage life, and, unlike carrots, they can be eaten only after cooking.[19] Roots are harvested 300–400 days after planting. The central root may have as many as 10 lateral roots aggregated around the central rootstock. Yields normally vary between 5 to 15 tons per hectare; one plant may produce 2–3 kg of edible lateral roots.[50] Roots are eaten boiled, baked, or fried, or added to stews. They have a crisp texture and a flavor characteristic of the family.[19] The young stems are used in salads or as a cooked vegetable. Although edible, the central root has a coarse texture and a strong flavor, and it is usually fed to livestock.[18] The nutritional value of arracacha is poorly studied, however, it is known that the dried roots have a starch content ranging from 10 to 25%. The starch is easily digested and can be used in foods for infants and invalids. This use has been commercialized in Brazil, where aracacha is known as mandioquinha salsa and grown in large acreages. During storage for several months, the roots increase in sweetness due to partial starch hydrolysis. All parts of the plant have high calcium content, and the roots are rich sources of vitamin A.[19]

Yacon (*Polymnia sonchifolia* P&P.) (Compositae), a species related to the sunflower, is an annual, herbaceous plant with dark green leaves. The aerial stems can reach 2 m in height and are hairy with purple color. Yacon's tuberous roots are irregular in shape, ranging from spindle-shape to round and varying considerably in size and sweetness.[18] On the outside they are tan to purplish brown, but inside they are white, yellow, purple, or orange. Yacon grows between 900 to 2750 m in the Andes, but has been grown at sea level in New Zealand and the United States.[50] The plant is tolerant of a wide range of temperatures and can grow in a variety of soil conditions. It does best in well-cultivated, rich, well-drained soil. Roots are eaten raw in salads and also can be consumed boiled and baked. Yacon might have potential as a forage crop. The foliage is luxuriant and the leaves have a protein content of 11 to 17% (on a dry weight basis). Fresh tubers have 70–83% moisture, 0.4–2.2 % protein, 0.4–1.3% fat, 4–6% fiber, and approximately 65% sugars.[10] The roots are high in potassium, and the dried root has 11–17% protein and 38–41% nitrogen Yacon is high in inulin, a fructan type non digestible carbohydrate used by diabetics, and with potential as a diet food.[48]

CONCLUSION

The phytochemistry and biochemistry of plant roots and tubers has been significantly less studied than other edible plant organs. A comprehensive and integrated effort to understand and utilize these fascinating underground chemical factories is bound to yield rewarding results in many respects. For example, a detailed understanding of the biological and cultural rationales for the many ethnobotanical uses known for underground storage organs should influence the direction of future biochemical and biotechnological approaches. The develop-

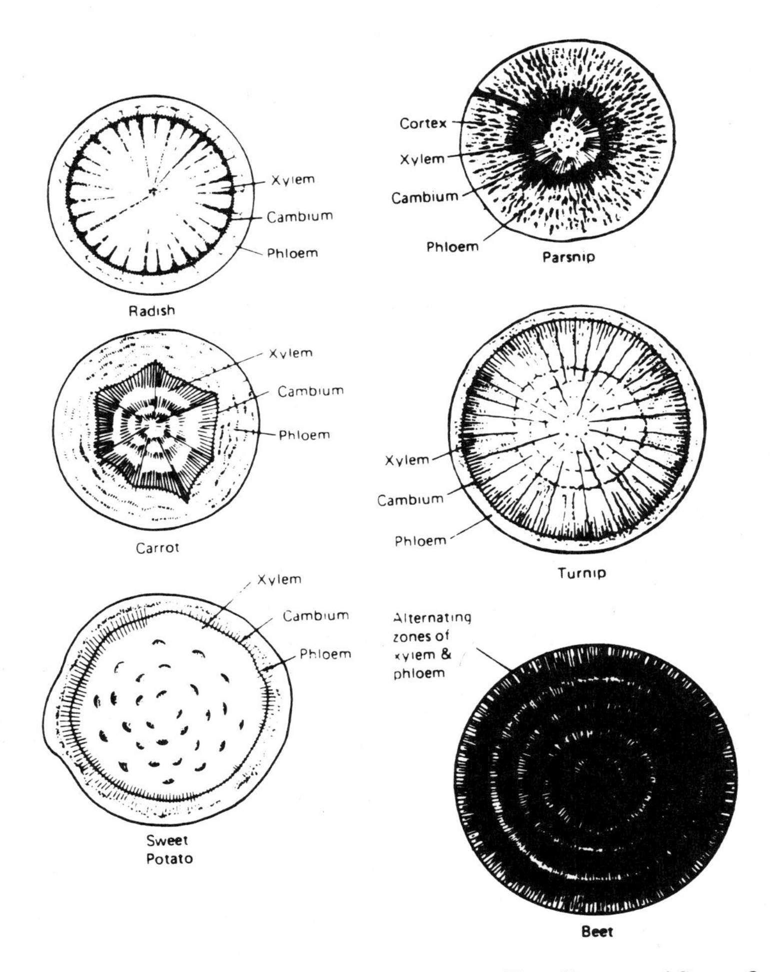

Figure 5. Basic developmental patterns in root storage organs (From Simpson and Conner-Ogorzaly, 1995).

mental aspects of storage organs, in particular roots, remain black boxes. For example, there are at least three basic patterns for storage root development, exemplified by carrots and radishes, beets, and sweet potatoes, respectively. We know basically nothing about the molecular events underlying the various patterns of cambium activation and subsequent parenchyma differentiation in carrots, radishes, beet, sweet potatoes, etc. (Fig. 5). Even for single root crops such as sweet potato, the variation in patterns of pigment accumulation present among common varieties suggests a rich diversity of metabolic capacities and a complex array of regulatory mechanisms. A sound understanding of secondary root formation and its underlying phyto- and biochemistry should lead to a better utilization of both common and neglected crops. We end this review with the hope that these intriguing questions will be resolved by young phytochemists willing to integrate their discipline with both traditional and modern approaches.

REFERENCES

1. LOPEZ, M., LARKINS, B. 1993. Endosperm origin, development and function. Plant Cell 5:1383–1399.
2. MARTIN, C., SMITH, A. 1995. Starch biosynthesis. Plant Cell 7:971–985.
3. SHEWRY, P.R. 1995. Plant storage proteins. Biol. Rev. Cambridge Physiol. Soc. 70:375–426.
4. HARBORNE, J. B. 1993. Introduction to Ecological Biochemistry. Academic Press, Great Britain, 318 pp.
5. FLORES, H.E., WEBER, C., PUFFETT, J. 1996. Underground metabolism: The biosynthetic potential of roots. In: Roots:The Hidden Half, (Y. Waisel, A. Eshel, U. Kafkafi, eds.), 2nd Ed., Marcel Dekker, New York. pp. 931–956.
6. SAUER, C.O. 1969. Agricultural Origins and Dispersals, the Domestication of Animals and Foodstuffs. 2nd Ed., Cambridge, MA., M.I.T. Press, 175 pp.
7. SIMPSON B.B, CONNER-OGORZALY, M. 1995. Economic Botany, Plants in Our World, 2nd Ed. McGraw-Hill, New York, 640 pp.
8. RICHARDSON, M. 1991. Seed storage proteins: The enzyme inhibitors. In: Methods in Plant Biochemistry, Vol. 5. Academic Press, New York, pp. 259–305.
9. LI, H.P. 1985. Potato Physiology. Academic Press, Orlando, FL, 586 pp.
10. BIETZ, J.A. 1983. Separation of cereal proteins by reversed-phase high-performance liquid chromatographic. J. Chrom. 255:219–238.
11. KASARDA, D.D., AUTRAN, J., LEW, E.J., NIMMO, C.C., SHEWRY, P. 1983. N-terminal amino acid sequences of ω-gliadins and ω-secalins: Implications of the evolution of prolamin genes. Biochem. Biophys. Acta 747:138–150.
12. PETERSON, D.M., SMITH, D. 1976. Changes in nitrogen and carbohydrates fractions in developing oat groats. Crop Sci. 16:67–71.
13. IRWIN, S.D., KEEN, J.N., FINDLAY, J.B., LORD, J.M. 1990. The *Recinus communis* 2S albumins precursor: a single pre-protein must be processed into two different heterodimeric storage proteins. Mol. General Gen. 222:402–408
14. MENENDEZ-ARIAS, L., MONSALVE, R.L., GAVILANEA, J.G., RODRIGUEZ, R. 1987. Molecular and spectroscopic characterization of a low molecular weight seed storage protein from yellow mustard (*Sinapsis alba*). Eur. J. Biochem. 19:899–907.
15. CASEY, R., DOMONEY, C., SMITH, A.M. 1986. Legume storage proteins and their genes. In: Surveys of Plant Molecular and Cell Biology, Vol. 3. Oxford University Press, Oxford, pp. 1–95.

16. ARGOS, P., NARAYANA, S.V., NIELSEN, N.C. 1985. Structural similarity between legumin and vicilin storage protein from legumes. EMBO J. 4:1111–1117.
17. HARLAN, J.R. 1992. Crops and Man. American Society of Agronomy, 2nd ed. Madison, WI, 284 pp.
18. NATIONAL RESEARCH COUNCIL. 1989. Lost Crops of the Incas: Little known Plants of the Andes with Promise for Worldwide Cultivation. National Academy Press, Washington, D.C., 407 pp.
19. HERNANDEZ, J.E., LEON, J. 1992. Cultivos Marginados: Otra Perspectiva de 1492. Organizacion de las Naciones Unidas para la Agricultura y la Alimentacion, Roma, 339 pp.
20. TAPIA, M.E. 1993. Semillas Andinas: El Banco de Oro. Consejo Nacional de Ciencia y Tecnologia (CONCYTEC), Lima, Peru, 76 pp.
21. MURRA, J.V. 1975. Formaciones Economicas y Politicas del Mundo Andino. Instituto de Estudios Peruanos, Lima, 339 pp.
22. MORRIS, C., VON HAGEN, A. 1993. The Inca Empire and its Andean Origins. Abbeville Press, New York, 251 pp.
23. CENTRO INTERNACIONAL DE LA PAPA. 1993. El Agroecosystema Andino: Problemas, Limitaciones, Perspectivas. Centro Internacional de la Papa, Lima, Peru, 8. pp.
24. KING, S.R. 1987. Four endemic andean tuber crops: promising food resources for agricultural diversification. Mountain Research and Development 7(1):33–39.
25. HODGE, W.H. 1954. The edible arracacha -a little known root crop of the Andes. Econ. Bot. 8:95–221.
26. TELLO, J., HERMANN, M., CALDERON, A. 1992. La maca (*Lepidium Meyenii* Walp) : Cultivo alimenticio potencial para las zonas alto andinas. Boletin de Lima 81:59–66.
27. JANICK, J., SIMON, J.E. 1990. New Crops. John Wiley and Sons, New York, 23. pp.
28. MONTALDO, A. 1977. Cultivo de Raices y Tuberculos Tropicales. IICA, San Jose, Costa Rica, 284 pp.
29. BRUSH, S.B., CARNEY, H.J., HUAMAN. Z. 1981. Dynamics of Andean potato agriculture. Econ. Bot. 35:70–88.
30. SANG-GO, S., PETERSON, J.E., STIEKEMA, W.J., HANNAPEL, D.J. 1990. Purification and characterization of the 22-kilodalton potato tuber proteins. Plant Physiol. 94:40–45.
31. MIGNERY, G.A., PIKAARD, C.S., PARK, W.D. 1984. Molecular characterization of patatin multigene family of potato. Gene 62:27–44.
32. ROCHA-SOSA, M., SONNEWALD, U., FROMMER, W., STRATMANN, M., SCHELL, J., WILLMITZER, L. 1989. Both developmental and metabolic signals activate the promoter of a class I patatin gene. EMBO J. 8:23–29.
33. PAIVA, E., LISTER, R., PARK, W. 1983. Induction and accumulation of major tuber proteins of potato stems and petioles. Plant Physiol. 71:161–168.
34. BRODGON, W.G., DICKINSON, J. 1983. A microassay system for measuring esterase activity and protein concentration in small samples and in high pressure liquid chromatography eluate fractions. Anal. Biochem. 131:499–503.
35. SPLITTSTOESSER, W.E. 1977. Protein quality and quantity of tropical roots and tubers. HortSci. 12:294–298.
36. DICKEY, L., COLLINS, W., YOUNG, C. 1984. Root protein and quality in a seedling population of sweet potatoes. HortSci. 19:689–692.
37. VARON, D., COLLINS W. 1989. Ipomoein is the major soluble protein of sweet potato storage roots. HortSci. 24:829–830.
38. JONES, D.B., GERSDORFF, C.E. 1931. Ipomoein a globulin from sweet potato, *Ipomoea batatas*. Isolation of a secondary protein derived from ipomoein by enzymatic action. J. Biol.Chem. 93:119–126.
39. MAESHIMA, M., SASAKI, T., ASAHI, T. 1985. Characterization of proteins in sweet potato tuberous roots. Phytochemistry 24:1899–1902.

40. HATTORY, T., MATSUOKA, K., NAKAMURA, K. 1988. Subcellular localization of the sweet potato tuberous root storage protein. Agric. Biol. Chem. 52:1057–1059.
41. OHTO, M., NAKAMURA-KITO, K., NAKAMURA, K. 1992. Induction of expression of genes coding for sporamin and β-amylase by poly galacturonic acid in leaf-petiole cuttings of sweet potato. Plant Physiol. 99:422–427.
42. ORBEGOSO, A.G. 1960. Estudios sobre la oca (*Oxalis tuberosa*) con especial reference a su estructura y variabilidad. Agron. 27:28–38.
43. CORTES, H.B. 1977. Avances en la investigacion de la oca. In: Advances en las Investigaciones sobre Tuberculos Alimenticios de los Andes. PISA. IICA, Lima, 144 pp.
44. CORTES, H.B. 1981. Avances de la investigacion en tres tuberculos Andinos. Curso Sobre Manejo de la Produccion Agraria en Laderas. IICA, Bogota, Colombia, 200 pp.
45. JOHNS, T., TOWERS, G.H.N. 1981. Isothiocyanates and thioureas in enzyme hydrolases of *Tropaeolum tuberosum*. Phytochemistry 20:2687–2689.
46. HODGE, W.H. 1957. Three native tubers of the high Andes. Econ. Bot. 5:185–201.
47. JOHNS, T., KITTS, W.D., NEWSOME, F., TOWERS, G.H.N. 1982. Anti-reproductive and other medicinal effects of *Tropaeolum tuberosum*. J. Ethnopharmacol. 5:149–161.
48. JOHNS, T. 1990. With Bitter Herbs They Shall Eat It: Chemical Ecology and the Origins of Human Diet and Medicine. The University of Arizona Press, Tucson, 356 pp.
49. KING, S., GERSHOFF, S. 1987. Nutritional evaluation of three underexploited Andean tubers: *Oxalis tuberosa* (Oxalidaceae), *Ullucus tuberosus* (Basellaceae), and *Tropaelolumetuberosum* (Tropalaceae). Econ. Bot. 41:503–511.
50. TAPIA, M.E. 1981. Los Tuberculos Andinos. PISA. IICA, Lima, 114 pp.
51. CASEDA, F., ROSSEL, J. 1985. Entomologia de los Cultivos Andinos: estudio de lascomunidades campesinas. Editorial Universitaria 85, Universidad Nacional Agraria, La Molina, Lima, 155 pp.
52. KING, S.R. 1988. Economic Botany of the Andean Tuber Crop Complex: *Lepidium meyenii, Oxalis tuberosa, Tropaeolum tuberosum* and *Ullucus tuberosus*. Ph.D. dissertation, Graduate Faculty in Biology, University of New York.
53. IBPGR. 1982. Plant genetic resources of the Andean region. Proceedings of meeting of IBGR, IICA, and JUNAC, Lima, Peru, 129 pp.
54. LEON, J. 1964. The "maca" (*Lepidium meyenii)*, a little-known food plant from Peru. Econ. Bot. 18:122–127.
55. CHACON, R.C. 1961. Estudio fitoquimico de *Lepidium meyenii*. Dissertation, Univ. Nac. Mayor de San Marcos, Peru.
56. STIRPE, F., BARBIERI, L., BATELLI, M.G., SORIA, M., LAPPI, D.A. 1992. Ribosome inactivating proteins from plants: present status and future prospects. Bio/Technol. 10:405–412.
57. ZARLING, J.M., MORAN, P.A., HAFFAR, O.J., SIAS, D., RICHMAN, D., SPINA, C.A., MYERS, D.E., KUELBELBECK, V., LEDBETTER, J.A., UCKUN, F.M. 1990. Inhibition of HIV replication by pokeweed antiviral protein targeted to CD4 cells by monoclonal antibodies. Nature 347:92–95.
58. VIVANCO, J.M. 1995. Efecto inhibitorio de los extractos de *Mirabilis jalapa* en contra de PVX, PVY y PSTVd. Tesis, Facultad de Agronomia-Universidad Nacional Agraria La Molina, Lima, Peru.
59. KUBO, S., IKEDA, T., IMAIZUMI, S., TAKANAMI, Y., MIKAMI, Y. 1990. A potent plant virus inhibitor foun in *Mirabilis jalapa* L. Ann. Phytopath. Soc. Japan 56:481–487.
60. SAVARY, B.J., FORES, H.E. 1994. Production of bioactive proteins in transformed root cultures of *Trichosanthes kirilowii* and related species. Plant Physiol. 106:1195–1204.
61. SAVARY, B.J., FLORES, H.E. 1997. Characterization of extracellular chitinases produced by root cultures of *Trichosanthes kirilowii*. Plant Physiol. Biochem. (in press).

Chapter Six

BEHAVIORAL DETERMINANTS FOR THE INGESTION OF FOOD PHYTOCHEMICALS

Timothy Johns

School of Dietetics and Human Nutrition
Macdonald Campus, McGill University
Ste Anne de Bellevue, QC, H9X 3V9, Canada

INTRODUCTION

People in industrial societies appear to be ingesting increasing quantities of phytochemicals in comparison to levels during most of the twentieth century, although quantitative data to support this supposition are scarce.[1,2] Certainly the public perception of the value of natural products has grown. Some of the reasons for alterations in ingestive patterns may reflect new scientific information, but the reasons for changes in consumer behavior are undoubtedly multifaceted. Phytochemicals may be ingested in food or as purified or semi-purified forms of the same types of chemical constituents taken as dietary supplements. Fresh fruits and vegetables and whole foods are recognized as important sources of a myriad of chemicals with a diversity of potential physiological effects.[3] Foods and food products, for which there is scientific information concerning the health benefits of their non-nutrient constituents, are the subjects of scientific and popular attention, and are variously referred to as nutraceuticals, functional foods or pharmafoods.[3] In addition, people in industrial countries are increasingly returning to the use of herbal medicines.[4–7]

The growing market for nutraceuticals and herbal supplements, notwithstanding the lack of substantiation of the efficacy and/or government approval of many of them,[8] is supported by considerable marketing and advertising by entrepreneurs. Although consumers relate to messages promoting health, humans are by nature conservative in their dietary habits and do not necessarily respond to nutrition education and advertising.[9] Explanations, therefore, for why phytochemicals and the entities that contain them are being ingested in increasing degree by at least a segment of the population should include other and varied perspectives.

Human consumptive patterns and the factors that determine them within populations and by individuals have been considered within the sciences of nutrition and behavior. Diet is typically examined as either the combination of nutrients or combinations of foods consumed by an individual or population.[10] Behavior can be viewed as a single outcome measure, such as total food consumption, or as a sum of components of that behavior, such as patterning of food and/or nutrient selection within a meal or throughout a longer time-frame.[9] Typically, approaches to this question have been divided into examination of psychosocial and cultural factors,[11] and physiological determinants.[12] The determinants of phytochemical ingestion, apart from clearly medicinal uses, are largely an unexplored field. This paper attempts to extend the methodological model of nutrient selection to include food phytochemicals. It draws insights from both the psychosocial and physiological perspectives.

PSYCHOSOCIAL DETERMINANTS OF BEHAVIOR AND CHANGING PARADIGMS IN BIOMEDICINE

To limit the determinants of consumer behavior to a consideration of only scientific developments and marketing campaigns suggests that the public is passive. It is more likely the case that science, economic activity, and values each influence each other, and that current profound changes in concepts of medicine and treatment have ties to all of these influences.

Etkin and Johns[13] have considered social, scientific and cultural factors together and suggest the emergence of a new biomedical paradigm that supports an increased ingestion of phytochemicals. On one level, this is a return to earlier forms of ingestive behavior. Prior to the twentieth century, western industrial societies, as most non-industrial societies, did not make strong distinctions between food and medicine. The importance of diet in influencing health and illness is particularly strong in the Hippocratic tradition.[14] In many cultures, foods were seen as health promoting, and medicines, principally herbal medicines, were viewed as agents for maintaining some kind of balance that is normally intrinsic to proper consumption of food. The Chinese system of medicine based on yin and yang provides a classic case of concepts of balance.[14] Similarly, Ayurvedic traditions make extensive use of food in disease prevention and healing.[15] An apparent feature of traditional subsistence systems was that people ingested appreciable levels of phytochemicals, and these played important roles in normal physiological ecology and health.[16]

Modern trends that increase phytochemical ingestion may represent a return step towards historical patterns of human behavior. Thus, the decreases in consumption of natural products in industrial societies as a result of manipulation of the food supply through plant breeding and food processing may in time be seen as somewhat of an historical aberration in human ecology. Modern pharmaceuticals, particularly those directed against infectious disease, have had profound effects on patterns and perceptions of disease. Diseases are seen as having specific causes, and treatments are targeted to their alleviation. In this model, food and medicine have been disconnected.[13]

Somewhat ironically then, the success of pharmaceutical-based medicine and of modern technology is also in large part responsible for the emergence of a modified biomedical paradigm that includes food and phytochemicals. Major diseases in industrial societies today, specifically cardiovascular disease, cancer, and diabetes, are not satisfactorily treated by targeted pharmaceuticals. They have complex etiologies, and diet is seen as an important contributor to the development of the disease and to its prevention and control. In particular, plant components of the diet are associated positively in epidemiological studies with health,[17] and natural products within plant foods are recognized as conferring important protective effects.[18]

Advances in analytical techniques mean that phytochemists are increasingly able to sort out the complex constituents of plant foods and to identify and test those that have specific properties. As well, biotechnologies such as tissue and cell culture, fermentation, and genetic engineering make it feasible to produce purified compounds that can be added to foods or other products.[3]

Public acceptance of phytochemicals encountered in foods considered as healthy, and in nutraceuticals and herbal supplements, seems to be related in part to the emergence of a synthetic homeostatic model of medicine. In this, traditional concepts of the relationship of food to medicine and of the importance of balance in health are included simultaneously with a continued acceptance of the value of modern pharmaceuticals. While the public shows certain disenchantment with drug-based medicine, it has not lost faith in the power of specific molecules. Embrace by scientists of homeostatic models in the understanding of complex diseases[13] can have a scientific basis but is in part directed by changing concepts and values in society as a whole. With the recognition of the benefits of molecular constituents within food and herbs, people have found a model that encompasses both tradition and modern science.

Entrepreneurs have seized on this convergence of values by promoting commodities that can be associated with plants and given a "natural" label. Identifying a perceived function with a specific constituent can strike a cord with consumers. Markets for herbal supplements and functional foods are growing at between 15–20% per year in North American and Europe.[3–7] Likewise, opportunities for research and for industrial development,[19] particularly involving technologies such as those discussed elsewhere in this volume,[20,21] are on the increase. While such developments have positive aspects, they present new challenges for government regulatory agencies.[3–5,22,23] Equally, they present interesting questions for phytochemists and other scientists concerned with the social impact of their research.

PHYSIOLOGICAL DETERMINANTS OF PHYTOCHEMICAL INGESTION

While individuals may or may not be able to articulate the cultural, social and psychological reasons for their choices, these factors are unlikely to be the sole determinants for their food selection. Attaining the nutrients essential to the body's metabolism is the primary purpose of eating, but few people would profess to eat for this reason. At a physiological level, nutrient appetite is only one of several physiological processes that affects intake. As determinants of behavior, such processes are mediated in some fashion through the nervous system,[12] although the mechanisms of this action are not necessarily clear.

In the most clear cut cases, phytochemicals have direct effects on nervous receptors. Indeed, many important neurological receptors, such as opioid, nicotinic and muscarinic receptors,[24] have been identified and named because of the actions of widely used plant products, *e.g.* morphine from *Papaver somnifera*, nicotine from *Nicotiana tabacum*, and muscarine from *Amanita muscaria*. It is conceivable that foods and other plants contain substances affecting receptors that have yet to be identified. However, in many cases the effects of phytochemicals likely are indirect.

Reward mechanisms, mediated through neurochemical changes in pleasure centers in the central nervous system (CNS), may provide the motivation for the ingestion of some compounds.[25] Conditioned learning, associated with improvement of well-being through relief of pain or other distress, may provide sufficient reward to encourage the consumption of a particular plant.[15,26]

This section will examine several processes linked with chemical constituents of plants that provide insights into how phytochemicals can determine behavior. Herbal remedies provide the most accessible examples. The list of potential herbs is vast, and this paper draws mainly on examples of plants that are in common use in Western industrial countries and for which tonic and psychodynamic action are their primary uses. Some of these plants, such as celery seed, chamomile, gentian, hops, peppermint, St. John's Wort, and valerian, while primarily used as herbal medicines, are approved as food flavourings or additives in certain countries.[27,28]

Tonics and Other Traditional Concepts

Apart from associating food and medicine as agents for health, people around the world recognize plants as having general effects on well-being. Tonics are used globally to energize, alleviate tiredness, and increase strength. In some cases they may restore a balance that is nutritional in nature.[29] We learned in our ethnobotanical work among the Batemi people of Tanzania that the stated purpose for the majority of plants used as routine additives in soup and other foods is as tonic, to counter pain, rheumatism and fever.[30] While these additives may make other positive contributions to health such as in modulating energy metabolism,[16,31] human behavior appears to by governed by those attributes that have more immediate impact on the nervous system.[30]

In the literature, tonic is usually related to effects on the nervous or immune systems. In the latter case they are largely synonymous with adaptogens[8] (see below). Modern naturopathic medicine recognizes as nerve tonics those plants with attributed properties in "repairing and nourishing the nervous system".[32] Such herbs with beneficial effects on the nervous system are referred to as nervines.[32] Among the attributes of plants equatable with nervines that are more generally recognized by pharmacologists are sedatives, stimulants, and antispasmodics. Plants may also be recognized for specific effects of pain alleviation, i.e. as analgesics. These terms correspond in general with concepts of neuropharmacology, and the mechanisms of action are recognizable.

The fundamental importance of plants with neuropharmacological and other behavioral effects in human affairs is underscored by the fact that the efforts of early organic chemists were in the isolation of compounds from such plants. The first, morphine, was isolated from *Papaver somnifera* in 1806 by Sertürner,[24] and others such as nicotine, caffeine, strychnine and cocaine drew early attention.[15,24] These compounds have contributed to our understanding of physiological mechanisms of drug action.

Neuropharmacological Mechanisms of Action

Natural products in food and herbals can potentially act through any of the many known mechanisms by which pharmaceuticals affect the nervous system.[24–26] They may be direct agonists or antagonists for neuroreceptors in the peripheral autonomic and central nervous systems, *e.g.* affecting cholinergic transmission involving acetylcholine (ACh) or adrenergic transmission involving the catecholamines dopamine, norepinephrine and epinephrine.[24] Alternately, they may affect transmission at other levels such as through stimulating or blocking ganglionic transmission, inhibition of the function of acetylcholinesterase in terminating the action of ACh, or as inhibitors of monoamine oxidase (MAO) in the oxidative deamination of catecholamines.[24] Compounds or precursors that affect synthesis of neurotransmitters can have effects on neurotransmission and subsequently behavior.

Both cholinergic and anticholinergic drugs cause behavioral effects in humans and animals.[24] Actions of compounds at the CNS level are likely to have the most direct effects upon behavior. In addition to ACh and the catecholamines, serotonin and the gamma-aminobutyrate (GABA) are important neurotransmitters that are affected by drugs.[25]

The effects of drugs at the neurophysiological and behavioral levels depend on the organization and function of particular neuorpathways. GABA is the main inhibitory neurotransmitter in the brain.[24] Mood and emotion-altering agents such as benzodiazepines, steroids, barbituates, and ethanol directly affect GABA levels and $GABA_A$ receptors. $GABA_A$ receptors[24,33] are coupled with chloride ion channels, and endogenous substances modulate chloride ion flow and neural responses. Multiple $GABA_A$ receptors exist in the brain and they differ in distribution and development patterns as well as in their effects on function.[34]

Other physiological mechanisms through which phytochemicals may affect well-being and behavior include pain relief and hormonal actions.[35] Naturally-occurring analgesics such as the salicylates act through the peripheral inhibition of the biosynthesis of the prostaglandins associated with the pain accomanying injury and inflammation.[24]

Food Selection and Behavioral Effects of Nutrients

Obtaining essential nutrients is the most fundamental form of chemically-directed behavior, and it has been hypothesized that humans and animals can inately select a balance of nutrients from available foods.[36] Appetite is primarily governed by energy need, and satiety is achieved by adequate intake of calories from fats, carbohydrate and protein. Specific appetites or hungers develop during nutritional deficiencies for sodium and iron, but appetites for other nutrients that might direct behavior are unclear.[37] The degree to which such behaviors are inate or conditioned is also uncertain. In addition, appetite is affected by anorexic agents in food and beverages, *e.g.*, methylxanthines such as caffeine and theobromine and small peptides with opioid activity called exorphins derived from wheat gluten and other sources.[38]

Carbohydrates, protein, and large neutral amino acids, most importantly tryptophan, have post-ingestional effects on behavior.[39] Tryptophan is a sedative, and carbohydrate consumption increases its peripheral concentration while

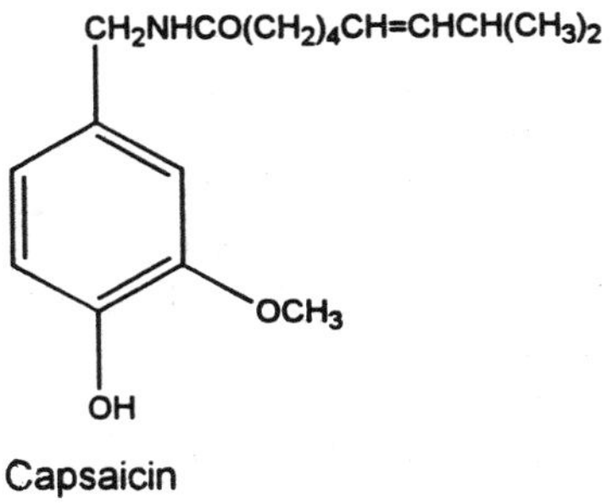

$CH_2NHCO(CH_2)_4CH{=}CHCH(CH_3)_2$

OCH_3

OH

Capsaicin

Figure 1. Structure of capsaicin from *Capsicum* spp.

protein meals lower it. This, in turn, affects the amount of tryptophan entering the brain. Sucrose has analgesic activity in newborns as determined by reduction in crying when administered prior to immunization,[40] and may similarly alleviate distress in older children.[41]

Flavor and Conditioned Effects

Whether as a response to sugar or other compounds, the perception of sweetness *per se* is pleasurable and as such can be a determinant of behavior.[15,42,43] Hedonic responses, those indicating pleasure,[43] to compounds perceived through taste and smell receptors are variable but are often shared among humans. On the other hand, liking of foods such as *Capsicum* spp.(chili pepper) and its constituent capsaicin (Fig. 1) varies from individual to individual as a reflection of experience and cultural associations.[44] Associations learned either through conditioned aversion or preference experiences that can be perceived through taste or smell may affect the behavior of individuals.

Pharmacological Effects

Apart from the specific neuropharmacological effects discussed below, phytochemicals in foods and herbal medicines have a wide range of potential pharmacological activities. Any positive contribution to health improves the sense of well-being and mood of a person and thereby mediates changes in behavior. Many common herbs have well-recognized therapeutic properties,[8,27] although the direct pharmacological effects of many of these, and certainly of staple food plants, have been poorly studied.[45]

Essential nutrients provide straight-forward examples of how a food constituent alleviates disease. Listlessness and behavioral abnormalities are symptoms of many deficiency states.[46] Nutritional anemia, characterized by loss of energy, tiredness, anorexia, behavioral abnormalities and insomnia, may be alleviated by iron, folic acid, cobalamin and other nutrients. Ascorbic acid, as well as common dietary condiments and beverage plants[47] that have a positive effect on iron uptake,[48] can act indirectly in this regard. Based on iron available through dialysis, we showed that plants used as tonics by rural women in Haiti are excellent sources of dietary iron,[49] and thus appear to be selected for their contribution to alleviating anemia. Among pharmacologically-active compounds that are widespread in plant foods, spices and natural flavors are salicylates.[50] These have antipyretic and analgesic activities.[24,50]

Adaptogens. A number of traditional herbal tonics and the natural products they contain have positive properties on well-being that, like the disease alleviating effects of nutrients, are not directly linked to nervous receptors. The reputed effects of these so-called adaptogens are subtle, long term, and sometimes controversial.[8] Included in this group are plants such as *Panax* spp. (ginseng) and *Eleutherococcus senticosus* (Rubr. & Maxim.) Maxim. (Siberian ginseng, syn. *Acanthopanax senticosus*). Wagner[51] provides a recent review of the diverse chemistry of adaptogens active in animal models. Typically, animals are subject to repeated episodes of emotional and physical stress, behavioral signs of fatigue are observed, and biochemical changes are measured. Administration of adaptogens is thought to normalize endocrine function along the pituitary-adrenal gland axis associated with stress and enhance the physiological adaptation to stress.

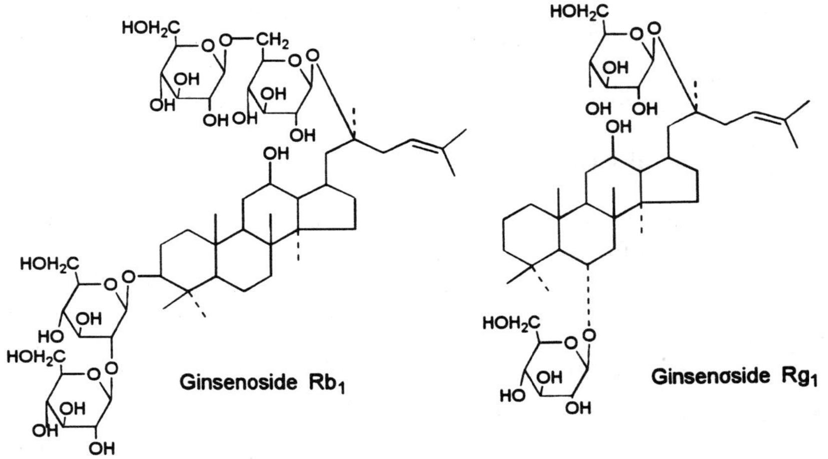

Figure 2. Examples of ginsenosides from *Panax ginseng* and other species.

The endocrine, nervous, and immune systems have links that, while not well-understood, are accepted as important in the actions of adaptogens.

In ginseng the compounds recognized as adaptogens are glycosides of the tetracyclic dammaran type, the so-called ginsenosides or panaxosides (Fig. 2).[51] In addition, these compounds are reported to have various pharmacological actions on the CNS[52,53] including anxiolytic action (the capacity to provide relief from symptoms of anxiety).

Stimulants. Caffeine (Fig. 3), the stimulant more regularly ingested than any other, is a constituent of coffee, tea, chocolate, and a number of other plants listed in Table 1. The large global trade in these products is an extension of the importance of caffeine-containing plants in many cultures through history. Along with other methylxanthines, caffeine has a number of physiological activities, but it and theophylline (Fig. 3), which is also found in tea, are the only two which act as CNS stimulants, apparently as antagonists of adenosine receptors.[54] The actions of these products in overcoming fatigue and drowsiness and increasing the capacity for intellectual effort[24] is the direct reason for the widespread ingestion of caffeine-containing products today. The mildly-addictive nature of caffeine further underlies the role of such stimulants in determining their ingestion from products that contain them.

Other than caffeine, there are no other stimulants that are generally recognized from food or common beverages. Plants such as coca (*Erythroxylum* spp.) and ephedra (*Ephedra sinensis*) are important sources of pharmacological and herbal stimulants. Coca is the source of cocaine[24] (Fig. 4) and ephedra contains ephedrine (Fig. 4).[8] There is anecdotal evidence also for the stimulant activity of rosemary (*Rosmarinus officinalis*) but experimental support for such effects is lacking, and a constituent of the plant responsible for such activity has not been identified. Rather, the volatile oil of rosemary has antispasmolytic activity.[27,55]

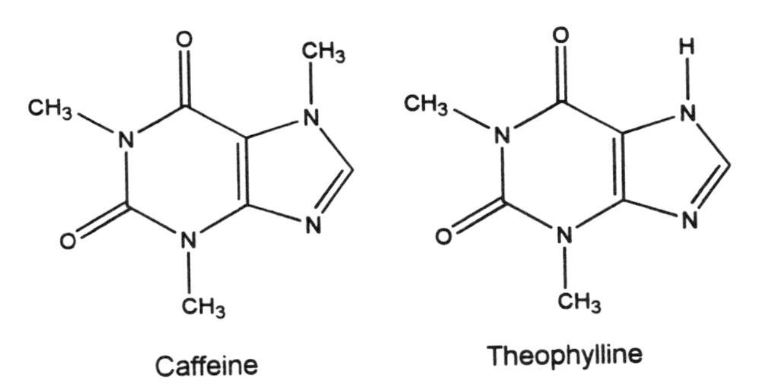

Figure 3. Methylxanthines active as CNS stimulants.

Table 1. Commonly consumed caffeine-containing plants

Species, family	Vernacular name	Plant part
Camellia sinensis (L.) O. Kuntze, Theaceae	tea	leaves, leaf buds
Coffea arabica L., Rubiaceae	coffee	seeds
Cola nitida (Vent.)Schott & Endl., Sterculiaceae	kola	cotyledons
Ilex paraguariensis St. Hil., Aquifoliaceae	maté	leaves
Paullinia cupana H.B.K., Sapindaceae	guarana	seeds
Theobroma cacao L., Sterculiaceae	cacao, cocoa	seeds

Digestive Stimulants and Antispasmodics. Disturbances of the gastrointestinal tract contribute to a general sense of malaise that can be alleviated by the same herbal agents that alleviate nausea, gastrointestinal extension and gas. These generally fall under the classes of stomachics (substances that stimulate appetite and promote functional activity of the stomach), carminatives (substances that relieve excess gas in the stomach or intestine) or gastrointestinal stimulants, and they act through a variety of mechanisms[8] such as facilitating digestion by increasing gut motility or bile flow. Antispasmodics quiet spasms that occur in smooth muscles, including those of the digestive system.

Chili peppers, as food or as condiment preparations, are important in the diets of people around the world, and they are cultural markers of particular cuisines.[44] Capsaicin (Fig. 1) and two closely related compounds are responsible for the pungency of chili and for its activity as a stomachic, carminative and gastrointestinal stimulant.[56]

The reputation of *Gentiana lutea* and related species as digestive stimulants is attributed to the secoiridoid glycosides that give it an extreme bitterness.[27]

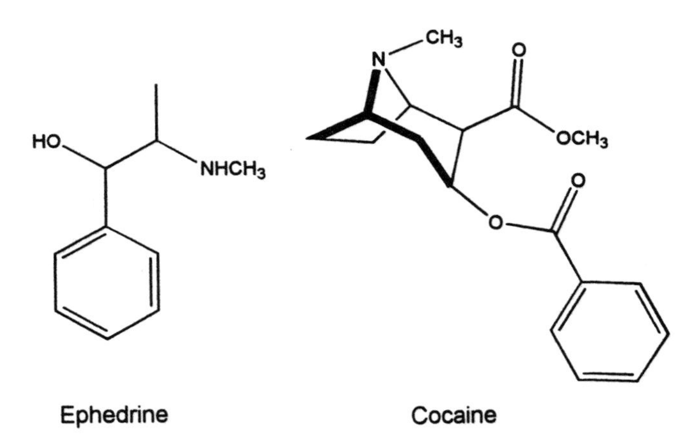

Figure 4. Examples of naturally-occurring stimulants of pharmacological importance.

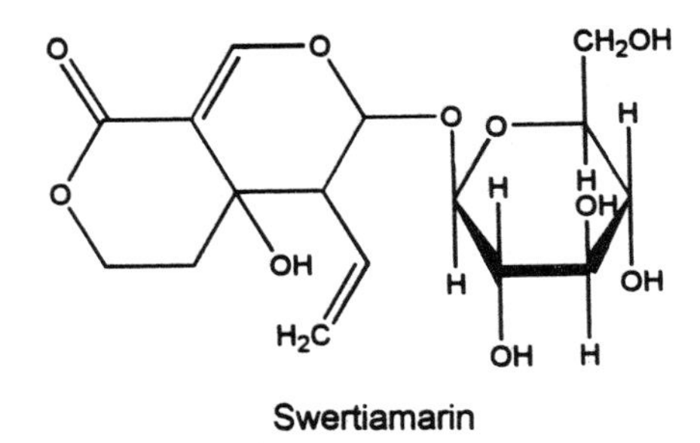

Figure 5. Example of secoiridoids from *Gentiana lutea*.

Bitters are often associated with such stimulation although data, certainly in the case of gentian, do not exist to support this effect.[56] Nonetheless, secoiridoids found in gentian such as swertiamarin (Fig. 5) are reported to act as CNS depressants in rodents.[57] Xanthines of the type present in gentian act as MAO inhibitors *in vitro*,[58] which could also account for the depressant activity of this plant.

The most widely used herbal antispasmodics are German chamomile (*Matricaria recutita* L.) and peppermint (*Mentha x piperita* L.). The former is consumed as a strong tea. Its volatile oil contains terpenes with anti-inflammatory and antispasmodic properties, principally (-)-α-bisabolol, bisabolol oxides A and B and matricin (Fig. 6).[27,56] As well, the flower heads contain flavones such as apigenin and luteolin (Fig. 7).[27] As discussed below, these have anxiolytic activity.

Peppermint also has demonstrated antispasmodic properties. These are attributable to menthol (Fig. 8), making up 50–78% of the essential oil of this plant, and other monoterpenes.[8] These act as antispasmodics in the upper gastrointestinal tract, as antibacterials, and as stimulants of bile and gastric secretions.

Sedatives. Herbal medicines that are used to alleviate sleep disorders and anxiety have related activities and are referred to under a number of names including sedatives, calmatives, anxiolytics, hypnotics, and soporifics. Considerable current investigation is focused on the activity of plant extracts and natural products from such plants as potentiators of the CNS inhibitory effects of gamma-aminobutyric acid. Specific binding sites for benzodiazepines have been identified in the brain; benzodiazepines and a number of unrelated natural products that have been shown to act as benzodiazepine receptor ligands enhance GABA activity through opening chloride ion channels.

Benzodiazepines were originally known as synthetic compounds. The first one chlordiazepoxide (LIBRIUM) was introduced for therapeutic use in the early 1960's; diazepam (VALIUM) is perhaps the best known of several benzodiazepi-

α-Bisabolol

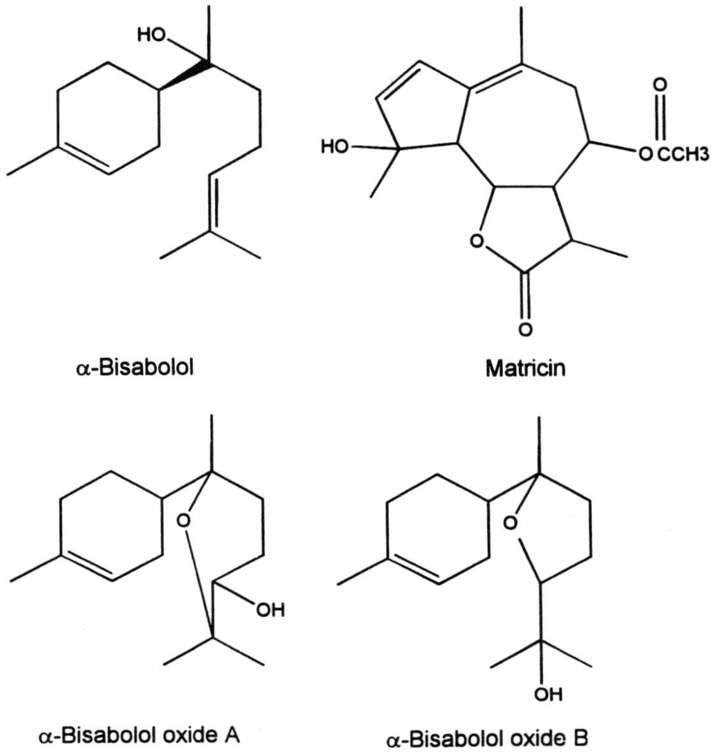

Matricin

α-Bisabolol oxide A

α-Bisabolol oxide B

Figure 6. Structure of terpenoids with antispasmodic and anti-inflammatory activity from *Matricaria recutita*.

nes in current use. In an interesting reversal of normal events where organic compounds are first discovered in nature and then synthesized, benzodiazepines have been found as apparently endogenous agents in food plants and other biological materials. In 1985, nordiazepam (Fig. 9) was identified in bovine, human, and rat brains,[59,60] and in 1987 Wildemann and colleagues reported the detection of diazepam (Fig. 9) and other pharmacologically-active benzodiazepines in wheat[61,62] and lormetazepam (Fig. 9) in potato.[62] This existence of naturally-occurring diazepines was met with scepticism but subsequent work detected diazepam in brain tissue from patients who had died prior to the introduction of benzodiazepines.[63] The demonstration that quantities of diazepam and lormetazepam increase in wheat and potato respectively during germination[64] supports the argument that these are naturally-occurring compounds. Although they appear to be synthesized by wheat, potatoes, and presumably other plants, the possibility that they are metabolites of microorganisms that are

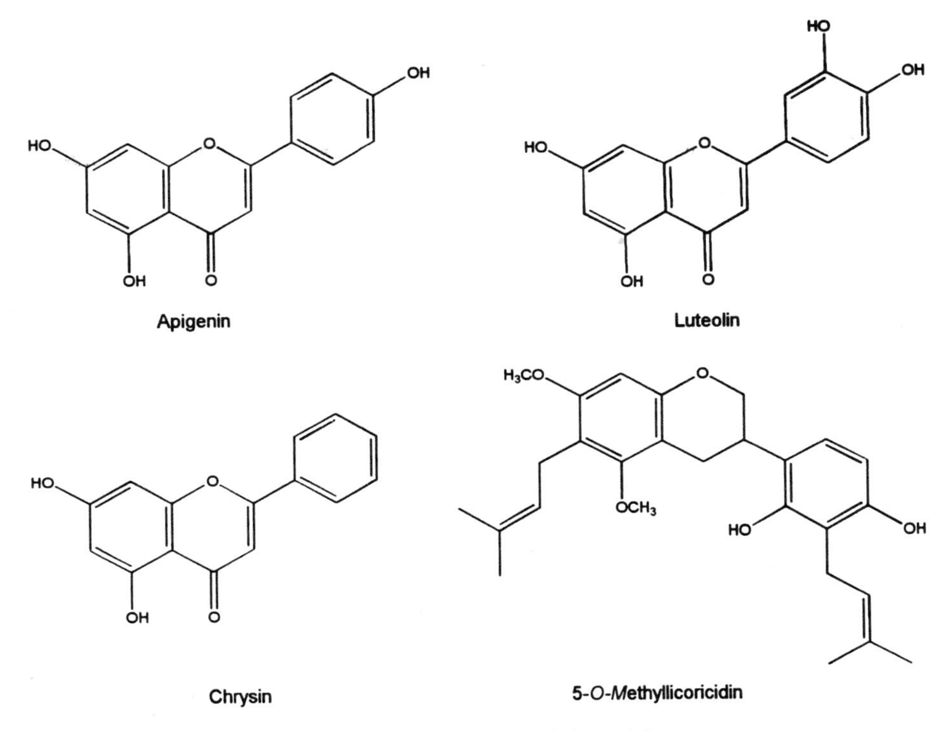

Figure 7. Flavonoids with sedative activity.

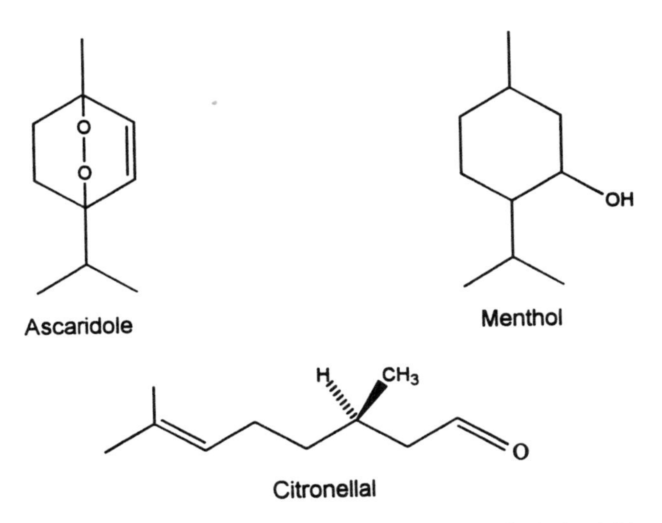

Figure 8. Examples of monoterpenoids with neuropharmacological activities.

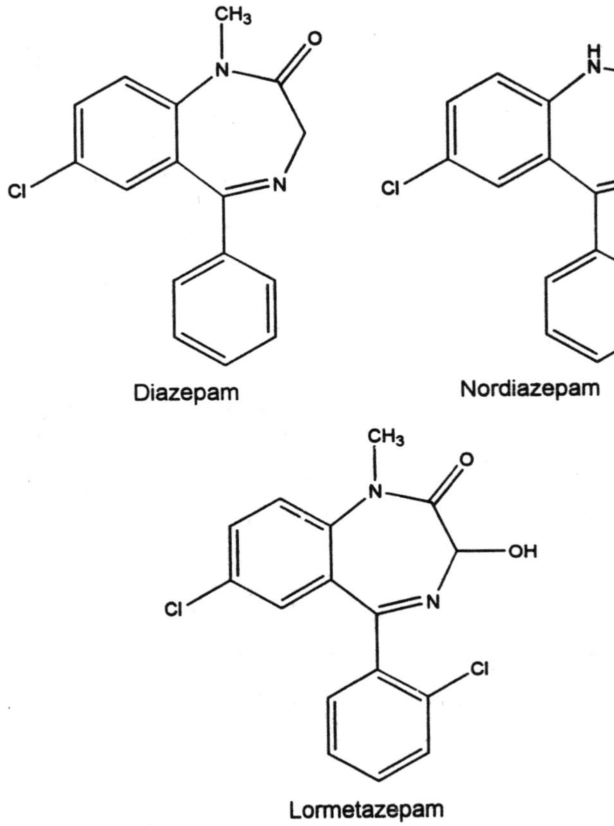

Figure 9. Examples of naturally-occurring benzodiazepines from plant foods.

taken up by plants has not been ruled out. In turn, the presence of benzodiazepines in mammalian, including human, tissues may reflect endogenous production, but it is considered more likely that they are obtained from ingested plants.

Benzodiazepines in food, in fact, occur only in the parts per billion (ppb) range.[61,62] In comparison with the doses of several milligrams per day needed for clinical activity,[24] they are unlikely to have an effect on receptors to a degree that would affect perception or behavior. Nonetheless, more subtle effects cannot be ruled out. If, in fact, benzodiazepines do accumulate in the brain, a role for them in psychophysiological processes is possible.

The activity of other benzodiazepine receptor ligands from herbal plants is more certain, and a range of compounds from different chemical classes have been demonstrated to have this activity in *in vivo* behavioral assays and *in vitro*.[65–68] *In vitro*, the ability of a phytochemical to competitively inhibit the binding of a radioactive ligand, such as flunitrazepam from receptors prepared from animals brains,[65] reflects the behavioral effects of benzodiazepines *in vivo*.

Paladini and colleagues[66] have recently identified benzodiazepine receptor activity in several plants that are widely employed in Western herbal medicine including extracts of *Tilia* spp. (lindin),[67] *Passiflora coerulea* (passion flower),[68] and chamomile.[65] From the latter two plants they have linked at least some of the activity with flavones, chrysin (Fig. 7) in the case of passion flower,[68] and apigenin (Fig. 7) in chamomile.[65] However, while these compounds competitively inhibit the binding of flunitrazepam to central benzodiazepine receptors, they lack some of the biochemical characteristics of benzodiazepines. In mice, they do not exhibit sedative and muscle relaxant effects at doses comparable to benxodiazepines, but rather have anxiolytic activity. Benzodiazepine receptor activity in *Glycyrrhiza uralensis* (licorice) is attributable to the isoflavonoid, 5–*0*-methyllicoricidin (Fig. 7).[69]

Various flavonoids have been reported to have other types of neuropharmacological activity.[70] Analgesic activity in flavones and flavone glucosides is opioid-mediated, and potency depends on the position of substitution in the flavone nucleus.[70] Hesperidin, a common flavanone in citrus peel as well as several isoflavonoids[71,72] have antiinflammatory activity, and flavonoids may have adaptogenic[73] and other physiological activities.

The sedative of greatest reputation in western and many other countries is valerian (*Valeriana officinalis* and related species). Germany's Commission E, a regulatory body providing leadership in the evaluation of phytomedicines,[8] approves 2–3 grams of valerian root, up to several times a day for restlessness and nervous disturbance of sleep. Although the sedative properties have been substantiated in studies in animals and humans, the identity of the active constituents remains unclear.[27,74] Valerenic acid (Fig. 10) and other sesquiterpenes in the volatile oil fraction inhibit the enzymes responsible for GABA catabolism,[27] and guaiane-type sesquiterpenoids such as α-kessyl alchohol (Fig. 10) are active in assays for antidepressant activity.[75] The iridoid valepotriates including valtrate (Fig. 10) have also been suggested as being responsible for sedative activity.

Sedative activity of unspecific mechanism has been detected in other plants. *Chenopodium ambrosoides* (American wormseed, 'paico' in Latin America) is used as a potent anthelminthic and a nervine in Latin America, but is also employed as a condiment in food.[76] Sedative and analgesic activities of this plant were associated in mice with the peroxide ascaridole (Fig. 8).[77] *Humulus lupulus* L. (hops) is a European herb of longstanding application as a sedative and which Germany's Commission E approves for anxiety, insomnia and restlessness. The sedative activity is associated with 2-methyl-3-buten-2-ol (a degradation product of humulone, lupulone, and other of the bitter principles in hops), although there is uncertainty that this compound occurs in sufficient quantity to provide signifi-

α-Kessyl alcohol

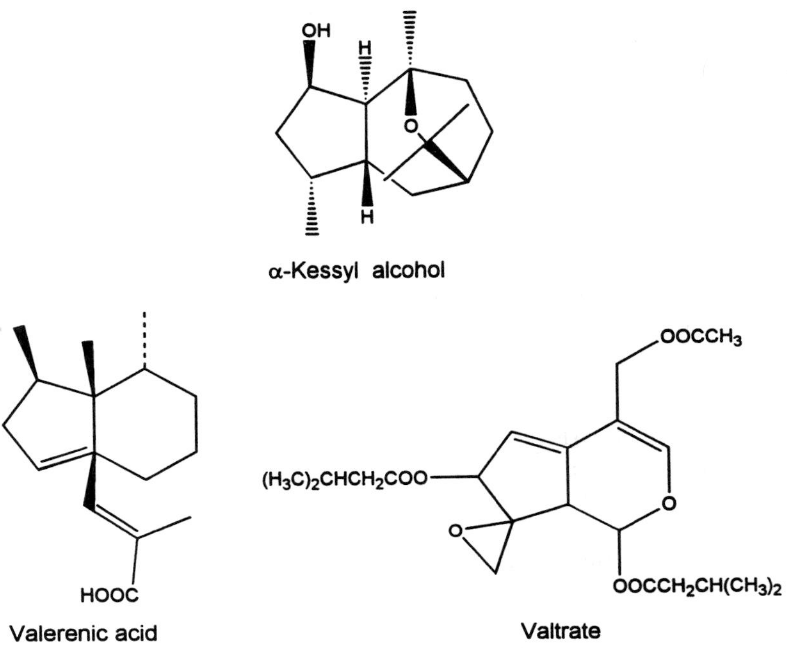

Valerenic acid

Valtrate

Figure 10. Structures of compounds thought to contribute to sedative activity of *Valeriana officinalis*.

cant sedative activity.[27,56] *Melissa officinalis* (lemon balm) is approved by Germany's Commission E as a carminative and sedative. The sedative action is attributed to a number of terpenes of which citronellal (Fig. 8) is thought to be the most important.[78] Sedative activity of the non-volatile residue from the essential oil of *Citrus bergamia* (bergamot) has been demonstrated in mice.[79] Bergamot oil is used to provide flavor to Earl Grey tea. Sedative activity of *Apium graveolens* L. (celery seed) has been documented for phthalide constituents.[27] Celery also contains apigenin. Among other plants with reputations as sedatives, but for which scientific data are limited or equivocal, are *Lavandula* spp. (lavender),[80,81] *Scutellaria lateriflora* (scullcap)[27] and *Nepeta cataria* (catnip).[56] The nepetalactones in catnip stimulant cats but there is disagreement on their effects in other mammals and humans.

Antidepressants

Plants used as sedatives may also serve some role as antidepressants, although they may not be specific in treating this disorder. Among the plants described above, valerian is sometimes recommended in this regard.[27] *Ginkgo biloba* L. (ginkgo) is a plant for which antidepressant activity is secondary to its well-established action as a vasodilator affecting cerebral blood flow.[27,56]

The antidepressant and sedative activities of L-tryptophan are directly related to the role of this amino acid as a precursor of serotonin.[56] It is also a precursor of the hormone melatonin. Another large neutral amino acid, tyrosine, is the precursor of the catecholamines. Although there is little evidence that increased ingestion of tyrosine has an effect on neural function in normal people, in individuals suffering from stress, tyrosine may increase their ability to function normally.[39]

In comparison to other plants, *Hypericum perforatum* L. (St. John's Wort) is the subject of the greatest amount of current research and discussion as an antidepressant.[82] This herb is sometimes used with valerian and also for anxiety and sleep disturbances. Although the activity is substantiated *in vivo*,[27] the mode of action is not well understood. St. John's Wort has MAO inhibitory activity which has sometimes been attributed to hypericin (Fig. 11) and related anthraquinone derivatives, but has more recently been associated with biflavonoids.[48]

CONCLUSION

The chemical complexity of food and other plants is being increasingly better documented as analytical techniques become more sophisticated. It should be expected that new psychodynamically-active substances will be discovered in food staples. While some effects can be extrapolated from known mechanisms,

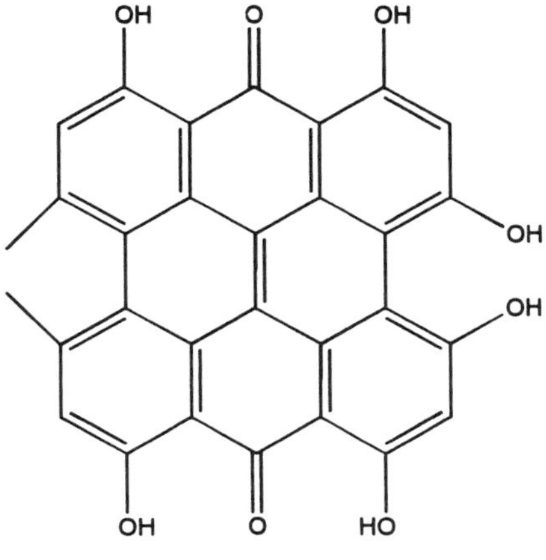

OH O OH
OH
OH
OH O HO
Hypericin

Figure 11. Structure of hypericin from *Hypericum perforatum*.

the activities of food constituents on nervous, hormonal, immune, and other systems, particularly in chronic doses, may be subtle and may be understood only gradually within the larger context of scientific investigation in food, nutrition, health, and behavioral sciences. Recent developments in the identification of food phytochemicals, such as benzodiazepines, suggest exciting opportunities for future discovery.

In many of the above examples, activity cannot be attributed to a single factor. In plants such as valerian or chamomile, single constituents are responsible for some of the activity, but not all. Where different constituents are independently demonstrated to have activity of the type seen in a whole extract, the activity of the plant may well be a result of their combined action. Synergistic actions involving two or more substances, especially if none of the ingredients show activity alone, are more difficult to elucidate.[83]

Furthermore, it is unlikely that single phytochemicals ingested from food and other plant sources determine the behavior of individuals or of populations. Just as more than one compound and mechanism of action may contribute to the properties of a plant, both psychosocial and physiological factors work together to determine the manner in which a food or herbal medicine is used. Capsicum, which has a physiologically-active constituent and a cultural identification with specific cuisines, or coffee and tea, which contain caffeine but also have important social roles, are obvious examples. The current trend in Western countries towards use of nutraceuticals and herbals has strong social reinforcement, but the scientific research behind many such products also supports their physiological role. The kind of psychodynamically-active phytochemicals discussed in this paper provide examples of such substances that can be expected to be sought after by individuals.

The ostensive reason why something is ingested may in fact be secondary to less recognizable factors. In the case of herbal medicines, for many of which a legitimate pharmacological role has not been shown, the primary function may be psychosomatic or important in supporting some culturally-defined criterium of how a medicine should act or of how a patient should be treated. On the other hand, ingestion in such cases, particularly where the user is able to self-medicate, may really be determined by previously unrecognized physiological effects of constituents within the plant. If the current trend of increased phytochemical ingestion continues, it can be expected that research into understanding the determinants of this phenomenon will proceed as well.

ACKNOWLEDGMENTS

I am grateful for financial support from the Natural Sciences and Engineering Research Council of Canada (NSERC). In addition, I thank N.L. Etkin, S.N. Young, H.K.M. Wagner, and E. Elisabetsky for useful insights as I prepared this paper.

REFERENCES

1. WOTEKI, C.E. 1995. Applications of antioxidants in physiologically functional foods. Consumption, intake patterns and exposure. Crit. Rev. Food Sci. Nutr. 35:143–147.
2. HERTOG, M.G.L., KROMHOUT, D., ARAVANIS, C., BLACKBURN, H., BUZINA, R., FIDANZA, F., GIAMPAOLI, S., JANSEN, A., MENOTTI, A., NEDELJKOVIC, S., PEKKARINEN, M., SIMIC, B.S., TOSHIMA, H., FESKENS, E.J.M., HOLLMAN, P., KATAN M.B. 1995. Flavonoid intake and long-term risk of coronary heart disease and cancer in the seven countries study. Arch. Intern. Med. 155:381–386.
3. GOLDBERG, I., (ed.) 1994. Functional Foods: Designer Foods, Pharmafoods, Nutraceuticals. Chapman & Hall, New York. 571 pp.
4. MARWICK, C. 1995. Growing use of medicinal botanicals forces assessment by drug regulators. JAMA 273:607–609.
5. COTTRELL, K. 1996. Herbal products begin to attract the attention of brand-name drug companies. Can. Med. Assoc. J. 155:216–219.
6. GRUNWALD, J. 1995. The European phytomedicines market: figures, trends, analyses. HerbalGram 34:60–65.
7. BREVOORT, P. 1995. The U.S. botanical market - an overview. HerbalGram 36:49–57.
8. TYLER, V.E. 1994. Herbs of Choice: The Therapeutic Use of Phytomedicines. Pharmaceutical Products Press, New York. 209 pp.
9. BOYLE, M.A., MORRIS, D.H. 1994. Community Nutrition in Action: An Entrepreneurial Approach. West Publishing Company, Minneapolis. 540 pp.
10. GREENWOOD, C.E. 1990. Methodologic considerations for diet and behavior studies: a nutrition's perspective. In: Diet and Behavior: Multidisciplinary Approaches, (G.A. Anderson, ed.), Springer-Verlag, London, pp. 193–207.
11. JOHNS, T., KUHNLEIN, H.V. 1990. Cultural determinants of food selection and behavior. In: Diet and Behavior: Multidisciplinary Approaches, (G.A. Anderson, ed.), Springer-Verlag, London, pp. 17–31.
12. SCHMIDT, H.J., BEAUCHAMP, G.K. 1990. Biological determinants of food preferences in humans. In: Diet and Behavior: Multidisciplinary Approaches, (G.A. Anderson, ed.), Springer-Verlag, London, pp. 33–47.
13. ETKIN, N.L., JOHNS, T. 1997. Pharmafoods and nutraceuticals: changing paradigms in biotherapeutics. J. Ethnopharmacol. In Press.
14. KAPTCHUK, T.J. 1983. The Web That Has No Weaver: Understanding Chinese Medicine. Congdon and Weed, New York. 402 pp.
15. DESAI, P.N. 1989. Health and Medicine in the Hindu Tradition. Crossroad Publishing Co., New York. 153 pp.
16. JOHNS, T. 1990. With Bitter Herbs They Shall Eat It: Chemical Ecology and the Origin of Human Diet and Medicine. University of Arizona Press, Tucson. 356 pp.
17. JOHNS, T., CHAPMAN, L. 1995. Phytochemicals ingested in traditional diets and medicines as modulators of energy metabolism. In: Phytochemistry of Medicinal Plants, (J.T. Arnason, R. Mata, J.T. Romeo, eds.), Plenum Press, New York, pp. 161–188.
18. PEZZUTO, J.M. 1995. Natural product cancer chemopreventive agents. In: Phytochemistry of Medicinal Plants, (J.T. Arnason, R. Mata, J.T. Romeo, eds.), Plenum Press, New York, pp. 19–45.
19. DUTTON, G. 1996. New moves attempt to boost research on nutraceuticals. Genet. Eng. News 16:1–3, 27.
20. CORMIER, F. 1997. Food colorants from plant cell cultures. In: Food Phytochemicals, (T. Johns, F. Cormier, J.T. Romeo, eds.), Plenum Press, New York, pp
21. VOELKER, T. 1997. Transgenic manipulations of edible oil seeds. In: Phytochemistry of Medicinal Plants, (T. Johns and J.T. Romeo, eds.), Plenum Press, New York, pp 223–236.

22. HATHCOCK, J.N. 1993. Safety and regulatory issues for phytochemical sources. Nutr. Today 28:23–25.
23. COCKBILL, C.A. 1994. Food law and functional foods. Br. Food J. 96:3–4.
24. HARDMAN, J.G., LIMBIRD, L.L., (eds.) 1996. Goodman and Gilman's The Pharmacological Basis of Therapeutics. McGraw-Hill, New York. 1905 pp.
25. STELLAR, J.R., RICE, M.B. 1989. Pharmacological basis of intracranial self-stimulation reward. In: The Neuropharmacological Basis of Reward, (J.M. Liebman, S.J. Cooper, eds.), Clarendon Press, Oxford. pp. 14–65.
26. CARR, G.D., FIBIGER, H.C., PHILLIPS, A.G. 1989. Conditioned place preference as a measure of drug reward. In: The Neuropharmacological Basis of Reward, (J.M. Liebman, S.J. Cooper, eds.), Clarendon Press, Oxford. pp. 264–319.
27. NEWALL, C.A., ANDERSON, L.A., PHILLIPSON, J.D. 1996. Herbal Medicines: A Guide for Health-Care Professionals. The Pharmaceutical Press, London. 296 pp.
28. LEUNG, A.Y. 1996. Encyclopedia of common natural ingredients used in food, drugs and cosmetics. 2nd Ed.. J. Wiley, New York, 649 pp.
29. ARNASON, T., HEBDA, R.J., JOHNS, T. 1981. Uses of plants for food and medicine by Native Peoples of eastern Canada. Can. J. Bot. 59:2189–2325.
30. JOHNS, T., MHORO, E.B., SANAYA, P., KIMANANI, E.K. 1994. Herbal remedies of the Batemi of Ngorongoro District, Tanzania: a quantitative appraisal. Econ. Bot. 48:90–95.
31. CHAPMAN, L., JOHNS, T., MAHUNNAH, R.L.A. 1997. Saponin-like *in vitro* characteristics of extracts from selected non-nutrient wild plant food additives used by Maasai in meat and milk based soups. Ecol. Food Nutr. In press.
32. HOBBS, C., CULLEN, S. 1990. Get on your nerves: give your nervous system herbs that can soothe, stimulate or heal. Veg. Times 155:73–74, 76, 90.
33. RABOW, L.E., RUSSEK, S.J., FARB, D.H. 1995. From ion currents to genomic analysis: recent advances in $GABA_A$ receptor research. Synapse 21:189–274.
34. LUDDENS, H., KORPI, E.R. 1995. Biological function of $GABA_A$/benzodiazepine receptor heterogeneity. J. Psych. Res. 29:77–94.
35. BAKER, M.E. 1995. Endocrine activity of plant-derived compounds: an evolutionary perspective. Proc. Soc. Exp. Biol. Med. 208:131–138.
36. STORY, M., BROWN, J.E. 1987. Sounding board: do young children instinctively know what to eat? New Eng. J. Med. 316:103–106.
37. ROZIN, P. 1976. The selection of food by rats, humans and other animals. Adv. Stud. Behav. 6:21–76.
38. WAHLQVIST, M.L. 1994. Functional foods in the control of obesity. In: Functional Foods: Designer Foods, Pharmafoods, Nutraceuticals. (I. Goldberg, ed.), Chapman & Hall, New York. pp. 71–86.
39. MEISELMAN, H.L., LIEBERMAN, H.R. 1994. Mood and performance foods. In: Functional Foods: Designer Foods, Pharmafoods, Nutraceuticals. (I. Goldberg, ed.), Chapman & Hall, New York. pp. 126–150.
40. BARR, R.G., YOUNG, S.N., WRIGHT, J.H., CASSIDY, K.L., HENDRICKS, L., BEDARD, Y., YAREMKO, J., LEDUC, D., TREHERNE, S. 1995. "Sucrose analgesia" and diphtheria-tetanus-pertussis immunizations at 2 and 4 months. J. Develop. Behavior. Ped. 16(4):220–225.
41. MILLER, A., BARR, R.G., YOUNG, S.N. 1994. The cold pressor test in children: methodological aspects and the analgesic effect of intraoral sucrose. Pain. 56:175–83.
42. SHALLENBERGER, R.S., ACREE, T.E. 1971. Chemical structure of compounds and their sweet and bitter taste. In: Handbook of Sensory Physiology. Vol. 4. (L.M. Deidler, ed.). Springer-Verlag, Berlin. pp. 221–277.
43. CABANAC, M. 1979. Sensory pleasure. Quart. Rev. Biol. 54:1–3.
44. ROZIN, P., SCHILLER, D. 1980. The nature and acquisition of a preference for chili pepper by humans. Motiv. Emot. 4:77–101.

45. ETKIN, N.L. 1991. Should we set a place for diet in ethnopharmacology? J. Ethnopharmacol. 32:25–36.
46. GIBSON, R.S. 1990. Principles of Nutritional Asssessment. Oxford University Press, New York. 691 pp.
47. EL-SHOBAKI, F.A., SALEH, Z.A., SAALEH, N. 1990. The effect of some beverage extracts on intestinal iron absorption. Zeit. fur Ernahrung. 29:264–269.
48. BODWELL, C.E., ERDMAN, J.W. 1988. Nutrient Interactions. M. Dekker, New York. 389 pp.
49. JEAN-BAPTISTE, J., ROUZIER, M.L., JOHNS, T. 1997. Ethnobotany of Haitian anti-anemic plants and their iron availability. Ecol. Food Nutr. In press.
50. PERRY, C.A., DWYER, J., GELFAND, J.A., COURIS, R.R., McCLOSKEY, W.W. 1996. Health effects of salicylates in foods and drugs. Nutr. Rev. 54:225–240.
51. WAGNER, H.K.M. 1995. Immunostimulants and adaptogens from plants. In: Phytochemistry of Medicinal Plants, (J.T. Arnason, R. Mata, J.T. Romeo, eds.), Plenum Press, New York, pp. 1–18.
52. ABE, K., CHO, S.I., KITAGAWA, I., NISHIYAMA, N., SAITO, H. 1994. Differential effects of ginsenoside Rb1 and malonylginsenoside Rb1 on long-term potentiation in the dentate gyrus of rats. Brain Res. 649:7–11.
53. BENISHIN, C.G. 1992. Actions of ginsenoside Rb1 on choline uptake in central cholinergic nerve endings. Neurochem. Int. 21:1–5.
54. NEHLIG, A., DAVAL, J.L., DEBRY, G. 1992. Caffeine and the central nervous system: mechanisms of action, biochemical, metabolic and psychostimulant effects. Brain Res. 17:139–170.
55. AQUEL, M.B. 1991. Relaxant effect of the volatile oil of *Rosmarinus officinalis* on tracheal smooth muscle. J. Ethnopharmacol. 33:57–62.
56. TYLER, V.E. 1993. The Honest Herbal, 3rd Edition. Pharmaceutical Products Press, New York. 375 pp.
57. BHATTACHARYA, S.K., GHOSAL, S., CHAUDHURI, R.K., SINGH, A.K., SHARMA, P.V. 1974. Chemical constituents of Gentianaceae XI: antipsychotic activity of gentianine. J. Pharmaceut. Sci. 63:1341–1342.
58. SCHAUFEBERGER, D., HOSTETTMANN, K. 1988. Chemistry and pharmacology of *Gentiana lactea*. Planta Med. 54:219–221.
59. SANNGMESWARAN, L., DE BLAS, A.L. 1985. Demonstration of benzodiazepine-like molecules in the mammalian brain with a monoclonal antibody to benzodiazepines. Proc. Nat. Acad. Sci. U.S.A. 82:5560–5564.
60. SANGAMESWARAN, L., FALES, H.M., FRIEDRICH, P., DE BLAS, A.L. 1986. Purification of a benzodiazepine from bovine brain and detection of benzodiazepine-like immunoreactivity in human brain. Proc. Natl. Acad. Sci. U.S.A. 83:9236–9240.
61. WILDMANN, J., MOHLER, H., VETTER, W., RANALDER, U., SCHMIDT, K., MAURER, R. 1987. Diazepam and N-desmethyldiazepam are found in rat brain and adrenal and may be of plant origin. J. Neural Transm. 70:383–398.
62. WILDMANN, J., VETTER, W., RANALDER, U.B., SCHMIDT, K., MAURER, R., MOHLER, H. 1988. Occurrence of pharmacologically active benzodiazepines in trace amounts in wheat and potato. Biochem. Pharmacol. 37:3549–3559.
63. UNSELD, E., FISCHER, C., ROTHEMUND, E., KLOTZ, U. 1990. Occurrence of 'natural' diazepam in human brain. Biochem. Pharmacol. 39:210–212.
64. WILDMANN, J. 1988. Increase of natural benzodiazepines of wheat and potato during germination. Biochem. Biophys. Res. Comm. 157:1436–1443.
65. VIOLA, H., WASOWSKI, C., LEVI DE STEIN, M., WOLFMANN, C., SILVEIRA, R., DAJAS, F., MEDINA, J.H., PALADINI, A.C. 1995. Apigenin, a component of *Matracaria recutita* flowers is a central benzodiazepine receptors-ligand with anxiolytic effects. Planta Med. 61:213–216.

66. MEDINA, J.H., PEÑA, C., LEVI DE STEIN, M., WOLFMAN, C., PALADINI, A.C. 1989. Benzodiazepine-like molecules, as well as other ligands for the brain benzodiazepine receptors are relatively common constituents of plants. Biochem. Biophys. Res. Comm. 165:547–553.
67. WOLFMAN, C., LEVI DE STEIN, M., WASOWSKI, C., PEÑA, C., MEDINA, J.H., PALADINI, A.C. 1994. Isolation of pharmacologically active benzodiazepine receptor ligands from *Tilia tomentosa* (Tiliaceae). J. Ethnopharmacol. 44:47–53.
68. WOLFMANN, C., VIOLA, H., PALADINI, A., DAJAS, F., MEDINA, J.H. 1994. Possible anxiolytic effects of chrysin, a central benzodiazepine receptor ligand isolated from *Passiflora coerulea*. Pharmacol. Biochem. Behav. 47:1–4.
69. LAM, Y.K.T., SANDRINO-MEINZ, M., HUANG, L., BUSCH, R.D., MELLIN, T., ZINK, D., HAN, G.Q. 1992. 5-*O*-Methyllicoricidin: a new and potent benzodiazepine-binding stimulator from *Glycyrrhiza uralensis*. Planta Med. 58:221–222.
70. THIRUGNANASSAMBANTHAM, P., VISWANATHAN, S., RAMASWAMY, S., KRISHNAMURTY, V., MYTHIRAYEE, C., KAMESWARAN, L. 1993. Analgesic activity of certain flavone derivatives: a structure-activity study. Clin. Exp. Pharmacol. Physiol. 20:59–63.
71. WU, E.S.C., LOCH, J.T., TODER, B.H., BORRELLI, A.R., GAWLAK, D., RADOV, L.A., GENSMAANTEL, N.P. 1992. Flavones. 3. Synthesis, biological activities, and conformational analysis of isoflavone derivatives and related compounds. J. Med. Chem. 35:3519–3525.
72. EMIM, J.A.D.S., OLIVEIRA, A.B., LAPA, A.J. 1994. Pharmacological evaluation of the anti-inflammatory activity of a citrus bioflavonoid, hesperidin, and the isoflavonoids, duartin and claussequinone in rats and mice. J. Pharm. Pharmacol. 46:118–122.
73. GHOSAL, S., JAISWAL, D.K., SINGH, S.K., SRIVASTAVA, R.S. 1985. Dichotosin and dichotosinin, two adaptogenic glucosyloxy flavans from *Hoppea dichotoma*. Phytochemistry 24:831–833.
74. LEUSCHNER, J., MULLER, J., RUDMANN, M. 1993. Characterisation of the central nervous depressant activity of a commercially available valerian root extract.-Arneimittel-Forschung 43:638–641.
75. OSHIMA, Y., MATSUOKA, S., OHIZUMI, Y. 1995. Antidepressant principles of *Valeriana fauriei* roots. Chem. Pharm. Bull. 43:169–170.
76. OKUYAMA, E., UMEYAMA, K., SAITO, Y., YAMAZAKI, M., SATAKE, M. 1993. Chem. Pharm. Bull. 41:1309–1311.
77. ARVIGO, R., BALICK, M. 1993. Rainforest Remedies: One Hundred Healing Herbs of Belize. Lotus Press, Twin Lakes, WI. 221 pp.
78. SOULIMANI, R., FLEURENTIN, J., MORTIER, F., MISSLIN, R., DERRIEU, G., PELT, J.M. 1991. Neurotropic action of the hydroalcoholic extract of *Melissa officinalis* in the mouse. Planta Med. 57:105–109.
79. OCCHIUTO, F., LIMARDI, F., CIRCOSTA, C. 1995. Effects of the non-volatile residue from the essential oil of *Citrus bergamia* on the central nervous system. Int. J. Pharmacogn. 33:198–203.
80. BUCHBAUER, G., JIROVETZ, L., JAGER, W., DIETRICH, H., PLANK, C. 1991. Aromatherapy: evidence for sedative effects of the essential oil of lavender after inhalation. Zeitschrift fur Naturforschung Section C. 46:1067–1072.
81. GUILLEMAIN, J., ROUSSEAU, A., DELAVEAU, P. 1989. Effets neurodepresseurs de l'huile essentielle de *Lavandula angustifolia* Mill. Annales Pharmaceutiques Francaises. 47:337–343.
82. SOMMER, H., HARRER, G. 1994. Placebo-controlled double-blind study examining the effectiveness of an hypericum preparation in 105 mildly depressed patients. J. Ger. Psych. Neur. 7:S9-S11.
83. ISMAN, M.B. 1996. Phytochemistry of the Meliaceae: So Many Terpenoids, So Few Insecticides. (J.T. Romeo, J.A. Saunders, and P. Barbosa eds.), Recent Advances in Phytochemistry, Plenum, New York. pp. 155–178.

Chapter Seven

PHYTOCHEMICALS AND WINE FLAVOR

Susan E. Ebeler

Department of Viticulture and Enology
University of California
Davis, California 95616

INTRODUCTION

Wine is a complex mixture containing aroma and taste compounds which are derived from grapes, produced or metabolized by yeast during fermentation, extracted from oak, and formed from chemical reactions that occur during processing and aging (Table 1). Aroma compounds, including terpenes, volatile esters, and alcohols, contribute significantly to the overall sensory impact of wines and will be discussed briefly at the end of this chapter. The tastes of sweetness (due largely to glucose and fructose) and sourness (due largely to tartaric acid with contributions by malic, lactic, and citric acids) also make important contributions to wine flavor, especially for white wines. However, wine, like coffee and tea, is largely characterized by distinct bitter and astringent sensory attributes, and, in the absence of these characters, may be considered

Functionality of Food Phytochemicals
edited by Johns and Romeo, Plenum Press, New York, 1997

Table 1. Typical gross composition (% by weight) of table wines

Component	White	Red
Water (by difference)	87	87
Ethanol	10	10
Other Volatiles	0.04	0.04
Sugar	0.05	0.05
Pectin and Related	0.30	0.30
Glycerol and Related	1.10	1.10
Acids	0.70	0.60
Ash	0.20	0.20
Phenols	0.01	0.20
Amino Acids and Related	0.25	0.25
Fats, Terpenes	0.01	0.02
Misc., Vitamins, etc.	0.01	0.01
Total	100	100

Reference: 76

"thin", "flat", or "insipid". This review will therefore focus primarily on the sensory properties of the bitter and astringent compounds found in wine.

Bitterness is a taste sensation perceived by receptors in the mouth and on the tongue and is elicited by a heterogeneous group of compounds including dipeptides, sugars, alkaloids, and phenols. Interestingly, the same compound can often produce both bitter and sweet tastes (*e.g.*, saccharin) indicating possible similarities in the mechanisms of perception which are thought to involve a weak interaction with lipophilic receptors. Astringency is a tactile sensation characterized by a "drying" or "puckering" sensation in the oral cavity, probably as a result of the precipitation of salivary proteins, especially proline-rich proteins.[1] Astringency is produced by a wide range of compounds including multivalent metals (alum, $Al_2(SO_4)_3$), acids, and phenols.

MEASUREMENT OF BITTERNESS AND ASTRINGENCY

Sensory Measurements

Sensory analysis is the systematic study of human responses to sensory stimuli (sight, smell, taste, and touch). Usually, analytical sensory tests are conducted under controlled conditions (light, temperature, time of day, etc.) and are designed to answer questions such as "is there a difference?", "how large is the difference?", and "what is the difference?". The types of tests that answer these questions generally include discrimination tests, intensity scaling, and descriptive analysis, respectively. The overall objective of analytical sensory testing is to define the sensory characteristics of a product such as wine. This is distinct from preference or affective sensory analyses where individual hedonic judgments, based on acceptance, preference, and liking, are obtained. Specific applications of sensory analysis methodologies to the study of wine and its sensory properties have been reviewed elsewhere.[2]

Sensory measurements of bitterness and astringency have generally focused on determining threshold concentrations and measuring perceived stimulus intensities at suprathreshold levels.[3–6] A threshold is defined as the minimum concentration of a stimulus that can be detected (absolute threshold), discriminated (just-noticeable-difference) or recognized (recognition threshold) a statistically significant percentage of the time.[7] Methods for determining thresholds vary but generally involve presenting sets of samples to judges in order of ascending or descending concentration; judges are asked to identify the level at which they can detect a difference in aroma or taste or the level at which they can recognize a specific aroma or taste.[2,7] Thresholds have been used to compare relative strengths of compounds, but the results are highly dependent on the test matrix (air, water, wine, etc.), compound purity, sensory methodology, and previous training and experience of the judges.[7] Threshold measurements cannot be used to predict taste (or aroma) intensity at higher concentrations because the rate of increase in perceived intensity is different for different compounds. Finally, many compounds can exhibit qualitative differences in taste (or aroma) properties at different concentrations.

Determining the relationship between the intensity of a perceived sensation and the physical concentration of a stimulus involves the use of scaling procedures. While there are many different specific procedures, category scales or magnitude estimation are commonly used. Category scales may be numerical (*e.g.*, ratings of 0–10; a sample receiving a rating of 10 would have a high intensity of a specified attribute) or graphical (a vertical or horizontal line anchored at the ends with terms such as "none" and "high" for specific attributes). However, judges do not necessarily use the intervals between category ratings equally, so a sample receiving a score of 4 is not necessarily double the intensity of sample receiving a score of 2. In general, category scales produce data which are linear with the log of the stimulus concentration.[7] This is also known as Fechner's law[7,8] where

$$\text{response} = k \bullet \log \text{concentration} \qquad (1)$$

In magnitude estimation, attribute intensities are evaluated in proportion to a reference sample (which may be defined by the judge or by the experimenter); a number is assigned corresponding to how many times lesser or how many times greater the sample's intensity is compared to the reference. This results in a relationship where equal ratio changes in concentration produce equal ratio changes in sensation magnitude (the sensation magnitude is expressed as the group geometric mean), also known as the psychophysical power law, or Stevens' Law:[7–9]

$$\log \psi = \beta \log \phi + \log k \quad (2)$$

where ψ = perceived magnitude
ϕ = stimulus concentration
β = the slope of the line
k = the intercept

The advantages and disadvantages of category scaling and magnitude estimation have been reviewed.[7] In general, for all sensory analyses there is a great deal of individual variability in the way subjects use scales. This is influenced in part by their training, background and familiarity with a given test and their sensitivity to specific attributes. All of these factors should be considered when choosing the appropriate sensory methodology for a given experimental objective. Specific applications of intensity scaling for the measurement of bitter and astringent properties of wine are discussed in greater detail throughout this review.

Astringency and bitterness are characterized by a distinct lingering nature in the mouth, therefore, monitoring changes in these sensory properties over time is essential for a complete evaluation of their sensory impact. The time course of these perceptions can be measured using time-intensity (T-I) procedures as described by Robichaud and Noble.[10] Using a computerized procedure to continuously rate perception, judges take the sample into their mouth (time 0) and are prompted by the computer to expectorate or swallow at a defined time. The perceived bitterness or astringency is rated continuously until extinction of the sensation. Intensity ratings are usually recorded on a category scale which is displayed on the computer screen. From a T-I curve (Fig. 1), the time to maximum intensity, maximum intensity, total duration, and rates of onset and decay can be obtained.

In general, T-I procedures correlate well with those produced by scalar methods.[10,11] Robichaud and Noble[10] observed a linear increase in astringency and bitterness perception as catechin, gallic acid, grape seed tannin, and tannic acid concentration increased using both scalar and T-I measurements of maximum intensity. However, T-I yielded additional information indicating that aftertaste, or total duration, of bitterness and astringency perception was also related to the maximum intensity[12] (Table 2). Hence, compound concentrations may need to be adjusted to yield samples of approximately equal maximum intensity for comparison of temporal patterns of bitterness and astringency of different compounds.

Using conditions that approximate those encountered during normal consumption, the cumulative effect of repeated ingestions on perception of astrin-

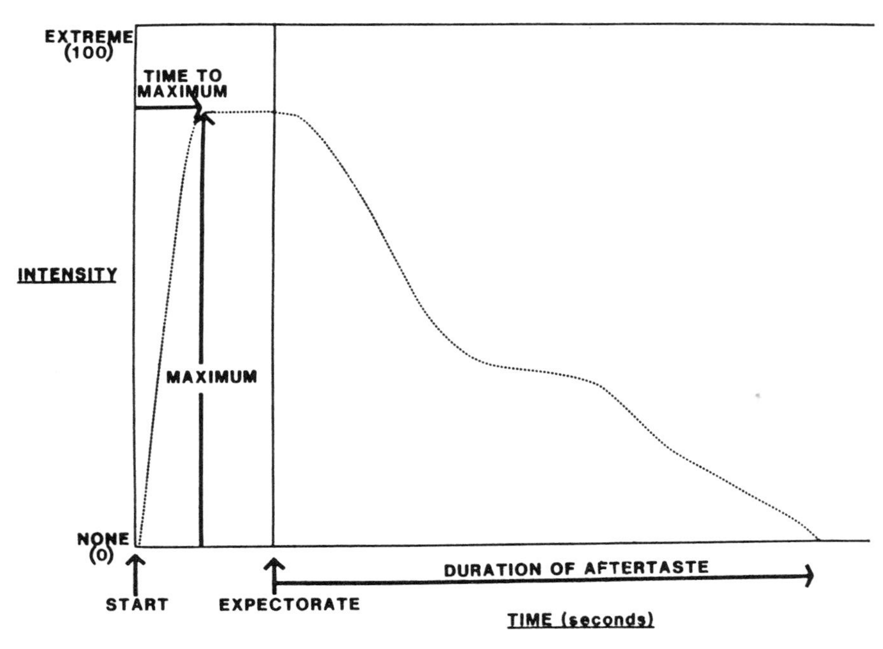

Figure 1. A typical Time-Intensity curve showing changes in perceived intensity (y axis) over time (x axis).

Table 2. Mean time-intensity measurements of maximum intensity and duration of aftertaste for bitterness and astringency of tannic acid added to a base white wine (n = 24)[1]

	Astringency		Bitterness	
Tannic acid conc (ppm)	Intensity (1–100)	Duration of aftertaste (sec)	Intensity (0–100)	Duration of aftertaste (sec)
0	30	32	28	32
375	56	46	47	44
1500	86	71	69	55

[1]The base wine was a neutral blended "dry" white wine made by standard practices in the University of California, Davis winery. The test samples were evaluated at room temperature by 12 judges trained in the use of Time-Intensity procedures and in the sensory measurement of astringency and bitterness. Each day, judges first evaluated a practice sample, then evaluated the test samples in random order. Two replications of the session were carried out on sequential days (n = 24; 12 Judges x 2 Replications).

Reference: 12

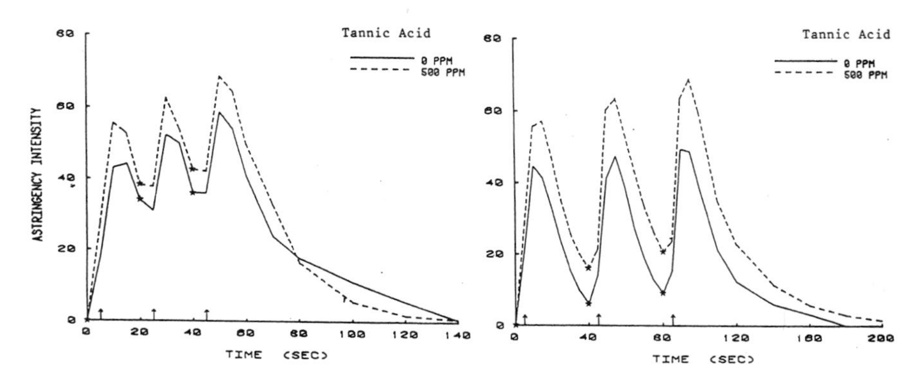

Figure 2. Average time-intensity curves for astringency in wine with 0 and 500 mg/L of added tannic acid upon three successive ingestions with 20 seconds between ingestions (left panel) and 40 seconds between ingestions (right panel). Sample uptake and swallowing are indicated by a star and an arrow, respectively. Samples (10 mL) were evaluated at 21°C in random order (n = 24; 8 Judges x 3 Replications). The base wine was a neutral blended "dry" white wine made by standard practices in the University of California, Davis winery. Reprinted with permission.[13] Copyright 1986 American Society for Enology and Viticulture.

gency in wine has been measured by T-I.[13] Wine samples with added tannic acid were presented to judges at specific time intervals (20 sec or 40 sec) and perception was monitored using a computerized T-I procedure. Astringency intensity increased upon repeated ingestion (Fig. 2). When sensory perceptions of astringency were allowed to return to 0 intensity prior to ingesting another sample, the total duration of astringency increased with repeated ingestion. Guinard et al.[13] theorized that these results are due to an increased time required for the mouth to return to normal lubrication and to remove tannin-protein precipitates in the saliva. Consistent with this, Smith et al.[14] found that increasing the viscosity of model solutions with carboxymethyl cellulose (from 2 to 45 centipoise) reduced the astringency of Grape Seed Tannin, possibly by restoring salivary lubrication.

Individual differences in perception of bitter and astringent tastes may be significant.[15] Individuals may have different thresholds for each of these attributes; for example, differential sensitivity to the bitter compounds phenylthiourea (PTC) and propylthiouracil (PROP) is well known.[16] Temporal responses may also be affected by individual differences in oral manipulation of the sample and in salivary composition[1] and flow rate. Fischer et al.[15] studied the effects of salivary flow rates on T-I responses to bitterness and astringency in model wine solutions. In general, for both attributes, individuals with low salivary flow took longer to reach maximum intensity and had a longer duration than high flow subjects. The decay rates followed first-order kinetics, suggesting that desorption from receptors, rather than diffusion, governed the perception. The decay curves for bitterness and astringency were also different, suggesting different mechanisms are responsible for bitterness and astringency perception. These studies are increasing our understanding of the role of saliva in sensory perception as well as providing mechanistic information about the chemical basis of the perception of bitterness and astringency.

Analytical or Chemical Measurements

As discussed in the following sections, bitterness and astringency in wine are elicited mainly by polyphenolic compounds. A number of methods are available for measuring total phenolic content, the Folin-Ciocalteu procedure being one of the most frequently used.[17] This method is based on the ability of phenolics to reduce the Folin reagent (a molybdotungstophosphate complex) to produce a blue color which absorbs at 765 nm. The structure, ionizability of the phenolic group, and number of hydroxyl groups present in a given phenol will affect pigment formation. For example, catechol reacts as if both hydroxyls were oxidized and gives double the molar color yield of phenol.[18] Therefore, a standard phenol (typically gallic acid) is used for reference purposes and results are generally reported in terms of Gallic Acid Equivalents (GAE). Any other wine components which can also be oxidized (SO_2, reducing sugars, etc.) will interfere with this reaction and must be accounted for during the analysis. More recently, HPLC methods for measuring phenolic levels have emerged which are more specific and allow quantitation of individual phenolics.[19–25] These methods do not provide complete separation of the polymeric fraction however.

In general, chemical measurements of "tannin" or polyphenolic concentration do not correlate well with sensory measurements of astringency. For example, the total polyphenol concentration of many fruits (*e.g.*, persimmons, bananas, raspberry, blackberry, grapes) increases or does not change as final ripening occurs, although the perception of astringency decreases.[26–29] The mechanisms for the decreased sensory astringency are not known but may involve polymerization reactions, interactions with other fruit components such as carbohydrates, or release of soluble pectin fragments from the cells of the fruit.[26–29] All of these mechanisms would involve alterations in the ability of polyphenols to interact with proteins in the mouth, leading to an altered perception of astringency. This concept of "active tannins" was first proposed by Barnell and Barnell in 1945 when they theorized that only "active tannins", and not the total amount of tannin, were directly related to perception of astringency.[26] They developed an amylase enzyme assay for quantifying "active tannins", which was based on the observation that amylase is precipitated or inactivated in the presence of tannins. Results from this enzyme assay showed that the "active tannin" concentration of bananas changed during ripening and was generally related to the perception of astringency. It should be noted that the sensory measurements of Barnell and Barnell were qualitative in nature, and the enzyme method was not validated with other standard polyphenolic compounds or with other foods and beverages, especially those which are highly astringent.

Other methods based on the ability of polyphenols to precipitate proteins (*e.g.*, hemoglobin, gelatin, BSA, lysozyme, β-glucosidase) have been proposed, and the chemistry of the tannin-protein interaction has been reviewed.[30–35] These methods are also limited in their ability to correlate with sensory perception of astringency, however. This is due to the fact that flavanol monomers such as catechin are astringent in sensory tests, however, only oligomers of three subunits or more are capable of precipitating proteins in the chemical tests.[36,37] Protein-catechin complexes which do not result in precipitates have been observed.[37] Therefore, further development of chemical methods which measure protein-polyphenol interactions, but do not rely on protein precipitation, may more closely correlate with the relative sensory astringency of monomers and polymers. For example, measurement of such interactions may involve assessing changes in protein functionality, shifts in UV absorption spectra, or protein binding.[38]

BITTER AND ASTRINGENT COMPOUNDS AND THEIR IMPACT IN WINE

Polyphenolics

Although a number of compounds can elicit bitter and astringent sensations in other foods and beverages, in wine, the polyphenolic fraction is largely responsible for these attributes. The total phenol content of wine is dependent on the grape used, skin contact time, oak storage, and use of fining agents.[39–41] Fining agents include materials such as casein, albumin, gelatin, bentonite, and polyvinylpolypyrrolidone (PVPP) which are added to aid in settling, precipitation, and removal of monomeric and polymeric phenols and proteins from the wine.[41] In general, red wines contain significantly more total phenols than whites due to extraction of phenols from seeds and skins during extended contact before pressing (Table 3).

Non-flavonoids. The non-flavonoid phenols are present at approximately the same concentrations in red and white wines (Table 4). Tyrosol (Fig. 3) is a bitter compound produced by yeast during fermentation from the deamination and decarboxylation of the amino acid tyrosine. As reviewed by Singleton and Noble,[42] the taste threshold for tyrosol in wine has not been determined, however, in beer, a threshold of 10–200 ppm has been reported. Typical tyrosol levels in

Table 3. Effect of skin contact time on phenolic content in Cabernet Sauvignon wines after 0 and 6 months of age

Days of skin contact	0 months age		6 months age	
	Total phenols[1]	% Polymeric[2]	Total phenols[1]	% Polymeric[2]
7	1827	34.0	1802	30.1
14	2753	35.1	2706	42.3
21	2943	36.3	2913	42.5
28	2901	36.7	2951	44.8

[1]Phenolic content expressed as mg/L gallic acid equivalent (GAE)
[2]Polymeric = Phenols with a molecular weight >3500
Reference: 39

red and white wine are 15–20 ppm, therefore, tyrosol may contribute to wine bitterness. The non-flavonoid hydroxycinnamates, caffeic and *p*-coumaric acids, and their tartaric acid esters (Fig. 3), also have bitter and astringent properties in aqueous model solutions.[43] However, Vérette et al.[43] concluded that they do not contribute to wine flavor at the levels found in wine (Table 5).

Aging in new oak barrels can contribute ~250 mg Gallic Acid Equivalents/L (GAE/L) of nonflavonoid phenols (*e.g.*, ellagitannins; Fig. 3) to the wine (Table 4). These compounds may contribute to bitterness and astringency of white wines, however, their contribution in red wine is probably negligible due to the high initial total phenolic content of the wine.

Table 4. Gross phenol composition estimated in mg Gallic Acid Equivalents/L (GAE/L) for typical table wines from *Vitis vinifera* grapes

	Source[1]	White	Red
Nonflavonoids, total		160–260	200–500
Volatile phenols	D, M, E	1	5
Tyrosol	M	14	15
Hydroxybenzoates (Gallic, etc.)	D, M, G, E	10	40
Hydroxycinnamates (Caffaric, etc.)	G, D	140	140
Hydrolyzable tannins (If aged in new oak)	E	0–250	0–250
Flavonoids, total		25–35	700–1,000
Catechins (flavan-3-ols)	G	25	75
Anthocyanins	G	0	400
Other monomeric flavonoids	G, D	Tr	25
Dimers and larger tannins	G, D	5	500
Total		190–290	955–1450

[1]D = Degradation Product; E = Environment, Cooperage; G = Grapes; M = Microbes, Yeast
Adapted from References 5, 76

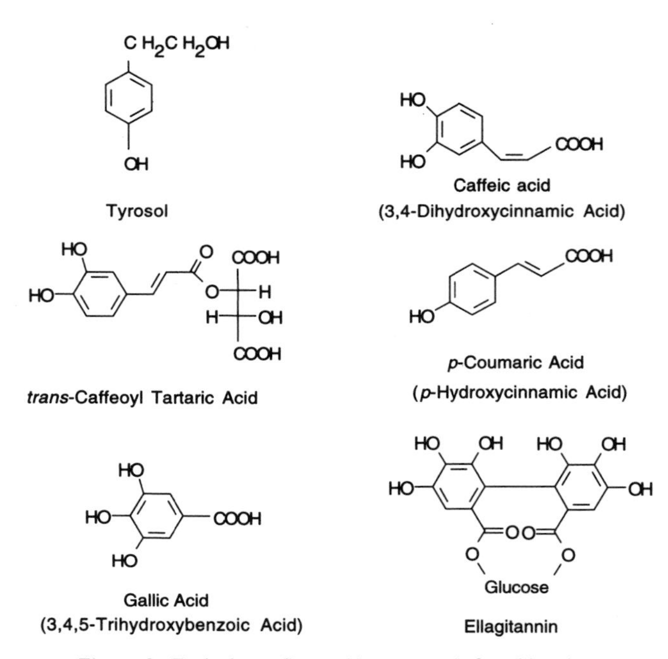

Figure 3. Typical non-flavonoid compounds found in wine.

Table 5. Evaluation of hydroxycinnamates in model solutions and in white wine. Duo-trio evaluations by 14 judges x 2 replications

Compound	mg/L	Model solution (1% aqueous ethanol, pH 3.85)	Base white wine (11.6% ethanol, pH 3.34)
p-Coumaric acid	30	NS	NS
Caffeic acid	120	**	NS
Caffeic & *p*-coumaric acids	120	**	NS
	30		
Caffeoyl tartaric acid	100	***	NS
	150	—	NS
2-S-Glutathionyl caffeoyl tartaric acid	50	NS	NS

NS, No significant difference; ** and ** significant at $p > 0.01$ and $p > 0.001$, respectively. Reprinted with permission.[43] Copyright 1988 Society of Chemical Industry.

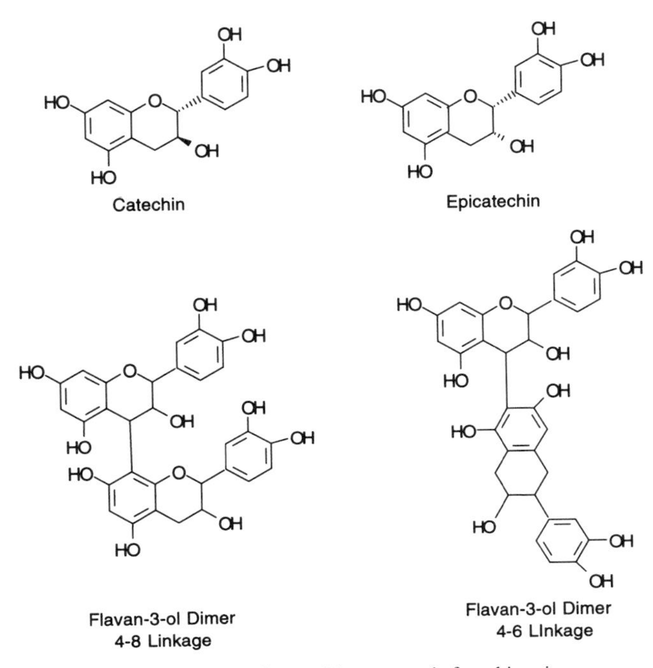

Figure 4. Typical flavonoid compounds found in wine.

Flavonoids. The flavonoid fraction is the predominant phenol fraction in red wine (Table 4) and is largely responsible for the perception of bitterness and astringency. Flavonoids can occur as monomers, *e.g.*, catechin and epicatechin, or as polymeric structures (it is the polymeric structures that are often referred to as "tannins" or procyanidins) (Fig. 4). Polymerization reactions are a result of enzymatic and chemical oxidation of the vicinal dihydroxy groups to yield electrophilic quinones. The quinones rapidly react with other nucleophilic species, including the original phenols, to yield the dimeric and polymeric products.[41]

In general, flavonoid monomers are more bitter than astringent[10] (Fig. 5). Using trained judges, the bitterness and astringency of catechin and grape seed tannin (GST; a mixture of oligomers containing mostly 4–8 flavonoid subunits) were evaluated as a function of stimulus concentration on a 9-point category scale. As concentrations of catechin and GST increased, perceived intensity of bitterness and astringency also increased. However, for catechin, bitterness increased at a faster rate than astringency; while for GST, astringency was more intense than bitterness at concentrations above ~400 mg/L. When individual phenol fractions extracted from grape seeds were evaluated in a model wine, the

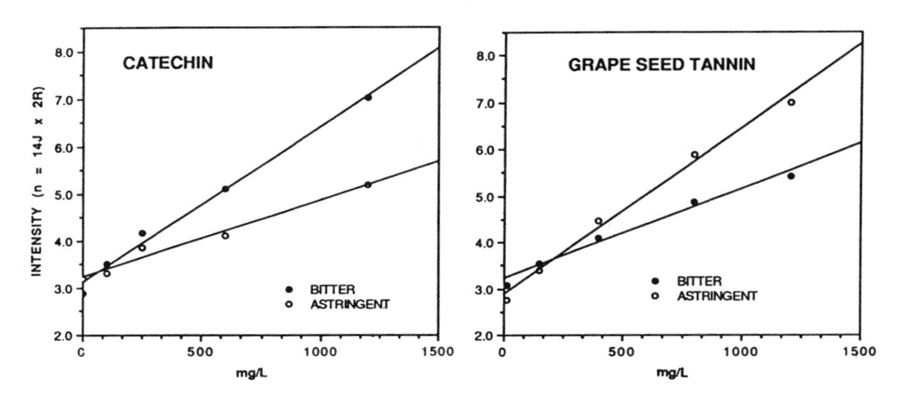

Figure 5. Mean intensity ratings for bitterness and astringency as a function of stimulus concentration. Samples were evaluated at room temperature and in random order (n = 28; 14 Judges x 2 Replications). The base wine was a neutral blended "dry" white wine made by standard practices in the University of California, Davis winery. Reprinted with permission.[10] Copyright 1990 Society of Chemical Industry.

relative intensity of both bitterness and astringency increased as the number of subunits increased[44] (Table 6). Again, the polymeric fractions were significantly more astringent than the monomer fraction.

These observations may have important implications in the taste perception of wines during aging. In general, as wines age, the fraction of total phenols occurring as polymers increases (Table 3) due to oxidative polymerization reactions. Noble[40] theorized that young red wines which contain larger amounts of flavonoid monomers and dimers may be perceived as more bitter than astringent, while older wines containing polymerized flavonoids may be perceived as less bitter and more astringent. If the oxidative reactions continue, the

Table 6. Relative astringency and bitterness of phenolic fractions isolated from grape seeds: Mean intensity ratings[1] per milligram of phenolics for bitterness and astringency and mean ratios of bitterness to astringency for four fractions

Fraction	Astringency rating per mg[2]	Bitterness rating per mg[2]	Bitter/astring[2]
I (monomer)	0.018[a]	0.032[d]	2.38[i]
II (dimer)	0.033[ab]	0.050[de]	1.89[gh]
III (tri and tetramer)	0.070[b]	0.110[e]	1.79[g]
IV (hexamer)	0.452[c]	0.913[f]	2.18[hi]

[1]Mean rating of three concentrations of each fraction by triplicate estimation of 11 judges (n = 99).
[2]Means with the same letter superscript are not significantly different at $p > 0.05$.
Reprinted with permission.[44] Copyright 1980 American Chemical Society

polymers can eventually reach a size where they will precipitate out of solution, resulting in a decreased astringency.

The absolute stereochemistry of the flavonoid molecule can also have important effects on the perception of bitterness and astringency. Catechin and epicatechin, diastereomers which differ only in configuration at the C3 position of the C ring (Fig. 4), have significantly different bitterness and astringency intensities in aqueous solution.[45] Time-intensity procedures were used to show that epicatechin is more bitter and more astringent than catechin, and the total duration of the bitterness and astringency is longer for epicatechin than for catechin (Fig. 6). The mechanism for these results is unclear and cannot be explained solely on the basis of chemical structure. Epicatechin, however, is more lipophilic than catechin, as demonstrated by its lower water solubility and

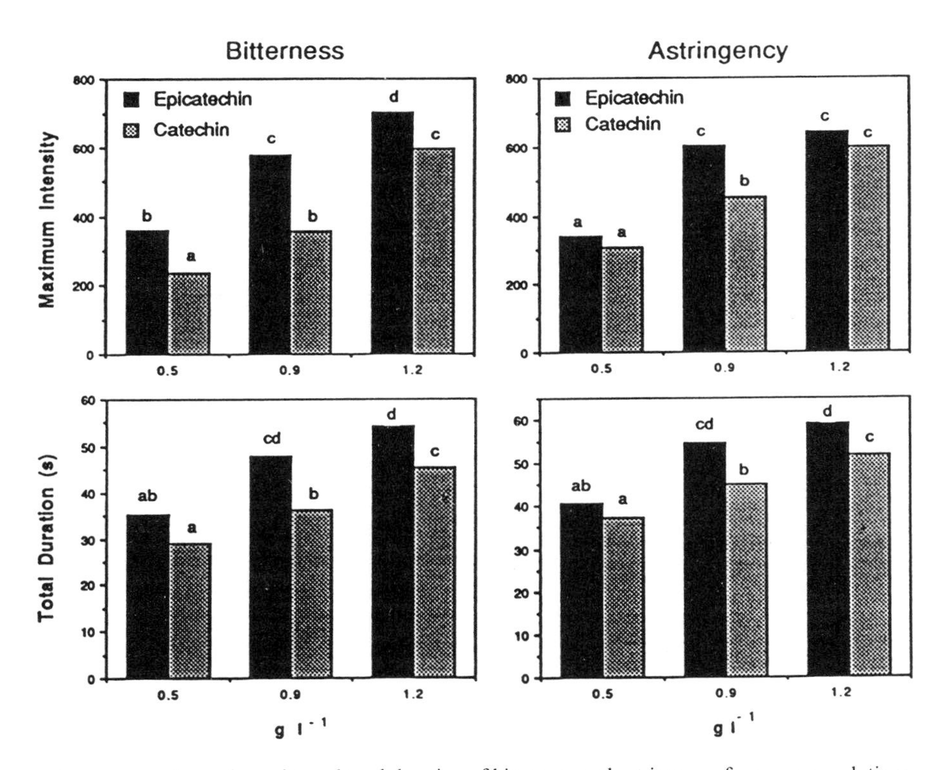

Figure 6. Maximum intensity and total duration of bitterness and astringency for aqueous solutions of epicatechin and catechin. For each T-I parameter, means which differ significantly ($p < 0.05$) have different superscripts. Samples were evaluated by trained judges under red light to eliminate visible color cues associated with flavonoid concentration. Sample presentation order was completely randomized within a session (n = 36; 12 Judges x 3 Replications). Reprinted with permission.[45] Copyright 1995 Society of Chemical Industry.

later elution on a reverse phase HPLC column.[45] This may allow for greater interaction with lipophilic receptors and promote hydrophobic interactions that drive the initial interaction with proteins responsible for the perception of astringency.

The effects of structure on sensory properties of other flavonoid monomers and dimers is largely unknown. For instance, flavonoid polymerization can occur at various positions (Fig. 4); over 20 isomers of dimeric flavonoids alone have been identified,[46] but their corresponding sensory properties are unknown. An understanding of the effects of chemical structure on perception of bitterness and astringency will be critical for a complete understanding of the effects of winemaking practices on sensory properties of wine.

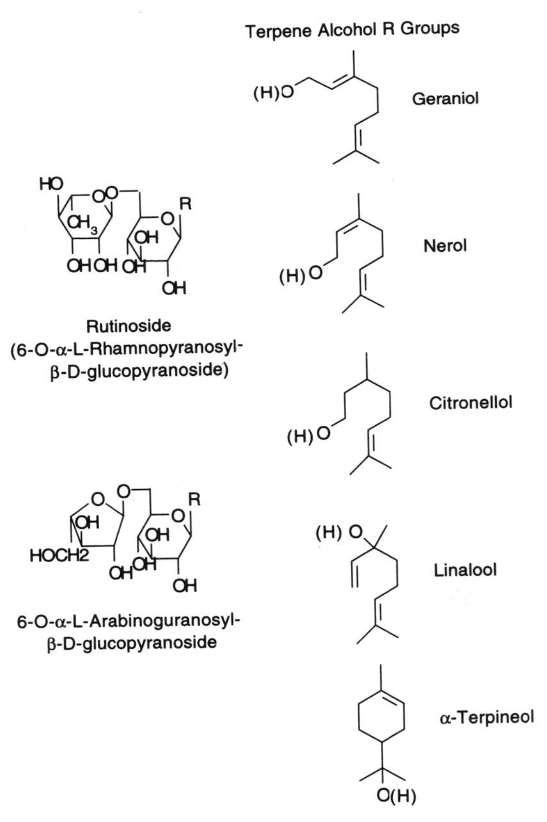

Figure 7. Typical terpene glycosides found in wine.

Glycosides

In the floral varieties, Gewürztraminer, Riesling, and Muscat, terpene glycosides (Fig. 7) make potentially important contributions to the volatile aroma of the wine upon hydrolysis of the monoterpenes. Many glycosides, *e.g.*, geranyl β-D-glucopyranoside and β-menthyl glucoside, also have bitter properties in aqueous solutions.[47,48] When glycosidic extracts from Chardonnay and Muscat of Alexandria wines were added to water at ten times the original concentration found in the wine, a faint bitterness was observed similar to that of 2 mg/L quinine sulfate, a commonly used non-flavonoid bitter standard. However, when the glycosidic extracts were added to model wines or to the original wine at the normal concentrations found in wine, the glycosides did not contribute to the sensory perception of bitterness.[49,50]

Interactions with Other Wine Components

As discussed, wine is a complex mixture containing many compounds which contribute to the overall sensory impact. The interaction of the tastes of sugars, acids, and ethanol in particular have been evaluated for their effects on the perception of bitterness and astringency.

After water, ethanol, a product of yeast fermentation, is the largest constituent of wine, comprising 10–14% of the total composition in table wines. Fischer and Noble[51] used T-I procedures to evaluate the bitterness of catechin in dealcoholized wine in the presence of different added ethanol levels. Ethanol had a significant effect on increasing the maximum bitterness intensity of catechin (Fig. 8). An increase of 3% v/v ethanol in 100 mg catechin per liter wine samples increased the maximum perceived bitterness intensity more than increasing catechin to 1400 mg per liter of the same wine. Total duration of the bitterness also increased at higher ethanol concentrations[52] (Fig. 9). Differences of 3% ethanol often occur in wines and could result in significant effects on bitterness perception. The mechanisms for the effects of ethanol on bitterness perception are unknown.

Acids, largely tartaric acid, are among the most prevalent nonvolatile wine constituents (Table 1). Recent studies have shown that acids, in addition to sourness, have astringent qualities (drying or puckering characteristics).[53–55] However, acids can also alter the bitterness and astringency of polyphenols. Varying the pH of model wine solutions from 2.9 to 3.8 had a small, non-linear effect on the maximum bitterness intensity[51] (Fig. 8). The perception of astringency, however, decreased significantly as pH increased above 3.[52]

In general, sweetness decreases the perception of bitterness in alcoholic beverages.[56–58] Such results support those observed in model solutions with sucrose and quinine sulfate, a non-flavonoid bitter compound.[59] The studies imply that residual sugar levels in wines may play an important role in masking bitterness perception.

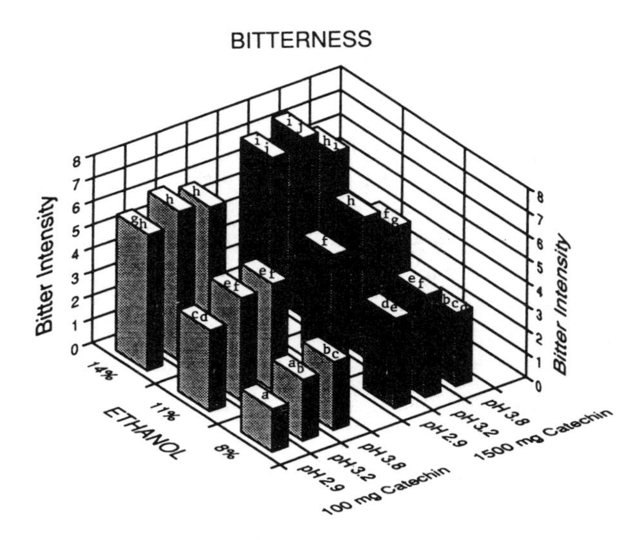

Figure 8. Mean bitterness intensity ratings of dealcoholized wine samples as a function of ethanol, pH, and catechin concentration. Time-intensity measurements were conducted at room temperature under red lights. Trained judges evaluated all samples in random order over three sessions with three replications (n = 60; 20 Judges x 3 Replications). Reprinted with permission.[51] Copyright 1994 American Society for Enology and Viticulture.

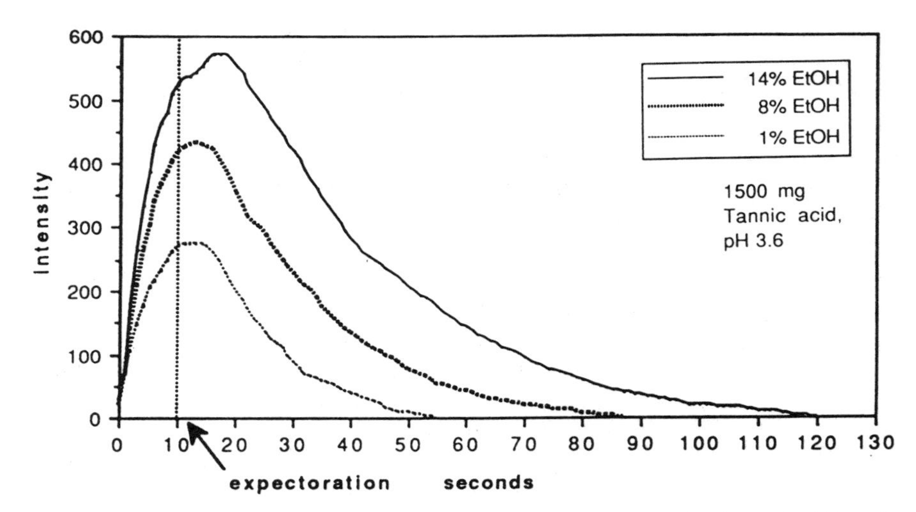

Figure 9. Average bitterness time-intensity curves as a function of ethanol in wine with 1500 mg/L tannic acid at pH 3.6. Arrow denotes time of expectoration (n = 24; Judges x 2 Replications). Reference: 52.

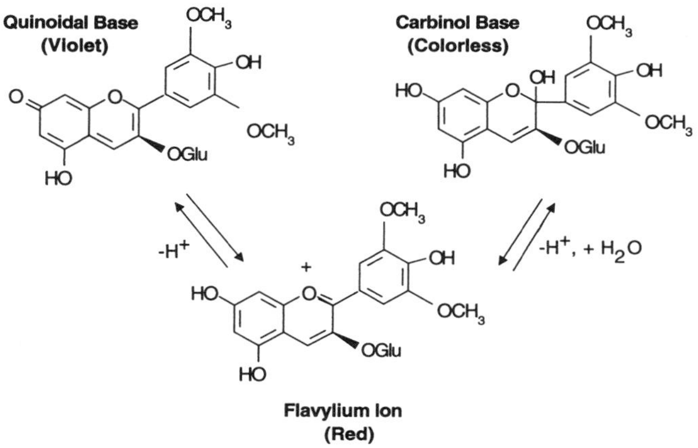

Quinoidal Base (Violet)

Carbinol Base (Colorless)

Flavylium Ion (Red)

Figure 10. Anthocyanin equilibria reactions affected by pH that influence wine color.

CONTRIBUTION OF PHENOLICS TO WINE COLOR

Anthocyanin pigments, extracted from grape skins during fermentations, are largely responsible for red wine color. The anthocyanins exist in several forms depending on pH: at low pH the red flavylium ion form is in equilibrium with the colorless pseudobase (carbinol) while at pH's greater than 4.25 the blue quinone form dominates (Fig. 10). At wine pH (~3.5), ~10% of the total anthocyanins are in the red, flavylium form.[60] As a result, most young red wines have an absorbance maximum at 520 nm and an absorbance minimum at 420 nm.

As wine matures, the monomeric anthocyanins tend to polymerize, forming oligomers and polymers. In addition, a shift in the absorption maximum to between 400 and 500 nm is observed as red wines age. This shift is probably due to the formation of brown pigments (quinones) which absorb at 420 nm.[61] Similarly, the oxidation of dihydroxyphenols (*e.g.*, caffeic acid, catechins, gallic acid) to quinones also results in formation of brown colors in white wines.

In addition to the monomeric and polymeric forms, anthocyanins exist in solution as copigmented forms. Copigmentation is defined as color enhancement resulting from complexing of an anthocyanin with a copigment (*e.g.*, flavonoids, phenolic acids, amino acids, sugars, alkaloids, and organic acids).[41,60] The specific nature of the copigments and their mechanisms of interaction with anthocyanin are unclear. As a result of copigmentation, the color intensity of red wines is higher than that obtained using model wines with similar anthocyanin concentration. Copigmentation also makes it difficult to predict the color of a red wine blend; as the wine is diluted, the equilibrium between the free anthocyanin and copigmented form can change resulting in less color than predicted.[60]

The effects of copigmentation on the bitterness and astringency of flavonoids is unknown. Several general reviews on the chemistry of anthocyanins, wine color, and copigmentation are available.[60–64] The reader is referred to these articles for detailed discussions.

WINE AROMA

Several hundred different volatile aroma compounds, comprising less than 1% of the total composition, have been identified in wines.[65,66] These aroma compounds can be considered to originate from the grape, from reactions occurring during fermentation (fermentation bouquet), and from chemical reactions occurring during maturation (maturation bouquet).

Considerable research effort has been expended to identify flavor compounds which can be used to characterize specific grape varieties. However, viticultural practices, climatic conditions, soil, and region of origin can also affect vine development and berry composition, exerting major influences on the distinctiveness of wine flavor.[67–71] Therefore, in only a few cases have specific compounds been identified which are responsible for characteristic varietal flavors. For example, the aroma of the floral varieties (Muscat, Gewürtraminer, Riesling) is dominated by monoterpene alcohols, particularly, linalool, geraniol, nerol, citronellol, and α-terpineol (Fig. 7). As discussed previously, these terpene alcohols occur largely as non-volatile glycosides. During processing and storage, chemical or enzymatic hydrolysis reactions increase the free terpene concentration with a corresponding increase in floral aroma.[72]

Alkoxypyrazinės in the vegetative or herbaceous cultivars (Sauvignon blanc, Cabernet Sauvignon) represent another example of "impact" compounds arising from grapes. The compound, 2-methoxy-3-isobutylpyrazine (Fig. 11), in particular, is responsible for the distinctive bell pepper aroma of these varieties.[73] This compound has an extremely low sensory threshold (2 ng/L) and is present at low part per trillion levels in Sauvignon blanc and Cabernet Sauvignon wines. Concentrations are dependent on fruit maturity with levels decreasing as fruit ripens.[73] In addition, high vine vigor and low light intensity in fruiting zone are strongly correlated with vegetative aroma and 2-methoxy-3-isobutylpyrazine concentration.[73]

The major aroma volatiles of wines are yeast fermentation products, including alcohols (other than ethanol) and esters. The alcohols (also called fusel oils) are formed anabolically from sugars or as transamination products of amino acids.[41] Isoamyl alcohol (3-methylbutanol; Fig. 11) is the most important of these compounds, making up as much as one-half of all the fusel alcohols, and contributing a characteristic "banana-like" aroma. The esters are also formed during yeast metabolism as a result of the reactions between Acyl CoA and alcohols (mainly ethanol). Ethyl acetate (Fig. 11) is the main ester found in wine

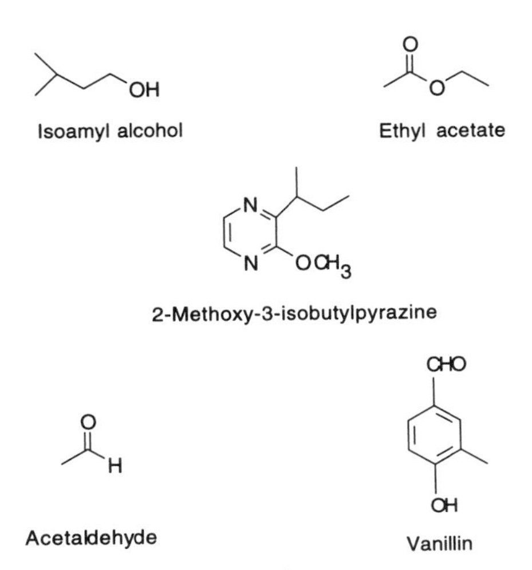

Figure 11. Common aroma volatiles found in wine.

contributing a "fruity" or "nail-polish remover" aroma, depending on concentration. Boulton et al.[41] and Cole and Noble[66] present excellent reviews on the formation and importance of the fusel oils and esters to wine aroma.

Finally, as reviewed by Rapp[65] and Cole and Noble,[66] a number of volatile aromas arise during aging and maturation of wines. These compounds are largely the result of oxidative reactions and volatile extraction during oak aging. For example, acetaldehyde (Fig. 11), which contributes a typical "green-apple" or "sherry-like" flavor to aged and heated wines, is formed from oxidative reactions involving polyphenols. In addition to the quinones discussed previously, hydrogen peroxide is a product of the reaction between molecular oxygen and the vicinal dihydroxy groups of polyphenolic compounds. Wildenradt and Singleton[74] have shown that hydrogen peroxide, a strong oxidant, readily reacts with ethanol forming acetaldehyde, resulting in an oxidized wine aroma. Oak components (*e.g.*, vanillin; Fig. 11) are extracted from oak barrels during barrel aging and can also contribute important sensory characteristics to the finished wine. The actual compounds extracted and their concentrations are dependent on storage time, oak species and source, cooperage practices, and age and size of the barrel.[66]

CONCLUSIONS

Bitterness and astringency are characterized by a persistent, lingering aftertaste. Time-intensity procedures have provided valuable insight into factors which affect bitterness and astringency perception, including the concentration and maximum intensity of the stimulus and the time between ingestions. However, the mechanisms of perception of these attributes are still not well understood, and individual subjects' differences in perception have not been well characterized (*e.g.*, effects of salivary flow or PROP status). Correlation of chemical and sensory measures of astringency have proven difficult, partially due to the variable and complex chemical nature of the stimuli, but also due to a poor understanding of the mechanisms of perception.

Flavonoid structure and degree of polymerization significantly affect the overall perception of bitterness and astringency in wines. However, sensory attributes of only a limited number of polyphenolic compounds have been evaluated, and no studies of the systematic effects of structure on bitterness and astringency have emerged. Such information will be critical for a complete understanding of the effects of winemaking practices on the sensory properties of wine. In addition, bitterness of Muscat, Gewürztraminer, and Riesling wines is not correlated with phenolic or glycoside content, indicating the presence of unidentified bitter fraction(s).

The effects of other wine components (sugars and acids) on the perception of bitterness and astringency has been largely unexplored. Recent studies have shown that acids not only alter the astringency perception of polyphenols but that they also have an astringent modality of their own. Such results point to the need for a greater understanding of the complex interactions which affect the overall taste and flavor perception of wine.

Phenolic compounds also make important contributions to wine color. Understanding the chemical nature of copigmentation and the mechanisms for interactions between anthocyanins and flavonoids (or phenolic acids) will be critical for stabilizing and maximizing the color of wines during processing and blending.

Finally, although this review focused largely on the sensory properties of phenolic compounds in wine, volatile aroma compounds also make significant contributions to wine flavor. Many of these aroma compounds are present at extremely low levels and require sensitive analytical and sensory methods for quantitation and identification. As a result, only a few flavor "impact" compounds characteristic of a given grape cultivar have been identified. Although varietally distinct flavor notes, such as berry aromas in Zinfandel,[75] Cabernet Sauvignon,[68] and Pinot Noir,[67] or black pepper notes in Zinfandel[75] have been identified in sensory tests, further work is needed to identify the chemical compounds which elicit these aromas.

ACKNOWLEDGMENTS

Special thanks to Dr. Ann Noble and John Ebeler for helpful comments and careful review during the preparation of this manuscript.

REFERENCES

1. CLIFFORD, M. N. 1996. Astringency. In: Proceedings of the Phytochemical Society of Europe, Vol. 41, (F.A. Tomas-Barbaran, ed.), Oxford University Press, London, pp. 87–107.
2. NOBLE, A.C. 1988. Analysis of wine sensory properties. In: Wine Analysis. Modern Methods of Plant Analysis, New Series Volume 6, (H. F. Linskens, J. F. Jackson, eds.), Springer-Verlag, Berlin, pp. 9–28.
3. HINREINER, E., FILIPELLO, F., BERG, H.W., WEBB, A.D. 1955. Evaluation of thresholds and minimum difference concentrations for various constituents of wines. IV. Detectable differences in wine. Food Techn. 9:489–490.
4. SINGLETON, V.L., SIEBERHAGEN, H.A., DE WET, P., VAN WYK, C.J. 1975. Composition and sensory qualities of wines prepared from white grapes by fermentation with and without grape solids. Am. J. Enol. Vitic. 26(2):62–69.
5. SINGLETON,V.L. 1982. Grape and wine phenolics: Background and prospects. In: Grape and Wine Centennial Symposium Proceedings, University of California, Davis, (A. D. Webb, ed.), UC Davis, Davis, CA, pp. 215–227.
6. NOBLE, A.C., WILLIAMS, A.A., LANGRON, S.P. 1984. Descriptive analysis and quality rating of 1976 wines from four Bordeaux communes. J. Sci. Food Agric. 35:88–98.
7. PANGBORN, R.M. 1981. A critical review of threshold, intensity and descriptive analyses in flavor research. In: Flavour '81, (P. Schreier, ed.), Walter de Gruyter, Berlin, pp. 3–32.
8. STEVENS, J.C. 1971. Psychophysics. In: Stimulus and Sensation: Readings in Sensory Psychology, Little, Brown & Co., Boston, pp. 5–18.
9. STEVENS, S.S. 1957. On the psychophysical law. Psychological Review 64:153–181.
10. ROBICHAUD, J.L., NOBLE, A.C. 1990. Astringency and bitterness of selected phenolics in wine. J. Sci. Food Agric. 53:343–355.
11. PANGBORN, R.M., LEWIS, M.J., YAMASHITA, J.F. 1983. Comparison of time-intensity with category scaling of bitterness of iso-α-acids in model systems and in beer. J. Inst. Brew. 89:349–355.
12. LEACH, E.J. 1984. Evaluation of astringency and bitterness by scalar and time-intensity procedures. M.S. Thesis, University of California, Davis, CA, 170 pp.
13. GUINARD, J.-X., PANGBORN, R.M., LEWIS, M.J. 1986. The time-course of astringency in wine upon repeated ingestion. Am. J. Enol. Vitic. 37(3): 184–189.
14. SMITH, A.K., JUNE, H., NOBLE, A.C. 1996. Effects of viscosity on the bitterness and astringency of grape seed tannin. Food Qual. Pref. 7(3/4):161–166.
15. FISCHER, U., BOULTON, R.B., NOBLE, A.C. 1994. Physiological factors contributing to the variability of sensory assessments: Relationship between salivary flow rate and temporal perception of gustatory stimuli. Food Qual. Pref. 5:55–64.
16. BARTOSHUK, L.M., DUFFY, V.B., MILLER, I.J. 1994. PTC/PROP tasting: Anatomy, psychophysics, and sex effects. Physiol. Behav. 56(6):1165–1171.
17. OUGH, C.S., AMERINE, M.A. 1988. Methods for Analysis of Musts and Wines. John Wiley & Sons, New York, 377 pp.
18. SINGLETON, V.L., ROSSI, J.A. 1965. Colorimetry of total phenolics with phosphomolybdic-phosphotungstic acid reagents. Am. J. Enol. Vitic. 16(3):144–158.
19. NAGEL, C.W. 1985. Appliction of high performance liquid chromatography to analysis of flavonoids and phenyl propenoids. Cereal Chem. 62(2): 144–147.
20. KANTZ, K., SINGLETON, V.L. 1990. Isolation and determination of polymeric polyphenols using Sephadex LH-20 and analysis of grape tissue extracts. Am. J. Enol. Viticul. 41(3): 223–228.
21. ROGGERO, J.-P., COEN, S., ARCHIER, P. 1990. Wine phenolics: Optimization of HPLC analysis. J. Liq. Chrom. 13(3): 2593–2604.
22. ROGGERO, J.-P., ARCHIER, P., COEN, S. 1991. Wine phenolics via direct injection: Enhancement of the method. J. Liq. Chrom. 14(3):533–538.

23. REVILLA, E., BOURZEIX, M., ALONSO, E. 1991. Analysis of catechins and proanthocyanidins in grape seeds by HPLC with photodiode array detection. Chromatographia 31:465–468.
24. McMURROUGH, I., BAERT, T. 1994. Identification of proanthocyanidins in beer and their direct measurement with a dual electrode electrochemical detector. J. Inst. Brew. 100(6):409–416.
25. LAMUELA-RAVENTOS, R.M., WATERHOUSE, A.L. 1994. A direct HPLC separation of wine phenolics. Am. J. Enol. Vitic. 45: 1–5.
26. BARNELL, H.R., BARNELL, E. 1945. Studies in tropical fruits. XVI. The distribution of tannins within the banana and the changes in their condition and amount during ripening. Ann. Botany, N.S. 9(3):77–99.
27. SINGLETON, V.L. 1966. The total phenolic content of grape berries during the maturation of several varieties. Am. J. Enol. Vitic. 15: 34–40.
28. RIBÉREAU-GAYON, P., GLORIES, Y. 1982. Structure of condensed phenolic compounds in *Vinifera* grapes and wine. Influence of ripening and infection by *Botrytis cinera* on phenolic content. In: Grape and Wine Centennial Symposium Proceedings, University of California, Davis, (A.D. Webb, ed.), UC Davis, Davis, CA, pp. 228–234.
29. HASLAM, E., LILLEY, T.H., WARMINSKI, E., LIAO, H., CAI, Y. MARTIN, R., GAFFNEY, S.H., GOULDING, P.N., LUCK, G. 1992. Polyphenol complexation. A study in molecular recognition. In: Phenolic Compounds in Food and their Effects on Health. Volume I. Analysis, Occurrence, and Chemistry, (C.-T. Ho, C.Y. Lee, M.-T. Huang, eds.), American Chemical Society, Washington, DC, pp. 8–50.
30. GOLDSTEIN, J.L., SWAIN, T. 1965. The inhibition of enzymes by tannins. Phytochemistry 4:185–192.
31. BATE-SMITH, E.C. 1973. Haemanalysis of tannins: The concept of relative astringency. Phytochemistry 12:907–912.
32. HAGERMAN, A.E., BUTLER, L.G. 1978. Protein precipitation method for the quantitative determination of tannins. J. Agric. Food Chem. 26(4):809–812.
33. HAGERMAN, A.E. 1989. Chemistry of tannin-protein complexation. In: Chemistry and Significance of Condensed Tannins, (R.W. Hemingway, J.J. Karchesy, eds.), Plenum Press, New York, pp. 323–333.
34. MARTIN, J.S., MARTIN, M.M. 1982. Tannin assays in ecological studies: Lack of correlation between phenolics, proanthocyanidins and protein-precipitating constituents in mature foliage of six oak species. Oecologia 54:205–211.
35. HASLAM, E., LILLY, T.H. 1988. Natural astringency in foodstuffs—a molecular interpretation. CRC Crit. Rev. Food Sci. Nutr. 27(1):1–40.
36. ASANO, K. OHTSU, K., SHINAGAWA, K., HASHIMOTO, N. 1984. Affinity of proanthocyanidins and their oxidation products for haze-forming proteins of beer and the formation of chill haze. Agric. Biol. Chem. 48(5):1139–1146.
37. YOKOTSUKA, K., SINGLETON, V.L. 1987. Interactive precipitation between graded peptides from gelatin and specific grape tannin fractions in wine-like model solutions. Am. J. Enol. Vitic. 38(3):199–206.
38. ITTAH, Y. 1991. Titration of tannin via alkaline phosphatase activity. Anal. Biochem. 192:277–280.
39. DYER, W.G. 1985. Evaluation of Commercial Practices in White Wine Fermentation in Barrels and Red Wine Pomace Contact. M. S. Research Report, University of California, Davis, CA, 54 pp.
40. NOBLE, A.C. 1990. Bitterness and astringency in wine. In: Bitterness in Foods and Beverages. Developments in Food Science 25, (R. Rousseff, ed.), Elsevier, New York, pp. 145–158.
41. BOULTON, R.B., SINGLETON, V.L., BISSON, L.F., KUNKEE, R.E. 1996. Principles and Practices of Winemaking. Chapman & Hall, New York, 604 pp.

42. SINGLETON, V.L., NOBLE, A.C. 1976. Wine flavor and phenolic substances. In: Phenolic, Sulfur, and Nitrogen Compounds in Food Flavors, (G. Charalambous, I. Katz, eds.), American Chemical Society, Washington, D.C., pp. 47–70.
43. VÉRETTE, E. NOBLE, A.C., SOMERS, T.C. 1988. Hydroxycinnamates of *Vitis vinifera*: Sensory assessment in relation to bitterness in white wines. J. Sci. Food Agric. 45: 267–272.
44. ARNOLD, R.A., NOBLE, A.C., SINGLETON, V.L. 1980. Bitterness and astringency of phenolic fractions in wine. J. Agric. Food Chem. 28:675–678.
45. THORNGATE, J.H., NOBLE, A.C. 1995. Sensory evaluation of bitterness and astringency of 3R(-)-epicatechin and 3S(+)-catechin. J. Sci. Food Agric. 67:531–535.
46. RICARDO DA SILVA, J.M., RIGAUD, J., CHEYNIER, V., CHEMINAT, A., MOUTOUNET, M. 1991. Procyanidin dimers and trimers from grape seeds. Phytochemistry 30:1259–1264.
47. FISCHER, E., HELERICH, B. 1911. Uber neue synthetische Glycosid. Ann. 383:68–74.
48. HASE, T., IWAGAWA, T., MUNESADA, K. 1982. A bitter monoterpene glucoside from *Viburnum phlebotrichum*. Phytochemistry 21:1435–1437.
49. NOBLE, A.C., STRAUSS, C.R., WILLIAMS, P.J., WILSON, B. 1987. Sensory evaluation of non-volatile flavour precursors in wine. In: Flavour Science and Technology, (M. Martens, G.A. Dalen, H. Russwurm, Jr., eds.), John Wiley & Sons, Ltd., pp. 383–390.
50. NOBLE, A.C., STRAUSS, C.R., WILLIAMS, P.J., WILSON, B. 1988. Contribution of terpene glycosides to bitterness in Muscat wines. Am. J. Enol. Vitic. 39(2):129–131.
51. FISCHER, U., NOBLE, A.C., 1994. The effect of ethanol, catechin concentration, and pH on sourness and bitterness of wine. Am. J. Enol. Vitic. 45(1):6–10.
52. FISCHER, U. 1990. The influence of ethanol, pH, and phenolic composition on the temporal perception of bitterness and astringency, and parotid salivation. M. S. Thesis, University of California, Davis, CA, 252 pp.
53. LEE, C.B., LAWLESS, H.T. 1991. Time-course of astringent sensations. Chem. Senses 16(3):225–238.
54. CORRIGAN THOMAS, C.J., LAWLESS, H.T. 1995. Astringent subqualities in acids. Chem. Senses 20:593–600.
55. SOWALSKI, R.A. 1996. Effect of the concentration, pH and anion species on the sourness and astringency of organic acids. M. S. Thesis, University of California, Davis, CA, 139 pp.
56. PYLE, H.J. 1989. Effect of ethanol, tartaric acid and sucrose on perceived bitterness of caffeine in model solutions. M. S. Thesis, University of California, Davis, CA, 149 pp.
57. BURNS, D.J.W., NOBLE, A.C. 1985. Evaluation of the separate contribution of viscosity and sweetness of sucrose to perceived viscosity, sweetness and bitterness of vermouth. J. Texture Studies 16: 365–381.
58. NOBLE, A.C. 1994. Bitterness in wine. Physiol. Behav. 56(6):1251–1255.
59. BIRCH, G.G., COWELL, N.D., YOUNG, R.H. 1972. Structural basis of and interaction between sweetness and bitterness in sugars. J. Sci. Fd. Agric. 23:1207–1212.
60. LEVENGOOD, J.S. 1996. A survey of copigmentation in Cabernet Sauvignon. M. S. Thesis, University of California, Davis, CA, 46 pp.
61. ZOECKLEIN, B.W., FUGELSANG, K.C., GUMP, B.H., NURY, F.S. 1995. Wine Analysis and Production. Chapman and Hall, New York, 621 pp.
62. SINGLETON, V.L. 1988. Wine phenols. In: Wine Analysis. Modern Methods of Plant Analysis, New Series Volume 6, (H.F. Linskens, J.F. Jackson, eds.), Springer-Verlag, Berlin, pp. 173–218.
63. SOMERS, T.C., EVANS, M.E. 1977. Spectral evaluation of young red wines: Anthocyanin equilibria, total phenolics, free and molecular SO_2, "Chemical Age". J. Sci. Food Agric. 28:279–287.
64. SOMERS, T.C., VÉRETTE. E. 1988. Phenolic composition of natural wine types. In: Wine Analysis. Modern Methods of Plant Analysis, New Series Volume 6, (H.F. Linskens, J.F. Jackson, eds.), Springer-Verlag, Berlin, pp. 219–253.

65. RAPP, A. 1988. Wine aroma from gas chromatographic analysis. In: Wine Analysis. Modern Methods of Plant Analysis, New Series Volume 6, (H.F. Linskens, J.F. Jackson, eds.), Springer-Verlag, Berlin, pp. 29–66.
66. COLE, V.C., NOBLE, A.C. 1994. Flavor chemistry and assessment. In: Fermented Beverage Production, (A.G.H. Lea, J.R. Piggott, eds.), Blackie Academic and Professional, Chapman and Hall, London, pp. 361–385.
67. GUINARD, J., CLIFF, M. 1987. Descriptive analysis of Pinot noir wines from Carneros, Napa, and Sonoma. Am. J. Enol. Vitic. 38: 211–215.
68. HEYMANN, H., NOBLE, A.C. 1987. Descriptive analysis of commercial Cabernet Sauvignon wines from California. Am J. Enol. Vitic. 38:41–44.
69. ABBOTT, N.A., COOMBE, B.G., WILLIAMS, P.J. 1991. The contribution of hydrolyzed flavor precursors to quality differences in Shiraz juice and wines: An investigation by sensory descriptive analysis. Am. J. Enol. Vitic. 42:167–174.
70. LACEY, M.J., ALLEN, M.S., HARRIS, R.L., BROWN, W.V. 1991. Methoxypyrazines in Sauvignon blanc grapes and wines. Am. J. Enol. Vitic. 42:103–108.
71. FRANCIS, L., SEFTON, M.A., WILLIAMS, P.J. 1992. Sensory descriptive analysis of the aroma of hydrolysed precursor fractions from Semillon, Chardonnay and Sauvignon blanc grape juices. J. Sci. Food Agric. 59:511–520.
72. WILLIAMS, P.J. 1993. Hydrolytic flavor release in fruit and wines through hydrolysis of nonvolatile precursors. In: Flavor Science. Sensible Principles and Techniques, (T.E. Acree, R. Teranishi, eds.), American Chemical Society, Washington, D.C., pp. 287–308.
73. NOBLE, A.C., ELLIOT-FIST, D.L., ALLEN, M.S. 1995. In: Fruit Flavors. Biogenesis, Characterization, and Authentication, (R.L. Rouseff, M.M. Leahy, eds.), American Chemical Society, Washington, D.C., pp. 226–234.
74. WILDENRADT, H.L., SINGLETON, V.L. 1974. Production of aldehydes as a result of oxidation of polyphenolic compounds and its relation to wine aging. Am. J. Enol. Vitic. 25:119–126.
75. NOBLE, A.C., SHANNON, M. 1987. Profiling Zinfandel wines by sensory and chemical analyses. Am. J. Enol. Vitic. 38:1–5.
76. SINGLETON, V.L. 1992. Wine composition. In: Potential Health Effects of Components of Plant Foods and Beverages in the Diet Symposium Proceedings, University of California, Davis, pp. 88–106.

Chapter Eight

FLAVOR BIOGENERATION

J. Crouzet

Laboratoire de Génie Biologique et Sciences des Aliments
Unité de Microbiologie et Biochimie Industrielles, Associée à l'INRA
Université de Montpellier II
34095 Montpellier Cedex 05 France

INTRODUCTION

The quality of food is largely dependent upon its organoleptic characteristics. Odorous compounds present in food reach the olfactory mucosa receptors in two different ways. By direct sniffing, the volatile compounds interact with the nose receptors and produce an odor sensation. When the volatile molecules are released in the mouth, the olfactory epithelium is reached by a retronasal route and the sensation perceived is known as aroma. The combined sensation of aroma and taste give the overall flavor of a food. A large number of volatile compounds responsible for fruit and vegetable aroma have been identified over the past 30 years by high resolution gas chromatography coupled to

Functionality of Food Phytochemicals
edited by Johns and Romeo, Plenum Press, New York, 1997

either mass spectrometry (HRGC/MS) or to Fournier Transformed Infra Red spectrometry (HRGC/FTIR), together with ^{1}H and ^{13}C NMR. These compounds are representative of different classes of organic compounds, several hundred volatiles having been found in the tissue of plants.

Until recently little was known about the biosynthesis and accumulation of volatile compounds in plants. Such knowledge is indispensable for control in the production of flavors by fruits and vegetables. Results obtained from study of biosynthetic pathways of aroma compounds can be applied through biotechnology. Plant tissue or cell culture, enzymes, or microrganismes can be used to produce natural flavor compounds from non volatile natural flavor precursors.

Food volatile compounds are generated by several routes. Sanderson and Graham[1] have distinguished four types of mechanisms involved in aroma formation in foods: biosynthesis, direct enzymatic action, indirect enzymatic or oxidative action, and pyrolytic action. Certain plant volatiles, produced by biogenetic pathways during ripening in the intact organ, are considered primary aroma compounds. In some fruits, the formation of volatile compounds is initiated by the climacteric phase, reaching a peak during the ripening phase. Several primary metabolites produced during the growing and preclimacteric stages such as fatty acids, amino acids and carbohydrates,[2,3] are involved in the biosynthesis of aroma compounds. The synthesis of the secondary metabolic compounds depends on such factors as climate, soil, nutrition, and water availability. Aroma is also dependent on the cultivar. Secondary aroma compounds also are formed during cell disruption.[2] In this case, non-volatile precursors act as substrates for enzymes involved in aroma formation. Recent studies have drawn attention to the importance of glycosidically bound volatile compounds as aroma precursors. More than 200 different aglycone structures have been found in about 150 botanical species belonging to 35 different families.[4]

This paper confines itself to those volatile compounds produced by biosynthesis and enzymatic action according to Sanderson and Graham criteria.[1] These correspond to primary and secondary aroma compounds as defined by Drawert.[2]

BIOSYNTHESIS

Ester Biosynthesis

Esters are qualitatively and quantitatively the most important class of volatile compounds in such fruits as apple, pear, banana, strawberry and pineapple. The biogenetic pathways of these compounds have been studied using peel and cortical tissue discs, banana tissue slices, and whole strawberries at several maturity stages, incubated with ^{14}C labeled precursors in a sucrose solution.[5–8] More recently, whole apple fruits treated with volatile precursors have been used.[9] Esters are produced enzymatically by combining acyl CoA and alcohol moieties according to the general reaction given in Figure 1.

Acyl CoA is produced by several pathways: fatty acid metabolism, activation of free fatty acids, decarboxylation and activation of α-keto acids, and from an amino acid degradative pathway for branched chain components. While short chain acid precursors are incorporated without further metabolization by banana discs to generate the corresponding esters, octanoic acid is metabolized to caproic and butyric acids by β-oxidation and to heptanoic acid by α-oxidation.[10]

Fatty acids are precursors of primary alcohols generated by reduction. In the case of banana, the results obtained using 8–^{14}C octanoate as precursor[7] can be explained by assuming the formation of the SCoA derivative of the acid to give caprylates. Octanal and 1-octanol are produced in a two step reduction, and octyl esters are the results of the transacylation reaction of the alcohol.

The role played by amino acid metabolism in ester biogenesis has been studied in such fruits as tomato,[11,12] banana,[6] and strawberry.[13,14] The high level of ethyl esters found in the Chandler strawberry variety is explained by the variation in alanine concentration during the ripening of this cultivar. The conversion of alanine, leucine and isoleucine to substitued propyl or butyl esters involves several enzymes: transaminase; decarboxylase -in some cases spontaneous decarboxylation has been reported-; alcohol dehydrogenase and alcohol acyl transferase; and α-keto acid dehydrogenase, which produces substitued propanoates and butanoates (Fig. 2).

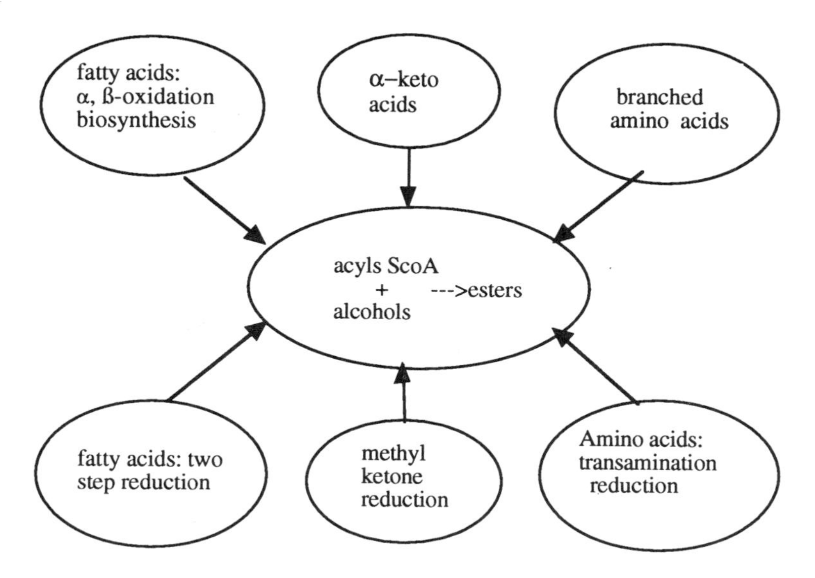

Figure 1. Ester biosynthetic pathways in plants.

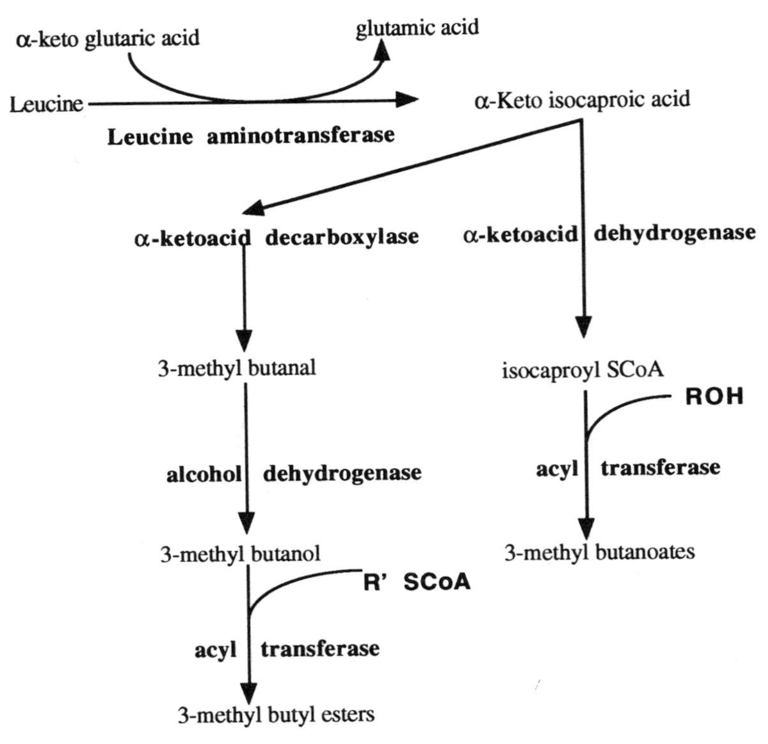

Figure 2. Ester biosynthesis from branched chain amino acids: Leucine.

It is generally accepted that the tissue needs to be intact for the development of the esterification reaction. Yamashita et al.,[8] for example, reported a slight increase in acetate ester concentration when a strawberry was cut into 2 to 4 pieces and acid and alcohol precursors were added. In contrast, ester formation decreased when the strawberry was cut into 8 pieces, and no ester formation occurred when a strawberry homogenate was used. In the latter case, the predominant reaction taking place was that of ester hydrolysis. Ester synthesis has, however, been detected after banana pulp extract containing several enzymes -transaminases, alcohol dehydrogenase, and polyphenol oxydase- were added to dearomatised banana powder.[15] The change in the nature and concentration of esters according to incubation time and temperature indicated a competition between synthesis and hydrolysis reactions.

Chiral secondary alcohols result from the reduction of methyl ketones, such as 2-pentanol and 2-heptanol, or from intermediates in the biosynthesis and β-oxidation of fatty acids such as 3-hydroxyderivatives. Different aroma characteristics are generally developed by stereoisomers, and information concerning the biosynthetic pathways can be obtained from the stereochemistry of these compounds. In the case of purple and yellow passion fruit, 2-pentanol, 2-heptanol and 2-nonanol and their esters have been identified. Free alcohols in the (S)(+) configuration are mainly found in the yellow variety, whereas the (R)(-) configuration is dominant in the purple variety.[16] These compounds originate through two different pathways involving two enzymes possessing different stereospecificity. One has the same specificity as yeast ADH (S configuration), the other has the same stereospecificity as dehydroxy-acetone reductase (R configuration). The corresponding esters are also detected in the purple variety.

3-Hydroxyacid esters, and 3- and 5-acetoxyacid esters have been identified in tropical fruits such as mango, pineapple, and passion fruit. The (S)(+) configuration predominates by more than 90% in compounds present in yellow passion fruit and pineapple, indicating that these compounds are generated by β-oxidation. In purple passion fruit and mango, where 3-hydroxybutanoates have an (R) (-) configuration, a different mechanism involving fatty acid biosynthesis or an abnormal hydration reaction of enoyl CoA in the β-oxidation reaction is suggested.[16] More recently it has been found that intact mature fruits such as apples are able to incorporate in their aged tissues volatile precursors such as alcohols added to the gas phase during storage.[17–19] This technique is known as Precursor Atmosphere (PA) storage, and has been proposed as a biotechnological process to increase fruit flavor.[20]

Methyl Ketones and Lactones

Methyl ketones possessing an odd number of carbon atoms, 2-pentanone, 2-heptanone, 2-nonanone and 2-undecanone, are considered to be characteristic of blue cheese aroma. They are, however, also present in several fruits: apricot,[21] mango,[22] passion fruit,[23] and quince.[24] These compounds are produced from the β-oxidation of fatty acids by decarboxylation of β-keto acids. While this pathway has been extensively studied using *Penicillium roqueforti*,[25] Tressl and Drawert [7] have shown using 8- ^{14}C octanoic acid added to banana slices or strawberries that small amounts of labeled 2-pentanone and 2-heptanone are produced. These results indicate that the same metabolic pathway is involved in fruits and fungi.

γ and δ-Lactones are present in several plants, and in particular in such fruits as peaches, apricots, mangoes, and strawberries. Lipid metabolism is clearly involved in lactone biogeneration, as indicated by results obtained with several micro-organisms.[26] Different patents describe the biotechnological production of lactones.[27–29] Some are also produced from amino acid precursors.

Findings concerning the biosynthesis of lactones in plants are scarce and sometimes speculative. Using deuterated precursors added to pineapple slices, Engel et al.[30] have shown that different pathways are involved in plant lactone biosynthesis. The first involves the reduction of keto acids precursors. δ-Octalactone is obtained after addition of 5-oxo-octanoic acid to pineapple tissue. On the other hand, the formation of this compound after incubation of 3-hydroxy-hexanoic acid $3d_1$ indicates that chain elongation and cyclization occur. This lactone is also detected after addition of (*Z*)-4-octenoic acid 2,2-d_2. Hydration of the double bond followed by cyclization of the intermediates 4-hydroxyoctanoic acid and 5-hydroxyoctanoic acid give respectively γ-octalactone and δ-octalactone. Hydroxylation of aliphatic acid has also occurred. Pineapple γ-octalactone mainly consists of the (R) enantiomer (85 %) whereas δ-octalactone possesses nearly a racemic composition (56 % (R): 44 % (S).[31] These results indicate that different pathways occur in the biogenesis of the two compounds.

Pyrazines

Alkyl pyrazines are widespread in food,[32] most of them being produced by non-enzymatic reactions during heating. Other compounds, in particular methoxypyrazines, have been detected in vegetables such as peas[33] and bell pepper,[34] and in fruits such as grape C.V. Cabernet Sauvignon,[35] and have a biosynthetic origin. Little information concerning the biosynthesis of these compounds is available. Some data concerning 3-isopropyl-2-methoxypyrazine have been obtained using bacteria, *Pseudomonas taetrolens*[36] and *Pseudomonas perolens*.[37] The results indicate that valine combined with glycine or glycine with methionine is involved in the biosynthesis. These mechanisms are, however, still speculative.

The biosynthetic pathways of methoxy-pyrazines in plants apparently have not yet been investigated. The only findings available are preliminary, where a physiological study has indicated that, in grape C. V. Cabernet Sauvignon, 3-isobutyl-2-methoxypyrazine is synthesized at the onset of ripening, decreasing as the fruit matures. This compound is light sensitive, and is dependent on climate and vine.[38]

BIOGENERATION OF AROMA COMPOUNDS BY ACTION OF ENZYMES ON PRECURSORS

Action of Lyases on Substituted L-Cysteine Sulfoxides

The characteristic odor of *Allium* plants (onion, garlic, leek, chives, and shallot) is due to a large number of volatile sulfur compounds. These compounds are not present in the intact organ and are only released when the tissue is damaged. They are produced after cell disruption by the action of specific enzymes: C-S lyases or alliim lyases or alliinases on precursors, S-alkyl and S-alkenyl L-cysteine sulfoxides or alliins.[39] Thiosulfinate and sulfenic acid are produced by two different mechanisms. In garlic, the precursors have been identified as methyl, methyl propyl, propyl, methyl allyl, propyl allyl and allyl derivatives, the last compound being the most important. Under the action of alliinase, thiosulfinate is the first compound released, and symmetrical or mixed mono, di and trisulfides, responsible for the fresh garlic aroma, are generated by a non-enzymatic disproportionation reaction (Fig. 3).[40]

Methyl, propyl, and especially propenyl cysteine sulfoxides are the major flavor precursors in onion and leek, with only traces of the allyl derivative detected. Propenylcysteine sulfoxide, the main precursor of important onion aroma compounds, is cleaved to give the hypothetical unstable intermediate 1-propenyl sulfenic acid. One of the important characteristics of these vegetables, apart from aroma compounds, is the release of a lachrymatory factor during cell disruption. Thiopropanal-S-oxide produced from sulfenic acid is generally recognized as being the lacrymatory factor.[41] Symmetrical and mixed thiosulfinate are produced from sulfenic acids, resulting from precursor enzymatic splitting.

Glucosinolate Hydrolysis

A pungent, sulfuraceous odor develops during the slicing, crushing and cooking of plants belonging to the Cruciferae family. These include cabbage, brussels sprouts, cauliflower, broccoli, black and white mustard, and radish. The compounds responsible for this odor are generated during enzymatic and non-enzymatic reactions from glucosinolate precursors. Under the action of a thioglycoside glucohydrolase or myrosinase, D-glucose, sulfate, and isothiocyanate are released. An intermediate, hydroxamic acid, results through a Lossen rearrangement to isothiocyanate, thiocyanate, or cyanide according to the reaction conditions. When the enzymatic reaction occurs at acid pH (pH : 3.0), a cyanide and sulfur are produced. These compounds probably result from an acid catalyzed

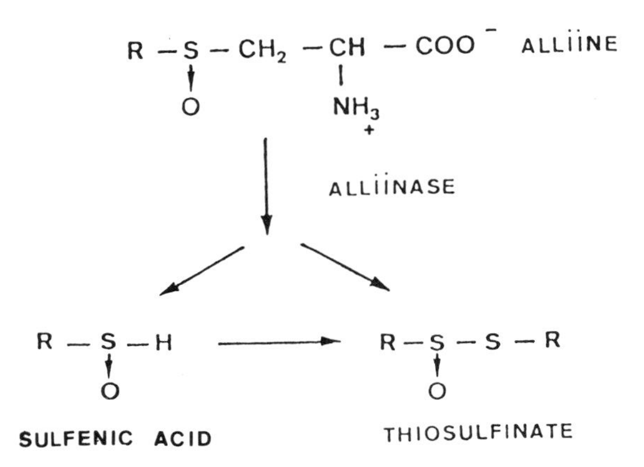

Figure 3. Action of alliinase on substituted L-Cysteine sulfoxides.

scission of the intermediate, although an enzymatic reaction is also possible. An enzymatic rearrangement is postulated to explain the thiocyanate formation, however, a direct formation from the precursor cannot be discarded.[42] In cabbage and cauliflower, 9-methyl cysteine sulfoxide is present as a precursor of dimethyl sulfide, whereas in radish volatiles, these compounds have been recognized as being produced by the transformation of isothiocyanates and nitriles.

Glycosidically Bound Aroma Compounds

Some volatile compounds are present in fruits and vegetables bound to glycosides, from which they can be released during fruit maturation, storage, pretreatment or processing by enzyme action or acid catalyzed reactions. The presence of these compounds was first reported by Bourquelot and Bridel,[43] who identified a geranyl-β-D-glucoside in *Pelargonium odoratissimum*. The increase of essential oil yield in peppermint during storage,[44] or after acidification of rose flower during steam distillation[45] has given rise to the hypothesis that bound monoterpenes exist in plants. The identification of terpenyl glycosides in rose petals[46] and in muscat grape[47] was an important and decisive step in this field. These bound compounds can be more important than free volatile compounds, for example, terpene derivatives in grape, apricot, mango and passion fruits (Table 1).

During the last 10 years, structural studies of glycosidically bound aroma compounds have attested to the recognized importance of structure with regard to enzymatic hydrolysis. Three different and complementary approaches have been used for this purpose: identification of compounds released from bound forms by acid and/or standard condition enzymatic hydrolysis; GC/MS study of

Table 1. Free and bound terpene alcohols in grapes, apricot, mango and passion fruit (mg/ kg, expressed as linalool)

Fruit	Free compounds	Bound compounds	Bound/free
Grapes (Muscat of Alexandria)	1.4	6.3	5.2
Apricot (Rouge du Roussillon)	1.0	5.2	5.2
Mango (African, ungrafted)	1.4	5.3	3.8
Purple passion fruit (Zimbabwe)	1.6	4.8	3.1
Yellow passion fruit (Cameroun)	1.5	1.5	1.0
Maracuja (Columbia)	0.5	1.1	2.2

derivatized compounds without previous separation; and MS, IR and NMR study of isolated compounds.

Structures. Information concerning the nature of the saccharides constituting the sugar moieties can be obtained by TLC and GC of trimethylsilyl derivatives of components released by acid hydrolysis of the whole heterosidic extract. For example, glucose, rhamnose and arabinose have been detected in passion fruit heterosidic extracts. GC and GC/MS of partially methylated alditol acetates have enabled the cyclic structure and the saccharidic units to be determined. In passion fruit, glucosides, rhamnofuranosyl glucosides (rutinosides), arabinopyranosyl glucosides (vicianosides), and glucopyranosyl glucosides (gentiobiosides) have been identified (Table 2).[48] Glycosidically bound compounds are generally present in plants as glucosides, diglycosides or triglycosides. A glucose unit is generally bound to the aglycone moity, and the terminal monosaccharide units which bind to glucose are shown in Figure 4. Trisaccharidic units, glucose-xylose-glucose,[49] glucose-glucose-glucose[50] have been also detected. Thymol and carvacrol galactosides were found in *Thymus vulgaris*.[51] Monoacetyl β-D-glucose bound to borneol has been detected in *Psilostrophe villosa*,[52] with malonated glycosides having been reported more recently.[53]

More than 200 different aglycones including aliphatic, terpene, and sesquiterpene alcohols, C_{13} norisoprenoïds, acids, hydroxyacids, and phenylpropane derivatives and related compounds have been identified. Aldehydes and phthalides are bound as acetates or hemiacetals. Polyols are of particular interest, these compounds being aroma compound precursors in their bound and free forms. They have been particularly studied in grape[54] and passion fruit.[5]

Gas chromatography coupled to negative chemical ionization mass spectrometry (NCI-MS) of TFA derivatives has been found to be an efficient method for the identification of glycosidically bound compounds.[55] The NCI-MS spectra of TFA derivatives of linalyl β-rutinoside are shown in Figure 5. The presence of specific fragment ions, molecular ion $M^- = 1038$, $[M\text{-}TFAO]^- = 925$, $[M + TFAO]^- = 1151$ in the experimental conditions used, unequivocally determined the mass of the glycoside. NCI-MS of glycosides gave some information concerning the sugar moiety. Ions at m/z 901, 882, 789 and 674 characterized

Table 2. Nature and composition of passion fruit glycosidic moieties

Glycoside	Relative molar percent
β-glucopyranoside	21
6-0-β-D-glycopyranosyl-β-D-glucopyranoside (gentiobioside)	28
6-0-α-L-arabinopyranosyl-β-D-glucopyranoside (vicianoside)	11
6-0-α-L-rhamnopyranosyl-β-D-glucopyranoside (rutinoside)	40

Chassagne et al. 1995 (48)

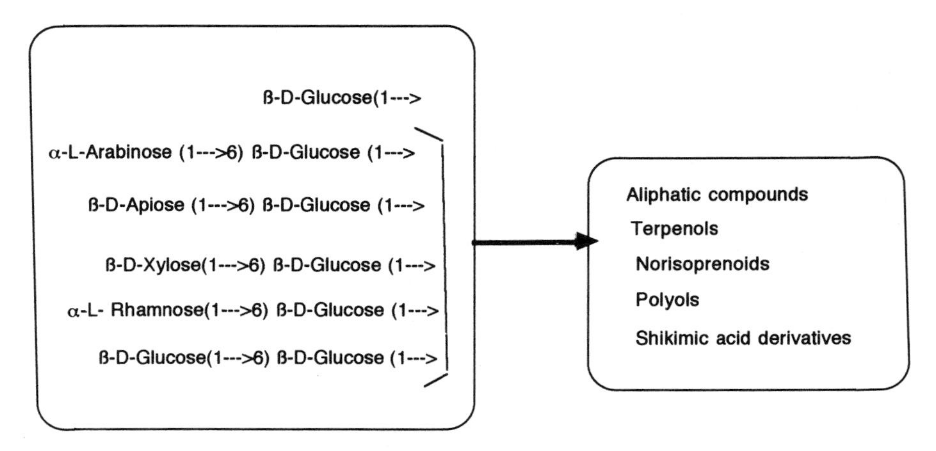

Figure 4. Main glycosidically bound aroma compounds found in fruits.

rutinosides, and an ion at m/z 451 was found to be specific to a glucose unit. Only tentative identification was obtained, however, authentic samples being needed for formal identification.

Several pure fractions have been obtained from grape and apricot heterosidic fractions by gel filtration on Fractogel TSK HW 40 S and preparative Over Pressured Liquid Chromatography (OPLC). Some information concerning the structure of isolated compounds has been obtained using soft ionization modes in mass spectrometry Field Desorption (FD), Fast Atom Bombardment (FAB), Chemical Ionization (CI) and Desorption-Chemical Ioinization (DCI) in positive and negative modes.[56] For example, the molecular weight and the arrangement of the constitutive units can be obtained by positive CI or DCI using ammonia as reagent gas, as in the case of terpene arabinoglucoside (Fig. 6): m/z 466 $(M + NH_4)^+$, m/z 334 $(\text{terpenylglucoside} + NH_4)^+$, m/z 312 $(\text{arabinosylglucose} + NH_4)^+$, m/z 180 $(\text{glucose} + NH_4)^+$, m/z 150 $(\text{arabinose} + NH_4)^+$. Using an authentic sample, this compound has been identified as linalyl arabinoglucoside. However, the distinction between isomers, aglycone, and saccharidic moities, could not be obtained. In contrast, isomers have been differentiated by tandem mass spectrometry, low energy Collisionally Activated Decomposition (CAD) spectra.[57] Use of deuterated ammonia in NIDCI has enabled the differentiation of isomeric compounds such as dienediol and linalol oxide glycosides. In the case of linalool oxide glucosides, the presence of four acidic protons gives a shift of 3 mass units, from m/z 331 to m/z 334, whereas a four mass unit shift, from m/z 331 to m/z 335, is obtained for dienediol glycosides due to the presence of five acidic protons.

6–0-α-L-Arabinopyranosyl-β-D-glucopyranosides (vicianosides) of linalool, benzylalcohol and 3-methylbut-2-en-1-ol have been isolated from passion fruit using partial enzymatic hydrolysis by the action of a glycosidase

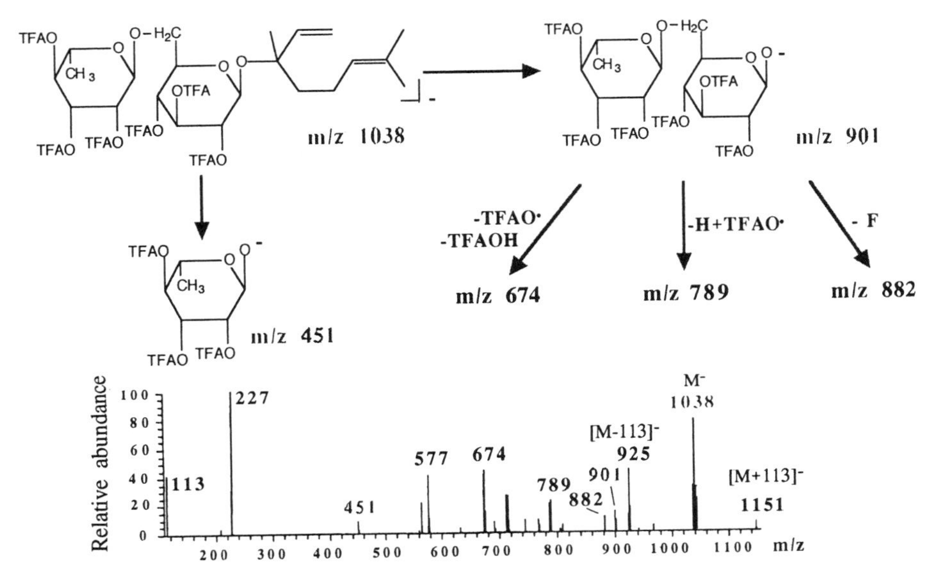

Figure 5. NCI mass spectrum of trifluroacetylated linalyl rutinoside.

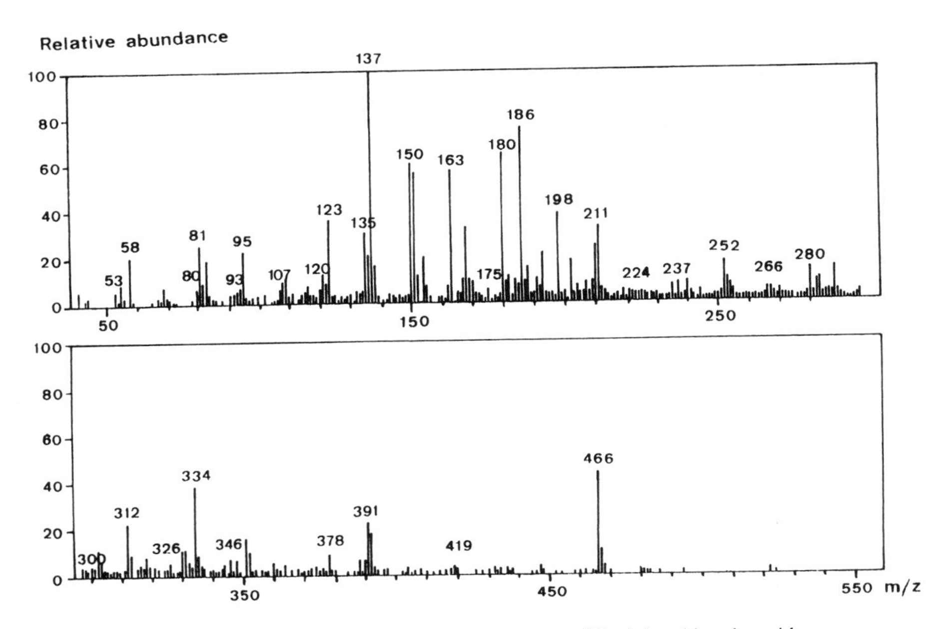

Figure 6. DCI in positive mode mass spectrum of linalyl arabinoglucoside.

preparation devoid of arabinopyranosidase activity, adsorption chromatography on XAD-2 resin, and semi-preparative HPLC on RP 18 phase. Their structures have been established by ^{1}HNMR and EIMS and NCIMS.[58]

Hydrolysis. Volatile compounds can be released from glycosidic fractions by acid hydrolysis at the natural pH of fruits and derived food products. Modifications in typical aroma have been reported during the aging of wine produced from non-floral cultivars.[59] The flavor modification occurring during storage of passion fruit juices and concentrates has been related to the acid hydrolysis of glycosides, more particularly linalyl glucoside and rearrangement reactions of polyhydroxylated compounds. These reactions increase at high temperatures, as for instance during heat treatment of mango pulp. In the latter, the tremendous increase in α-terpineol content, from 1800 μg/kg to 55000 μg/kg, can be explained by the hydrolysis of α-terpineyl glucoside, arabinoglucoside and rutinoside identified in this fruit.[60] Whereas vanillin is the most characteristic volatile compound in vanilla extract, this compound is only present in small quantities in mature green vanilla pods. Glucovanillin, present in the fresh pod, is hydrolyzed by endogenous β-glucosidase during the curing process, and vanillin is released.[61]

Aryan et al.[62] have pointed out the problem related to the action of plant enzymes on glycosides. *Vitis vinifera* endogenous glycosidases have an optimum pH at 5.0, and are strongly inhibited by the presence of glucose, 50 % at 50 mM. Almond and papaya glucosidases have the same optimum pH as grape enzymes, and are also glucose inhibited, 10 % at 50 mM for almond[63] and 50 % at 10 mM for papaya.[64] Exogenous microbial glucosidases are also inhibited by glucose, with the exception of *Saccharomyces cerevisiae* enzyme.[65] Other characteristics of these enzymes are their ethanol tolerance[66] and their substrate specificity.

According to Gunata et al.,[67] glycoside hydrolysis is a sequential process, the terminal unit of the saccharidic sequence being released by the action of a specific glycosidase, and the corresponding glucoside then being produced. Aglycone is then liberated by the action of a β-glucosidase. However, an endo β-glucosidase, not glucose inhibited, has been isolated and used in immobilized form for the flavor enhancement of muscat wine and passion fruit juice.[68] Aroma compounds can be released from the glycosidic fraction isolated from fruits by the use of *Aspergillus niger* preparations, containing several glycosidase activities. However, the efficiency in the release of monoterpene alcohols from passion fruit glycosides varies considerably from one preparation to another.[48]

Oxidative Pathways

Formation of Aldehydes and Alcohols from Unsaturated Fatty Acids. Aliphatic aldehydes and alcohols have been isolated from the volatile fraction of most fruits and vegetables. Among them are C_6 and C_9 compounds: hexanal,

(*Z*)-3-hexenal, (*E*)-2-hexenal (leaf aldehyde), (*Z*)-3-hexenol (leaf alcohol), (*Z*)-3-nonenal, (*Z,Z*)-3,6-nonadienal. All are generated from polyunsaturated fatty acids, linoleic and linolenic acids (Fig. 7). Four enzymes, acyl hydrolase, lipoxygenase, hydroperoxide cleavage enzyme, and alcohol dehydrogenase are involved in this pathway.[69–71] The presence of an hydroperoxide isomerase is sometimes involved.[69] Polyunsaturated fatty acids, the main constituents of plant membranes, are released by the action of an acyl hydrolase, for example, during wounding or tissue disruption. Lipoxygenase or linoleate oxygen oxidoreductase, first isolated from soya bean, then from different leguminous seeds, is present in numerous plants.[72] Several isoforms have been identified, for example six isoenzymes have been isolated from soya bean seeds, leaves and cotyledons.[72, 73]

Hydroperoxides are the primary products resulting from the catalyzed oxidation, by molecular oxygen, of fatty acids possessing a (*Z,Z*)-1,4-pentadiene unit. The reaction mechanism was described by Hamberg and Sammuelson in 1967.[75] 13-(S)-Hydroperoxides and 9-(R)-hydroperoxides are the main compounds initially produced by the reaction. The ratio of these compounds is dependent mainly

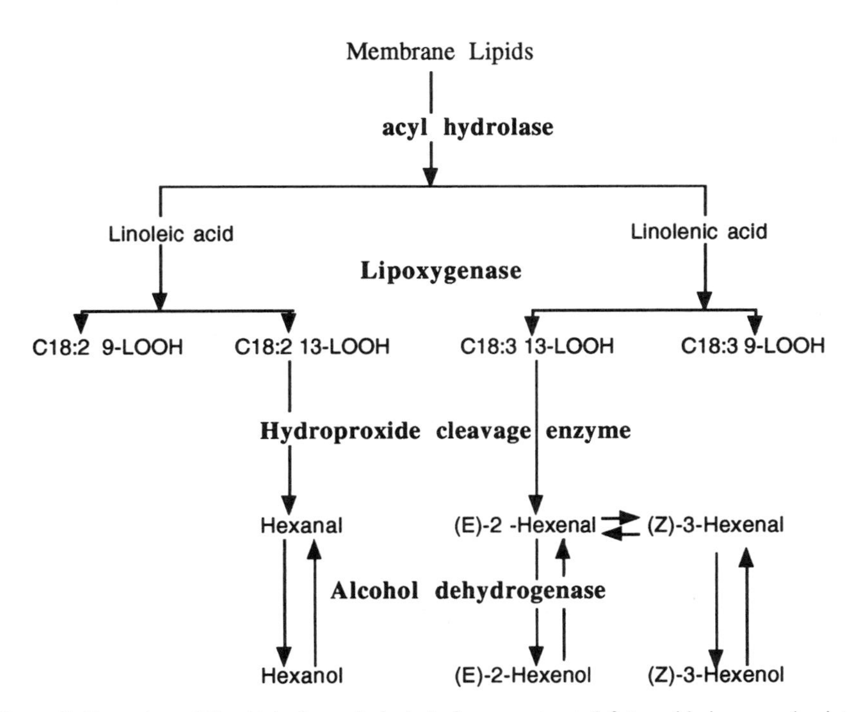

Figure 7. Formation of C_6 aldehydes and alcohols from unsaturated fatty acids in grape (variety Carignane).

on the enzyme specificity, although it can be modified by the action of pH, Ca++ ions, and temperature.[76] Furthermore, isomerization can take place during experiments.[77] For example, 9-(R)-hyperoxide is the most abundant compound produced by tomato or potato lipoxygenases, whereas 13-(S)-hydroperoxide represents 90% of the mixture resulting from the action of grape lipoxygenase.[78]

Hydroperoxide cleavage enzyme, improperly designated as hydroperoxide lyase, was first identified in watermelon seedlings,[79] and then in etiolated and green tissues.[80] It is a membrane bound enzyme located in chloroplasts and microsomes. This enzyme cleaves the carbon chain between carbon 12 and 13 or 9 and 10 to produce volatile aldehydes, hexanal or (*Z*)-3-nonanal from linoleic acid hydroperoxides and a non-volatile oxo acid. Three types of hydroperoxide cleavage enzymes can be distinguished according to the substrate specificity.[71] Some of them are non-specific and cleave both 9- and 13-hydroperoxides, *e.g.* cucumber fruit enzyme.[80] A second type is 9-hydroperoxide specific, *e.g.* pear hydroperoxide cleavage enzyme.[81] Tomato fruit and grape enzymes cleave exclusively 13-hydroperoxides.[82,83] One particular hydroperoxide cleavage enzyme isolated from mushrooms is involved in the 1-octen-3-ol formation and is specific for 10-hydroperoxides.[84]

The identification among fruit and vegetable volatile compounds of alcohols having the same carbon chain as aldehydes indicates the presence in plant tissues of an alcohol dehydrogenase. This enzyme, initially detected in pea,[84] was ultimately isolated and characterized from such plants as cucumber,[86] apple,[87] strawberry,[88] tomato,[89,90] and grape.[91,92] In comparison with most plant alcohol dehydrogenases studied, grape enzyme has a broad substrate specificity (Table 3), although some aromatic aldehydes and alcohols are not substrates. Aldehydes are generally better substrates than the corresponding alcohols, as indicated by catalytic efficiency values.

Co-Oxidation Reactions. The co-oxidation of β-carotene or crocine, enzymatic bleaching by lipoxygenase in the presence of unsaturated fatty acid, has been well studied in vitro.[93–96] Several volatile compounds are produced during enzymatic degradation of β-carotene by soya bean lipoxygenase: 2,6,6-trimethylcyclohexanone, 2-hydroxy-2,6,6-trimethylcyclohexanone, β-ionone, 5,6-epoxy-β-ionone, and dihydroactinidiolide (DHA).[97–99] As indicated in Figure 8, β-ionone and DHA are the first compounds produced, and 5,6 epoxy-β-ionone can be the result of the cleavage of luteochrome and of β-ionone oxidation. Other enzymes such as polyphenol oxidase,[100] lactoperoxidase,[101] and xanthine oxidase,[102] are able to promote degradative oxidation of β-carotene.

Carotenoid Cleavage. While little is known about the *in vivo* cleavage of carotenoids in plants, and the mechanisms proposed are somewhat speculative, several points have nevertheless been made. The first is that the observed decrease in carotenoid concentration during ripening is associated with an

Table 3. Substrate specificity of grape alcohol dehydrogenase

Substrate	K'M (mM)	V'max (mM x min^{-1})	V'max/ K'M x 10^3 (min^{-1})
Ethanol	10.0	90	9
n-Hexanol	9.1	9.6	1.05
(*E*)-2-Hexenal	6.2	106	17.1
Benzyl alcohol	not substrate	not substrate	not substrate
Acetaldehyde	1.02	590	578
Hexanal	6.3	370	58.7
(*E*)-2 Hexenal	3.1	37	11.9
Benzaldehyde	0.23	1.2	5.2

Molina et al. 1987 (92)

increase in C_{13} norisoprenoid concentration.[103] Additional carotenoid fragments, C_{27} apocarotenoids and C_{14} diapocarotenoids have been identified.[104] In addition to C_{13} norisoprenoid compounds, C_{10} and C_{15} carotenoid metabolites have also been characterized as constituents of saffron, quince and starfruit (Fig.9).[104–106] Finally, in all cases examined, the stereochemistry of 5-hydroxyl is the same in the C_{13} norisoprenoid fragment, and the parent carotenoid, neoxanthin.[104]

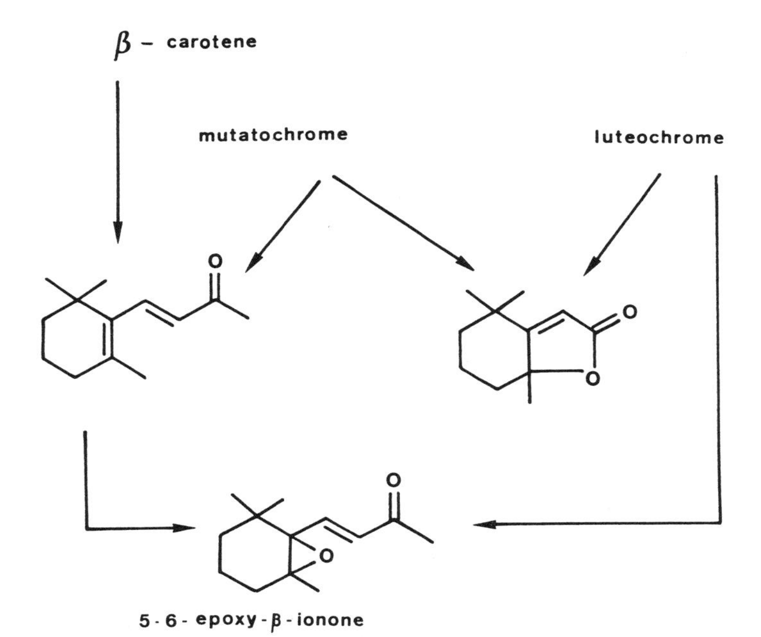

Figure 8. Formation of volatile compounds during enzymatic degradation of β-carotene.

According to Winterhalter and Schreier[104] the regioselective cleavage of starfruit carotenoids, lutein, violaxanthin, neoxanthin, catalyzed by a 9–10 (9'-10') dioxygenase, gives a C_{13} end group and C_{27} fragment which is ultimately cleaved by a second dioxygenase to C_{13} and C_{14} compounds. A highly specific 11–12 (11'-12') dioxygenase cleaves epoxycarotenoids, neoxanthin or violanxanthin, to produce an abcissic acid (ABA) derivative, xanthoxin, according to the proposed biosynthetic pathway called the "apo-carotenoid" pathway. Xanthoxin is then metabolized to ABA via ABA aldehyde under the action of xanthoxin oxidase and ABA-aldehyde oxidase.[104] An alternative pathway involves the action of cytochrome P450 monooxygenase on ABA alcohol.[104]

Compounds possessing the ABA structure represent 57% of starfruit free compounds and 32% of the glycosidically bound compounds in this fruit.[104] 2,7-Dimethyl-8-hydroxy-4(*E*),6(*E*)-octadienoic acid, isolated for the first time from quince,[107] derives from the central portion of the carotenoid chain. The primary degradation products obtained after carotenoid cleavage, with the exception of α and β-ionone generated from α and β-carotene, are generally odorless, and several steps involving enzymatic and acid catalyzed reactions are required for the biosynthesis of aroma compounds. For example, in the postulated pathway of β-damascenone from neoxanthin,[104] the first intermediate, the grasshopper ketone, is reduced to a triol, and an acetylenic diol is also formed. These two compounds yield β-damascenone and 3-hydroxy-β-damascone. ABA was found, using co-oxidation *in vitro* in the presence of soya bean lipoxygenase, to be a precursor of dehydrovomifoliol.[108] It can be assumed that oxidation occurring *in vivo* produces this compound, and several aroma precursors such as 7,8-dihydrovomifoliol.

Acid hydrolysis has revealed norisoprenoid derivatives to be important flavor precursors, in such fruits as quince and starfruit, particularly during processing or storage. For example, β-ionyl glucoside is the precursor of

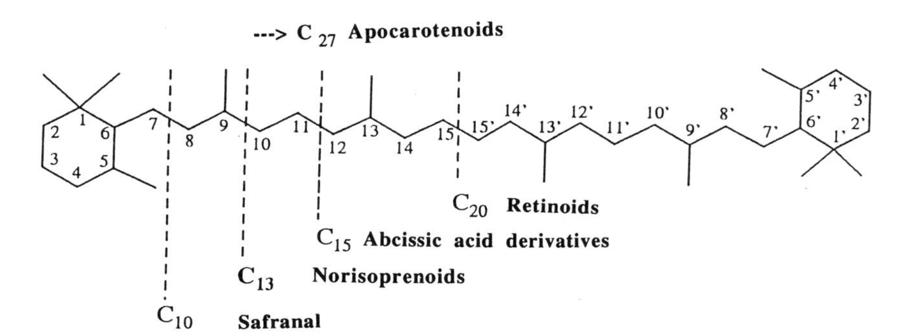

Figure 9. Regioselective cleavage of carotenoids under the action of 9–10 and 11–12 dioxygenase [adapted from Winterhalter and Schreier (104)].

isomeric megatisma-4,6,8, trienes, and it plays an important role in purple passion fruit aroma.[109,110] 7,8-Dihydrovomifoliol glycoside[109] is considered as the precursor of theaspirones, key components in black tea aroma.[110] Glycosidically bound 2,7-dimethyl-8-hydroxy-*4(E)-6(E)* octadienoic acid is the precursor of marmelo lactones isolated from quince and starfruit.[107,108]

CONCLUSION

Our knowledge concerning the mechanisms involved in aroma generation have significantly progressed during the last 20 years, mainly for the compounds obtained by hydrolysis or cleavage reactions such as sulfur compounds in *Allium* or C6 and C9 aldehydes produced from unsaturated fatty acids. However, progress remains to be realized in several fields. Proposed biosynthetic pathways of lactones and pyrazines in food are generally speculative. The use of cell cultures or of fungi can certainly be useful for establishing these. Another field needing study remains the glycosidically bound compounds and more particularly the compounds produced by the *in vivo* cleavage of carotenoids. Two different approaches must be used to increase our knowledge of plant volatiles and their biogenesis - structure determination and isolation of enzymes involved in the biogeneration process. The first approach is presently being developed in our laboratory by the use of immunoaffinity chromatography for the isolation of free and bound compounds resulting from the carotenoid cleavage.

REFERENCES

1. SANDERSON, G.W., GRAHAM, H.N. 1973. On the formation of black tea aroma. J. Agric. Food Chem. 21: 576–585.
2. DRAWERT, F. 1985. Formation des arômes à différents stades de l'évolution du fruit. Enzymes intervenant dans cette formation. In: Facteurs et Régulation de la Maturation des Fruits. Colloques Internationaux CNRS n° 278, Paris, pp. 309–319.
3. TRESSL, R., HOLZER, M., APETZ, M. 1985. Biogenesis of volatile in fruit and vegetables. In: Aroma Research (H. Maarse, P.J. Groenen, eds.), Pudoc, Wageningen, pp. 41–62.
4. STAHL-BISKUP, E., INTERT, F., HOLTHUIJZEN, J., STENGELE, M., SCHULTZ, G. 1993. Glycosidically bound volatiles. A Review 1986–1991. Flav. Fragr. J. 8: 61–80.
5. PAILLARD, N. 1985. Biosynthèse de produits volatils à partir d'acides gras, par des disques de tissu de pomme. In: Facteurs et Régulation de la Maturation des Fruits. Colloques internationaux CNRS n° 278, Paris, pp. 321–326.
6. MYERS, M.J., ISSEMBERG, P., WICK E. 1970. Leucine as a precursor of isoamyl alcohol and isoamyl acetate, volatile aroma constituents of banana fruit discs. Phytochemistry 9: 1693–1700.
7. TRESSL, R., DRAWERT F. 1973. Biogenesis of banana volatiles. J. Agric. Food Chem. 21: 560–565.
8. YAMASHITA, I., NEMOTO, Y., YOSHIKAWA, S. 1975. Formation of volatile esters in strawberries. Agr. Biol. Chem. 39: 2303–2307.

9. De POOTER, H.L., DIRINCK, P.J., WILLAERT, G.A., SCHAMP, N.M. 1981. Metabolism of propionic acid by golden delicious apples. Phytochemistry 20: 2135–2138.
10. STUMPF, P.K., CONN, E.E. 1980. The Biochemistry of Plants. A Comprehensive Treatise. Vol 4. Lipids Structure and Function (P.K. Stumpf ed.), Acad. Press, New York. 693pp.
11. YU, M.H., OLSON, L.E., SALUNKE, D.K. 1967. Precursors of volatile components in tomato fruit. Phytochemistry 6: 1457–1465.
12. RECH, J., CROUZET, J. 1974. Partial purification and initial sudies of the tomato L.alanine : 2-oxoglutarate aminotransferase. Biochim. Biophys. Acta. 350: 392–399.
13. DRAWERT F., BERGER, R.I. 1981. Possibilities of the biotechnological production of aroma substances by plant tissue culture. In: Bioflavor'81 (P. Schreier, ed.), W. de Gruyter, Berlin, pp. 509–527.
14. PEREZ, A., RIOS, J.J.,SANS, C., OLIAS, T.M. 1992. Aroma components and free amino acids in variety Chandler during ripening. J. Agric. Food Chem. 40: 2232–2235.
15. CROUZET J., NGALANI J., SIGNORET, A. 1980. Formation de composés volatils par action d'extraits végétaux sur des précurseurs, régénération enzymatique de l'arôme des fruits. Bull. Soc. Chem. II:108–109.
16. TRESSL, R., ALBRECHT, W. 1986. Biogenesis of aroma componds through acyl pathways. In: Biogeneration of Aromas (T.H. Parliment R. Croteau eds.), Am. Chem. Soc., Washington D.C., pp. 114–133.
17. WILLAERT, G.A., DIRINCK, P.J., POOTER, H.L., SCHAMP, N.M. 1983. Objective measurement of aroma quality of Golden Delicious apples as a function of controlled-atmosphere storage time. J. Agric. Food Chem. 31: 809–813.
18. BERGER, R.G., DRAWERT, F. 1984. Changes in the composition of volatiles by post-harvest application of alcohols to red delicious apples. J. Agric. Food Chem. 35: 1318–1325.
19. BARTLEY, I.M., STOKER, P.G., MARTIN, A.D.E., HATFIELD, S.G.S, KNEE, M. 1985. Synthesis of aroma compounds by apples supplied with alcohols and methyl esters of fatty acids. J. Agric. Food Chem. 36: 567–574.
20. BERGER, R.G. 1990. The biogenesis of fruit flavors. A continuing story. Perfumer and Flavorist. 15: 33–39.
21. GUICHARD, E., SOUTY, M. 1988. Comparison of the quantities of aroma compounds found in fresh apricot (*Prunus armeniaca*) from six varieties. Z. Lebensm. Unters. Forsch. 186: 301–307.
22. ENGEL, K.H., TRESSL, R. 1983. Studies on volatile components of two mango varieties., J. Agric. Food Chem. 31: 796–801.
23. MURRAY, K.E.J., SHEPTON, J., WHITFIELD, F. B. 1972. The chemistry of food flavor. I.Volatile constituents of passion fruit, *Passiflora edulis*. Aust. J. Chem. 25: 1921–1933.
24. SCHEYEN L., DIRINCK P., SANDRA, P., SCHAMP, N. 1979. Flavor analysis of quince. J. Agric. Food Chem. 27: 872–876.
25. DARTEY C.K., KINSELLA, J.E. 1973. Oxidation of sodium U-^{14}C palmitate into carbonyl compounds by *Penicillium roqueforti* spores. J. Agric. Food Chem. 21, 721–726.
26. DUFOSSE, L., LATRASSE, A., SPINNLER, H. E. 1994. Production de lactones par les microorganismes. Sci. Aliments 14: 17–50.
27. FARBOOD, M.I., WILLIS, B.J. 1985. Production of g-decalactone. American patent n° 4.560.656.
28. CHEETAM, P.S.J., MAUME, K.A., de ROAY, J.F. 1988. Production of lactones. European Patent EP 0258.993.
29. CARDILLO, R., FUGANTI, C., BARBENI, M., CABELLA, P., GUARDA, P.A., GIANNA, A. 1991. Procédé de production microbiologique des gamma et delta lactones. European Patent EP 0412.880.
30. ENGEL, K.H., HEIDLAS, J., ALBRECHT, W., TRESSL, R. 1989. Biosynthesis of chiral flavor and aroma compounds in plants and microorganisms. In: Flavor Chemistry. Trends and

Developments.(R. Teranishi, R.G. Buttery, F. Shahidi, eds.), Am. Chem. Soc., Washington, D.C., pp. 8–22.

31. TRESSL, R., ENGEL, K H., ALBRECHT, W., BILLE-ABDULLAH, H. 1985. Analysis of chiral aroma components in trace amount. In: Characterization and Measurement of Flavor Compounds.(D.D. Bills, C.J. Mussinan eds.), Am. Chem. Soc., Washington, D.C., pp. 43–60.
32. MAGA, J. 1982. Pyrazine in flavour. In: Food Flavours. Part A. Introduction, (I.D. Norton, A.J. Macleod, eds.), Elsevier. Amsterdam, pp. 283–321.
33. MURRAY, K.E., SHIPTON, J., WHITEFIELD, F.B. 1970. 2-Methoxypyrazines and the flavor of green peas (*Pisum sativum*). Chem. Ind.(London), pp. 897–898
34. BUTTERY, R.G., SEIFERT, R.M., GUADAGNI, D.G., LING, L.C. 1969. Charaterization of some constituents of bell peppers. J. Agric. Food Chem. 17: 1322–1327.
35. BAYONOVE, C., CORDONNIER, R., DUBOIS, P. 1975. Etude d'une fraction caractéristique de l'arôme du raisin de la variété Cabernet-Sauvignon; mise en évidence de la 2-methoxy-3-isobutylpyrazine. C. R. Acad. Sci., Series D. 281: 75–78.
36. GALLOIS, A., KERGOMARD, A., ADDA, J. 1988. Study of the biosynthesis of 3-isopropyl-2-methoxypyrazine by *Pseudomonas taetrolens*. Food Chem. 28: 299–309.
37. CHENG, T.B., REINECCIUS, G.A., BJORKLUND, J.A., LEETE, E. 1991. Biosynthesis of 2-methoxy-3 isopropylpyrazine in *Pseudomonas perolens* J. Agric. Food Chem. 39: 1009–1012.
38. NOBLE, A.C., ELLIOT-FISK, D.L., ALLEN, M.S. 1995. Vegetable flavor and methoxypyrazines in Cabernet Sauvignon In: Fruit Flavors, Biogenesis, Characterization and Authentication, (R.L. roussef, M.M. Leahy, eds.), Am. Chem. Soc., Washington, DC, pp. 226–234.
39. BLOCK, E. 1997. Organosulfur and selenium phytochemicals in garlic, onion and other genus *Allium* plants. In: Recent Advances in Phytochemistry Vol 31
40. SHANKARANARAYANA, M.L., RAGHAVAN, B., ABRAHAM, K.O., NATARAJAN, C.P. 1982. Sulfur compounds in flavours. In: Food Flavours Part A. Introduction, (D. Norton, A.J. Macleod, eds.), Elsevier, Amsterdam, pp. 169–281.
41. BRODNITZ, M.H., PASCALE, J.V. 1971. Thiopropanal-S-oxide a lachrymatory factor in onion. J. Agric. Food Chem. 19: 269–272.
42. FENWICK, G.R., HEANY, R.K., MULLIN, J.W. 1982. Glucosinolates and their breakdown products in food and food plants. In: CRC Critical Reviews in Food Science and Nutrition, (Van Etten, C.H. Referee), pp. 123–201.
43. BOURQUELOT, E., BRIDEL, M. 1913. Synthèse du geranylglucoside β à l'aide de l'emulsine; sa présence dans les végétaux. C.R. Acad. Sci. 157: 72–74.
44. ESDORN, I. 1950. Etheral content of wilting plants. Pharmazie 5: 481–488.
45. NAVES, Y.R. 1974. Technologie et Chimie des Parfums Naturels. Masson, Paris, 535 pp.
46. FRANCIS, M.J.U., ALLCOCK, C.1969. Geraniol β-D-glucoside; occurrence and synthesis in rose flowers. Phytochemistry 8: 1339–1347.
47. CORDONNIER, R., BAYONOVE, C. 1974. Mise en évidence dans la baie de raisin, variété muscat d'Alexandrie, de monoterpenes liés, révélables par une ou plusieurs enzymes de fruit. C.R. Acad. Sci. serie D, 278: 3387–3390.
48. CHASSAGNE, D., BAYONOVE, C., CROUZET, J., BAUMES, R. 1995. Formation of aroma by enzymic hydrolysis of glucosidically bound components of passion fruit In: Bioflavour 1995, (P. Etievant, P. Schreier, eds.), INRA Paris, pp. 217–222.
49. HERDERIC, M., FESER, W., SCHREIER, P. 1992. Vomifoliol 9-O-β-glucopyranosyl-4O-β-D-xylopyranosyl-6-o-β-glucopyranoside: a trisaccharide glycoside from apple fruit. Phytochemistry 31: 895–897.
50. OKAMURA, N., YAGI, A., NISHIOKA, I. 1981. Study on the constituents of *Zizyphi fructus*. V. Structures of glycosides of benzyl alcohol, vomifiliol and naringenin. Chem. Pharm. Bull. 29: 3507–3514.
51. SKOPP, K., HÖrster, H. 1976. An Zucker gebundene reguläre monotermene. Terl I Thymol und carvacrolglykoside in *Thymus vulgaris*, Planta Med. 29: 208–215.

52. BOHLMANN, F., JAKUPOVIC, J., DUTTA, L., GOODMAN, M. 1980. Neue, abgewandelte pseudoguajanolide ans *Psilostrophe villosa*. Phytochemistry 19: 1491–1494.
53. MOON, J.H., WATANABE, N., SAKATA, K., INAGAKI, J., YAGI, A., INA, K., LUO, S. 1994. Linalyl β-D-glucopyranoside and its 6'-O-malonate as aroma precursors from *Jasminum sambuc*. Phytochemistry 36: 1435–1437.
54. WILLIAMS, P.J., STRAUSS, C.R., WILSON, B. 1980. Hydroxylated linalool derivatives as precursors of volatile monoterpenes of muscat grape. J. Agric. Food Chem. 28: 766–771.
55. CHASSAGNE, D., CROUZET J., BAUMES, R., LEPOUTRE, J.P., BAYONOVE, C.L. 1995. Determination of trifluoracetylated glycosides by gas chromatography coupled to methane negative chemical ionization mass spectrometry. J. Chromatogr. A. 694: 441–451.
56. SALLES, C., ESSAIED, H., CHALIER, P., JALLAGEAS, J.C., CROUZET J. 1988. Evidence and characterization of glucosidically bound volatile components in fruits. In: Bioflavor'87, (P. Schreier, ed.), W. de Gruyter, Berlin, pp. 145–160.
57. COLE, R.B., TABET J.C., SALLES, C., JALLAGEAS, J.C., CROUZET, J. 1989. Structural "memory effects" influencing decomposition of glucose alkoxide onions produced from monoterpene glycoside isomers in tandem mass spectometry. Rapid Commun. Mass Spectrom. 3: 59–63.
58. CHASAGNE, D., CROUZET J., BAYONOVE, C., BRILLOUET, J.M., BAUMES, R. 1996. 6-O-α-L-arabinopyranosyl-β-D-glucopyranosides as aroma precursors from passion fruit. Phytochemistry 41: 1497–1500.
59. WILLIAMS, P.J., SEFTON, M.A., FRANCIS, I.L. 1992. Glycosidic precursors of varietal grape and wine flavor. In: Flavor Precursors. Thermal and Enzymatic Conversions. (R. Teranishi, G.R., Takeoka, M. Güntert, eds.), Am. Chem. Soc., Washington, D.C., pp. 74–86.
60. SAKHO, M., CROUZET, J., SECK, S. 1985. Evolution des composés volatils de la mangue au cours du chauffage. Lebensm. Wissen. u-Technol. 18: 89–93.
61. ARANA, F. 1943. Action of β-glucosidase in the curing of vanilla. Food Research 8: 343–351.
62. ARYAN, A.P., WILSON, B., STRAUSS, C.R., WILLIAMS, P.J. 1987. Properties of glycosidases of *Vitis vinifera* and a comparison of β-glucosidase activity with that of exogenous enzymes. Assesment of possible applications in enology. Am. J. Enol. Vitic. 38: 182–188.
63. GUNATA, Y.Z., BAYONOVE, C.L., BAUMES, R.L., CORDONNIER, R.E. 1985. The aroma of grapes. I. Extraction and determination of free and glycosidically bound fractions of some grape aroma components. J. Chromatogr. 331: 83–90.
64. HARTMANN-SCHREIER, J., SCHREIER P. 1987. Properties of β-glucosidase from *Carica papaya* fruit. Food Chem. 26: 201–212.
65. DUBOURDIEU, D., DARRIET, P., OLLIVIER, C., BOIDRON, J.N., RIBEREAU-GAYON, P. 1988. Role de la levure *Saccharomyces cerevisiae* dans l'hydrolyse enzymatique des hétérosides terpéniques du jus de raisin. C.R. Acad. Sci., Série III, 306: 489–493.
66. GUNATA, Z., DUGELAY, I., SAPIS, J.C., BAUMES, R., BAYONOVE, C. 1993. Role of enzyme in the use of flavour potential from grape glycosides in winemaking In: Progress in Flavour Precursor Studies - Analysis, Generation, Biotechnology. (Schreier, P., Winterhalter, P. eds.), Allured Publ., Carol Stream, pp. 219–234.
67. GUNATA, Z., BITTEUR, S., BRILLOUET, J.M., BAYONOVE, C., CORDONNIER, R. 1988. Sequential enzymic hydrolysis of potentially aromatic glycosides from grape. Carbohydr. Res. 184: 139–149.
68. SHOSEYOV, O., BRAVAS, B.A., SIEGEL, D., GOLDMAN, A., COHEN, S., SHOSEYOV, L., IKAN, R. 1990. Immobilized endo-β-glucosidase enriches flavor of wine and passion fruit juice. J. Agric. Food Chem. 38: 1387–1390.
69. GAILLARD, T. 1979. The enzymic degradation of membrane lipids in higher plants. In: Advances in the Biochemistry and Physiology of Plant Lipids, (L. Å. Applequist, C. Ligenberg, eds.), Elsevier/North Holland Biomedical Press, Amsterdam. pp. 121–131.
70. WHITAKER, J.R. 1991. Lipoxygenases. In: Oxidative Enzymes in Foods, (D.S. Robinson, A.M. Eskin, eds.), Elsevier Applied Science. Amsterdam, pp. 175–215.

71. HATANAKA, A., KAJIWARA, T., SEKIYA, J. 1986. Fatty acid hydroperoxide lyase in plant tissues. In: Biogeneration of Aromas, (T.H. Parliment, R. Croteau, eds.), Am. Chem. Soc. Washington, D.C., pp. 167–175.
72. SEKIYA, J., KAJIWARA, T., MUNECHIKA, K., HATANAKA, A.1983. Distribution of lipoxygenase and hydroperoxide lyase in the leaves of various plant species. Phytochemistry 22: 1867–1869.
73. KATO, T., OHTA, H., TANAKA, K., SHIBATA, D. 1992. Appearance of a new lipoxygenase in soybean cotyledons after germination and evidence for expression of a major new lipoxygenase gene. Plant Physiol. 98: 324–330.
74. KATO, T., TERAO J., SHIBATA, D.1992. Partial amino acid sequences of soybean lipoxygenase L-G isolated from cotyledons. Biosc. Biotech. Biochem. 56: 1344.
75. HAMBERG, M., SAMUELSON B. 1967. On the specificity of the oxygenation of unsatured fatty acids catalysed by soybean lipoxidase. J. Biol. Chem. 242: 5329–5335.
76. CHRISTOPHER J.P., PISTORIUS, E.K., REGNIER, F.E., AXELROD, B. 1972. Factor influencing the positional specificity of the soybean lipoxygenase. Biochem. Biophys. Acta 289: 82–87.
77. CHAN, H.W.S., CISTARAS, C.T., PRESCOTT, F.A.A., SWOBODA, P.A.T. 1975. Specificity of lipoxygenases: thermal isomerisation of linoleate hydroperoxides, a phenomenon affecting the determination of isomeric ratios. Biochem. Biophys. Acta 398: 347–350.
78. CAYREL, A., CROUZET J., CHAN, H.W.S., PRICE, K.R. 1983. Evidence of the occurrence of lipoxygenase activity in grapes (variety Carignane). Am. J. Enol. Vitic. 34: 77–82.
79. VICK B.A., ZIMMERMAN, D.C. 1976. Lipoxygenase and hydroperoxide lyase in germinating water-melon seedlings. Plant Physiol. 57: 780–788.
80. GAILLARD, T., PHILLIPS, D.R., REYNOLDS, J. 1976. The formation of *cis*-3-nonenal, *trans*-2-nonenal and hexanal from linoleic acid hydroperoxide isomers by hydroperoxide cleavage enzyme system in cucumber (*Cucumis sativus*) fruit. Biochem. Biophys. Acta 441: 181–192.
81. KIM, I.S., GROSCH, W. 1981. Partial purification and properties of a hydroperoxide lyase from fruits of pear. J. Agric. Food Chem. 24: 938–942.
82. SCHRIER, P., LORENTZ, G. 1981. Formation of "green-grassy" notes in plant tissues : characterization of the tomato enzyme system. In: Flavour'81, (P. Schreier, ed.), W. de Gruyter, Berlin, pp. 495–507.
83. VALENTIN, G., CROUZET J. 1990. L'enzyme de clivage des peroxides du raisin. In: Actualités Oenologiques 1989, (P. Ribereau-Gayon, A. Lonvaud, eds.), Dunod, Paris, pp. 133–138.
84. WURZENBERGER, M.H., GROSCH, W. 1984. The formation of 1-octen-3-ol from the 10-hydroperoxide isomer of linoleic acid by a hydroperoxide lyase in mushrooms (*Psalliota bispora*). Biochem. Biophys. Acta 794: 25–30.
85. ERICKSSON, C.E. 1968. Alcohol NAD oxidoreductase E.C. 1–1–1–1- from peas. J. Food Sci. 33: 525–532.
86. LEBLOVA, S., MANCAL, P. 1975. Characterization of plant alcohol dehydrogenase. Physiol. Plant. 34: 246–249.
87. BARTLEY, I.M., HUNDKEY, S.J. 1980. Alcohol dehydrogenase of apple. J. Exp. Botany 31: 449–459.
88. YAMASHITA, I., NEMOTO, Y., YOSHIKAWA, S. 1976. NAD dependent alcohol dehydrogenase and NADP dependent alcohol dehydrogenase from strawberry seeds. Agric. Biol. Chem. 40: 2231–2235.
89. NICOLAS, M., CROUZET J. 1980. Purification de l'alcool desydrogensase de tomate par chromatographie d'affinité. Phytochemistry. 19: 15–18.
90. NICOLAS, M., CROUZET, J., AYMARD, C. 1980. L'alcool desydrogensase de tomate (*Lycopersicon esurlentum*). Etude cinétique et mécanisme d'action. Phytochemistry. 19: 395–398.

91. MOLINA, I., NICOLAS, M., CROUZET J. 1986. Grape alcohol dehydrogenase I. Isolation and characterization. Am. J. Enol. Vitic. 37: 169–173.
92. MOLINA, I., SALLES, C., NICOLAS, M., CROUZET, J. 1987. Grape alcohol dehydrogenase II. Kinetic studies : mecanism, substrate, and coenzyme specificity. Am. J. Enol. Vitic. 38: 60–64.
93. ZINSOU, C. 1971. Dégradation enzymatique et autoxydation du β-carotène. Physiol.Veget. 9: 149–167.
94. BEN-AZIZ, A., GROSSMAN, S., BUDOWSKI, P., ASCARELLI, I. 1971. Enzymatic oxidation of carotene and linoleate by alfalfa: Properties of active fractions. Phytochemistry 10: 1823–1830.
95. WEBER, F., GROSCH, W. 1976. Co-oxidation of carotenoid by the enzyme lipoxygenase : influence of the formation of linoleic acid hydroperoxides. Z. Lebenm. Unters. Forsch. 161: 223–230.
96. CABIBEL, M., NICOLAS, J. 1991. Lipoxygenase from tomato fruit (*Lycopersicon esculentum*). Partial purification, some properties and *in vitro* cooxidation of some carotenoid pigments. Sci. Aliments. 11: 277–290.
97. GROSCH, W., WEBER, F., FISHER, K.H. 1977. Bleaching of carotenoid by the enzyme lipoxygenase. Ann. Technol. Agric. 26: 133–137.
98. DRAWERT, F., SCHEIER, P., BHIWAPURKAR, S., HEINDZE, I. 1981. Chemical-technological aspects for concentration of plant aromas. In: Flavour'81 (P. Schreier, ed.), W. de Gruyter, Berlin, pp. 649–663.
99. DAWERITZ-BRABET, I., CROUZET, J. 1993. Enzymatic degradation of beta-carotene. In: Proceedings of the First International Symposium on Natural Colorants, (F.J. Francis, ed.), University of Massachussetts, Amherst. 335 pp.
100. SANDERSON, G.W., CO, H., GONZALES, J.G. 1971. Biochemistry of tea fermentation : the role of carotenes in black tea aroma formation. J. Agric. Food Chem. 36: 231–236.
101. EKSTRAND, B., BJÖRCK, L. 1986. Oxidation of β-carotene by bovine milk lactoperoxidase-halide-hydogen peroxide system. J. Agric. Food Chem. 34: 412–415.
102. BOSSER, A., BELIN, J.M. 1994. Synthesis of β-ionone in an aldehyde/ xanthine oxidase/β-carotene system involving free radical formation. Biotechnol. Prog. 10: 129–133.
103. RAZUNGLES, A., GÜNATA, Z.Y., PINATEL, S., BAUMES, R., BAYONOVE, C. 1991. Etude quantitative de composés terpéniques norisoprénoïdes et de leurs précurseurs dans diverses varietés de raisins. Sci. Aliments 13: 59–72.
104. WINTERHALTER, P., SCHREIER, P. 1995. The generation of norisoprenoid volatiles in starfruit (*Averrhoa carambola* L.) : A review. Food reviews International. 11: 237–254.
105. PFANDER, H., SCHUTENBERGER, H. 1982. Biosynthesis of C20. carotenoids in *Crocus sativus*. Phytochemistry 21: 1039–1042.
106. WINTREHALTER, P., LUTZ, A., SCHEIER, P. 1991. Isolation of a glycosidic precursor of isomeric marmelo lactones from quince fruit. Tetrahedron Lett. 32: 3669–3670.
107. LUTZ, A., WINTERHALTER, P. 1992. Isolation of additional carotenoid metabolites from quince (*Cydonia oblonga* Mill.) J. Agric. Food Chem. 40: 1116–1120.
108. WINTERHALTER, P. 1997. Carotenoid-derived aroma compounds : biogenetical and biotechnological aspects. In: Biotechnology of Foods and Flavors, (R. Teranishi, ed.), Am. Chem. Soc., Washington, D.C., (in press).
109. SKOUROUMOUNIS, G.K., MASSY-WESTROPP, R.A., SEFTON, M.A., WILLIAMS, P.J. 1992. Precursors of damascenone in fruit juices. Tetrahedron Lett. 24: 3533–3536.
110. INA, K., SAKAMATO, Y., FUKAMI H. 1968. Isolation and structure elucidation of theaspirone, a components of tea essential oil. Tetrahedron Lett. 2777–2780.

Chapter Nine

FOOD COLORANTS FROM PLANT CELL CULTURES

François Cormier

Food Research and Development Centre
Agriculture and Agri-Food Canada
St. Hyacinthe, Quebec, Canada J2S 8E3

INTRODUCTION

Food Colorants

Food colorants are often considered simply cosmetic in nature, but the role they play in our food supply is actually very significant. Color is the first sensory quality by which foods are judged, and food quality and flavor are closely associated with color. Consumers are conditioned to expect foods of certain colors and to reject any deviation from their expectations. The physiological basis of the need for food colors is well established.[1]

Colorants are added to foods and beverages for numerous manufacturing reasons: (i) to alleviate damage to the appearance caused by processing and to preserve the product's identity, (ii) to ensure color uniformity in the market place of products which naturally vary in color, (iii) to intensify the color of manufactured foods such as sauces and soft drinks, and to make colorless foods such as gelatine-based jelly more attractive, (iv) to help protect flavor- and light-sensitive vitamins during shelf storage by a sun screen effect, and (v) to serve as a visual indication of quality.

The colorants that can be added to foods are regulated in most countries of the world. In the United States, there are two classes of color additives permitted for use: colorants exempt from certification and colorants subject to certification. Colorants exempt from certification are those that are obtained from vegetable, animal, or mineral sources or are synthetic duplicates of naturally occurring colorants. These colorants are improperly referred to as "natural colors" or "nature identical colors". The Food and Drug Administration does not recognize the description "natural" for these colors. Certified food colors include chemically-synthesized pigments which have no equivalent in nature. Noncertified food colorants from botanical sources are listed in Table 1.

The world market for food colorants is worth more than $500 million.[2] Colors of natural origin are increasing in use and quantity; this is catalyzed by concerns over the safety of artificial colorants. The market for natural food colorants has an annual growth rate of 4 to 6% compared to 1 to 2% for artificial colors.[3] Food pigments from natural sources are important to our well being.

Table 1. Examples of colorants from botanical sources

Colorant	Source	Color
Anthocyanins	Fruits, mainly grape; red cabbage	Red to violet to blue
Betalains	Beetroot	Red to red-purple
Carotenoids:		
bixin	Seeds of *Bixa orellana*	Yellow to orange
β-carotene[a]	Carrot and palm oil	Yellow to orange
capsanthin, capsorubin	Red pepper	Orange to red
crocetin, crocin	Saffron stigmas; gardenia fruits	Yellow to orange-red
lycopene	Tomatoes	Orange to red
lutein	Alfalfa, marigold petals	Yellow to orange
zeaxanthin	Corn, marigold petals	Yellow to orange
Chlorophylls	Plants	Green
Santalan	Red sandlewood tree heartwood	Red
Tumeric	Root of *Curcuma longa*	Yellow to yellow-green

[a]Mainly produced synthetically as "nature-identical".

They are not important because they are colors, but because of their biochemical properties.[4] There are seven major groups of food pigments found in biological materials. Some play a significant physiological role in mammals, but there are still doubts as to whether others have any such function. Table 2 lists the groups, together with their function or suspected function in plants, and provides an indication of their value to herbivores and carnivores further along the food chain. Animals must obtain vitamins A, B2, K1 and K2 from plants, as they are unable to synthesize them. Hence, to man, their value in biological terms is indisputable. Artificial colorants have no known beneficial biological function.

Plant Cell Bioprocesses

Extracted natural colors are subject to problems with consistency, reproducibility, and seasonal variations in plant or animal sources. Bioprocesses are production processes which are based on the *in vitro* use of living organisms such as plant cell cultures. Some advantages of plant cell-based bioprocesses over traditional sources of pigments include (i) independence from environmental factors such as climate, pathogens and geographical and seasonal constraints, (ii) a well-defined production system which enables a modulation of production in accordance with market demand, (iii) a greater consistency in quality and quantity of products and, (iv) independence from political interference *e.g.* production cartel, embargo, etc.

Perhaps because of their 'obvious' nature which facilitates the selection of productive cell lines, many metabolites which have been produced successfully in plant cell cultures are pigments. In a recent review, Dörnenburg and Knorr list the numerous plant species from which pigment producing callus and cell suspension cultures have been obtained.[5] The pigments include anthocyanins,

Table 2. Physiological functions of natural pigments (Reproduced with permission from Hutchings[4])

Pigment groups	Function in plant	Value to herbivores and carnivores
Carotenoids	Photosynthesis Pollination	Vital, precursors of vitamin A which cannot be synthesized.
Porphyrins	Photosynthesis	None known, haemoglobin can be synthesized
Flavonoids	Growth control Pollination	No physiological function known
Quinones	Respiratory enzymes	Vital, includes vitamins K1, K2 which cannot be synthesized
Indoles	Unknown	None known, melanins can be synthesized
Flavins	Quench molecular oxygen	Vital, vitamin B2 cannot be synthesized
Betalains	Pollination? Virus protection?	No physiological function known.

betalains, carotenoids such as β-carotene, lycopene and crocin, quinones such as anthraquinones and naphthoquinones, and others. Plant cell cultures are used for the production of valuable secondary metabolites in two ways: *de novo* synthesis and bioconversion. *De novo* synthesis of metabolites starts out with the establishment of a heterogeneous cell culture followed by the selection of productive cell lines *i.e.* cytodifferentiated cultures. Then, two distinct culture stages, *i.e.* biomass production and metabolite production stages, are optimized. Two culture stages are usually required since there is an inverse relationship between growth and metabolite production. Metabolites are synthesized from the carbon and nitrogen sources in the nutrient medium. Bioconversions are carried out using an undifferentiated cell culture as a catalyst. Exogenous sbustrates which are fed to the culture undergo a limited number of reactions. In this chapter, an example of each mode of metabolite production will be presented: the first and second parts will respectively consider the *de novo* biosynthesis of anthocyanins in a grape cell culture and the bioconversion of crocetin catalyzed by a saffron cell culture.

ANTHOCYANIN PRODUCTION USING A GRAPE (*Vitis vinifera* L.) CELL CULTURE

Most of the red and blue colors of flowers, leaves and fruits are due to the presence of anthocyanins. These are intensely colored water-soluble glycosides of 2-phenylbenzopyrylium salts. Most anthocyanin-based colorants are obtained from grape press residues or pomace. Grape pomace is the least expensive and most available source of anthocyanins. This is because grapes are the most important fruit crop in the world and because much of the grape is processed into wine and juices which yield a sizeable amount of grape press pomace. Plant cell cultures have been considered as potential sources of anthocyanins as demonstrated by numerous patent applications.[6] However, to date no patent application has dealt with a true commercial process. A description of a process which produces anthocyanins using a grape (*Vitis vinifera*) cell suspension culture along with a preliminary economic analysis has been previously reported.[6] This section focuses mainly on anthocyanin composition as regulated by culture conditions, enzymatic activities, and ultrastructural features.

Cell Line Selection and Anthocyanin Composition

A cell culture of *V. vinifera* L. cv Gamay Fréaux was established at the Plant Biotechnology Laboratory of the École Nationale Supérieure Agronomique (Toulouse, France) from fragments of immature berries. The level of anthocyanins in callus cultures was maintained at a moderate level by pooling together and subculturing pigmented portions of the culture at the end of each culture cycle. A typical anthocyanin content of a suspension culture derived from this callus culture was approximately 0.26 $mg \cdot g^{-1}$ fresh weight.[7] The level of anthocyanins in callus cultures can be progressively improved during the course of repeated subculturing of separate minute, intensely pigmented portions *i.e.* cell-aggregate cloning.[8,9] As an example, the cell-aggregate cloning technique produced, after 20 months of selection, cell line No. 5.4 which contained 0.44 $mg \cdot g^{-1}$ fresh weight on average, and after 37 months, cell line No. 13.1 which contained 1.02 $mg \cdot g^{-1}$ fresh weight.[10] Prolonged selection of anthocyanin-producing cell lines not only improves the anthocyanin content but also modifies the anthocyanin composition.

Anthocyanins found in the grape cell culture derive from the anthocyanidins cyanidin and delphinidin (Fig. 1). The major anthocyanins of cell line No. 5.4 derive from cyanidin while those of cell line No. 13.1 derive from both cyanidin and delphinidin (Table 3). The ratio of cyanidin- and delphinidin-derived anthocyanins may be regulated by the activities of the cytochrome P450, flavonoid 3'- and 3',5'-hydroxylases as is the case in *Petunia hybrida*.[11] The predominant anthocyanin of cell lines No. 5.4 and No. 13.1 are peonidin 3-glucoside and peonidin 3-*p*-coumaroylglucoside respectively, indicating that selection favors the establishment of cell lines which contain more metabolically-evolved anthocyanins.

Anthocyanin-Promoting Culture Conditions and Anthocyanin Composition

Anthocyanin accumulation in plants is influenced by many environmental factors such as light, temperature, nutritional effects, plant hormones, mechanical damage, and pathogen attack.[12] The effects of several of these factors have been studied in plant cell cultures of various species.[13] In cell cultures of *Vitis vinifera*, the anthocyanin content is greatly affected by nutritional and physical factors. Indeed, osmotic stress, brought about by a high concentration of carbohydrates in the culture medium, inhibits growth, promotes the accumulation of anthocyanins, and increases the percentage of pigmented cells.[14,15] The positive effect of osmotic stress on anthocyanin accumulation also has been observed in cell cultures of other species.[16–21] The reduction of the level of nitrates has similar effects on cell growth, anthocyanin accumulation and cytodifferentiation, in *Vitis vinifera*[22] and other cultures.[23–26] The effect of nitrate reduction is 'switchlike' rather than gradual. Osmotic stress and reduction of the level of nitrates have a synergistic effect when applied together.[22]

The kinds of anthocyanins being synthesized under anthocyanin-promoting conditions differ from one cell line to another (Table 3). Cell line No. 5.4 accumulates mostly peonidin 3-glucoside while cell line No. 13.1 accumulates mostly peonidin 3-*p*-coumaroylglucoside followed by malvidin 3-glucoside. There is a correlation between the synthesis of more biosynthetically-evolved

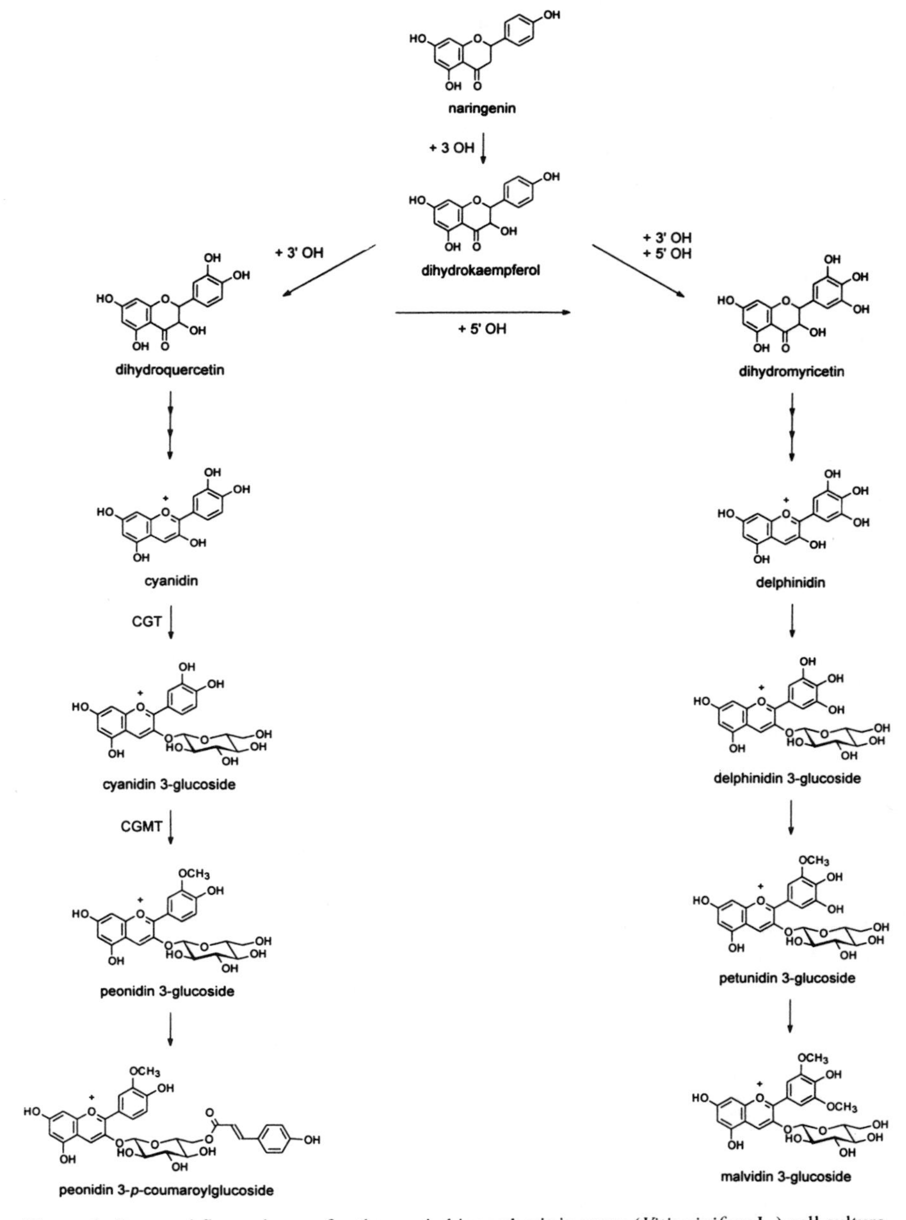

Figure 1. Proposed flow scheme of anthocyanin biosynthesis in grape (*Vitis vinifera* L.) cell culture.

Table 3. Anthocyanin composition of two selected cell lines of grape (*Vitis vinifera* L.) cell culture grown in a maintenance and an anthocyanin-promoting medium (Reproduced with permission Cormier et al.[10])

	Anthocyanin concentration (mg / g FW)*			
	Cell line No. 5.4		Cell line No. 13.1	
	Stage M	Stage A-P	Stage M	Stage A-P
Cyanidin 3-glucoside	0.07	0.54	0.02	0.10
Peonidin 3-glucoside	0.22	1.17	0.37	0.44
Peonidin 3-p-coumaroylglucoside	0.14	0.37	0.40	2.17
Malvidin 3-glucoside	tr	tr	0.09	0.94
Total:	0.53	2.36	1.09	4.93

Cell line No. 5.4 was obtained after 20 months of selection;
Cell line No. 13.1 was obtained after 37 months of selection;
Culture stage: M, maintenance; A-P, anthocyanin-promoting;
*determined at the end of culture cycle

anthocyanins and prolonged selection. In both cell lines, the anthocyanins that are enhanced, are methylated in the B ring. Accumulation of methylated anthocyanins under anthocyanin-promoting conditions also has been observed in *Aralia cordata* cell cultures.[27] The section that follows look into the regulation of anthocyanin accumulation at the enzymatic level.

Enzyme Regulation of Anthocyanin Composition

Glycosyltransferase. In most plant sources, glycosylation, usually a glucosylation, in the 3-position of anthocyanidin is an obligatory reaction leading to the formation of the first stable anthocyanin.[28] Because of the instability of the aglycones, lack of 3-O-glucosyltransferase activity would be expected to prevent accumulation of anthocyanins, and it is assumed that the anthocyanidin formed is immediately glycosylated at the 3-hydroxyl group.[29] An enzyme catalyzing the transfer of the glucosyl moiety of UDP-glucose to the 3-hydroxyl group of cyanidin has been isolated from *Vitis vinifera* cell suspension culture and purified 75-fold.[30] Some characteristics of this enzyme are presented together with glucosyltransferases from other plant sources (Table 4). To date, all glycosyltransferases involved in the synthesis of an anthocyanin monoglycoside are specific to the 3-hydroxyl group of the aglycone. Most have a preference for cyanidin as the acceptor but can also utilize other anthocyanidins and flavonols in varying degrees. Other glycosyltransferases that are involved in the synthesis of di- or tri-glycosides have been reported.[34,38,39,41,43,44] Most use uridine-diphospho-glucose as the sugar donor. Glycosylation of anthocyanins can also occur via the transfer of 1-O-acylglucosides.[31] Most glycosyltransferases have a molecular weight between 49 and 56 KDa. The maximal pH of the reaction can vary greatly. Most glycosyltransferases are not stimulated by divalent cations.

Table 4. Properties of anthocyanidin and anthocyanin glycosyltransferases

Source (reference)	Position of glucosylation	Substrate (Km)			MW (KDa)	pH	pI	Remarks
		Consumed	Sugar donor[a]	Not used[b]				
Brassica oleracea—seedlings								
(32)	3	cyanidin (0.4 mM), pelargonidin, peonidin, malvidin, kaempferol, quercetin, isorhamnetin, myrcetin, fisetin	UDPGlu (0.51 mM)	cyanidin 3-glucoside, cyanidin 3-sophoroside, cyanidin 3,5-diglucoside, apigenin, luteolin, naringenin, dihydroquercetin		8.0		
Daucus carota—cell suspension								
(33)	3	cyanidin (35 μM), delphinidin, pelargonidin, quercetin, kaempferol	UDPGlu (1 mM)	naringenin eriodictyol dihydroquercetin		8.0		no stimulation by divalent cations; total inhibition by 0.4 mM *p*-chloromercuribenzoate
(34)	3	cyanidin (1.5 mM)	UDPGal (0.4 mM)		52	6.8		optimal temperature, 23°C; no stimulation by divalent cations
(34)	2″	cyanidin 3-galactoside (0.1 mM)	UDPXyl (0.2 mM)		52	6.9		optimal temperature, 32°C; no stimulation by divalent cations
Gentiana triflora—petals								
(35)	3	in decreasing order: delphinidin, cyanidin, pelargonidinlittle activity with: kaempferol, quercetin, myricetin	UDPGlu	dihydrokaempferol, dihydroquercetin, apigenin, delphinidin 3-glucoside				

Source (reference)	Position of glucosylation	Substrate (Km)			MW (KDa)	pH	pI	Remarks
		Consumed	Sugar donor[a]	Not used[b]				
Happlopappus gracilis—cell suspension								
(36)	3	cyanidin (0.33 mM), much lower activity with:pelargonidin, peonidin, malvidin, kaempferol, quercetin, isorhamnetin, fisetin	UDPGlu(0.5 mM)	apigenin, luteolin, naringenin, dihydroquercetin, cyanidin 3-glucoside, cyanidin 3,5-diglucoside		8.0		
Hippeastrum—petals								
(37)	3	in decreasing order of activity:kaempferol (12.5 μM), quercetin (70 μM), dihydroquercetin, dihydrokaempferol; low activity with: pelargonidin, cyanidin (70 μM), malvidin	UDPGlu(1 mM)	naringenin apigenin	49 (24.5[c])	5.0 (8–8.5)	5.6	
Matthiola incana—flower								
(38)	2′	cyanidin 3-*p*-coumaroylglucoside and -caffeoylglucoside; 3-glucoside of cyanidin, pelargonidin, delphinidin, quercetin, kaempferol	UDPXyl	UDP Glu, UDPMan, UDPGal, UDPGluc, CDPGlu, GDPGlu, ADPGlu, cyanidin 3,5-diglucoside; 7-glucosides of apigenin, naringenin, eriodictyol, dihydrokaempferol		6.5		slight stimulation by Ca^{2+}; inhibition by EDTA

(Continued)

Table 4. *(Continued)*

Source (reference)	Position of glucosylation	Substrate (Km)			MW (KDa)	pH	pI	Remarks
		Consumed	Sugar donor[a]	Not used[b]				
(39)	5	the following pelargonidin- and cyanidin- derivatives (in decreasing order of activity): 3-sambubioside acylated with *p*-coumarate-3-sambubioside-3-glucoside acylated with *p*-coumarate	UDPGluTDP Glu	ADPGluCDPGluGDPGlu		7.5		inhibited by EDTA, divalent cations and *p*-chloromercuribenzoate; optimal temperature, 40°C
Petunia hybrida- flower								
(40)	3	cyanidin, delphinidin	UDPGlu(1.7 mM)	ADPGlu		7.5–8		cyanidin glucosylation inhibited by delphinidin and *vice versa*
(41)	5	delphinidin- and petunidin-3-(*p*-coumaroyl)-rutinoside (3 μ M)	UDPGlu(0.22 mM)	delphinidin- and cyanidin-3-rutinoside delphinidin 3-glucoside	52	8.3	4.75	no stimulation by Mg^{2+} or Ca^{2+}; not inhibited by EDTA
Silene dioica—flower								
(42)	3	cyanidin (40 μM) and at a lower rate: pelargonidin and delphinidin	UDPGlu (0.41 mM)	kaempferol quercetin ADPGlu	125 (60*)	7.5		no stimulation by divalent cations; inhibition by *p*-chloromercuribenzoate and $HgCl_2$
(43)	5	cyanidin 3-rhamnosylglucoside (3.6 mM), pelargonidin 3-rhamnosylglucoside; low activity with cyanidin 3-glucoside	UDPGlu (0.5 mM)	ADPGlu	55	7.4		stimulation by divalent cations mainly by Ca^{2+}; inhibition by EDTA, $HgCl_2$

Source (reference)	Position of glucosylation	Substrate (Km)			MW (KDa)	pH	pI	Remarks
		Consumed	Sugar donor[a]	Not used[b]				
(44)	6″	3-glucosides of cyanidin (2.2 mM), pelargonidin and delphinidin; lower affinity for the 3,5-diglucoside of cyanidin and pelargonidin	UDPRha		45	8.1		stimulation by Co^{2+}, Ca^{+2}, Mn^{+2} and Mg^{+2}; inhibition by $HgCl_2$, *p*-chloromercuribenzoate, *N*-ethylmaleimide
Vitis vinifera—cell suspension								
(30)	3	in decreasing order of activity: cyanidin (18 μM) delphinidin (28 μM) pelargonidinpeonidin malvidin	UDPGlu (1.2 mM)	quercetin kaempferol	56	8.0	4.55	no stimulation by divalent cations

[a]UDPGlu, uridine-diphospho-glucose; UDPGal, uridine-diphospho-galactose; UDPXyl, uridine-diphospho-xylose; TDPGlu, thymidine-diphospho-glucose;
[b]ADPGlu, adenosine-diphospho-glucose; UDPMan, uridine-diphospho-mannose; UDPGluc, uridine-diphospho-glucuronic acid; CDPGlu, cytidine-diphospho-glucose; GDPGlu, guanosine-diphospho-glucose;
[c]active monomer.

In *Vitis vinifera* cell cultures, a single enzyme is responsible for the glucosylation of the 3'-hydroxyl anthocyanidin cyanidin and the 3'- and 5'-hydroxyl anthocyanidin delphinidin. The enzyme activity is lower with a substrate lacking a 3'-hydroxyl group *i.e.* pelargonidin. Even lower activity is obtained with a substrate with a 3'-methoxyl group *i.e.* peonidin, and 3'- and 5'-methoxyl goups *i.e.* malvidin. The highest affinity towards cyanidin and delphinidin suggests that the enzyme may accept a hydrogen bond from the substrate's 3'-hydroxyl group. The slower activity observed with anthocyanidins methoxylated in 3' and 5' compared with the protonated substrate may be attributed to steric hindrance. The enzyme does not use flavonols. The affinity is slighly greater for cyanidin than for delphinidin. However, such differences might not explain the predominance of cyanidin-derived anthocyanins over delphinidin-derived anthocyanins in cell culture.

Methyltransferase. Once glycosylated, anthocyanins can undergo further transformations such as additional glycosylation, methylation, and acylation of sugar residues.[39] Since anthocyanin-promoting culture conditions favor the accumulation of methylated anthocyanins in *Vitis vinifera* cell cultures, it is presumed that methylation might play an important role in determining the type of anthocyanins being accumulated. An enzyme catalyzing the transfer of the methyl moiety of S-adenosyl-L-methionine to the 3'-hydroxyl group of cyanidin 3-glucoside has been isolated from *Vitis vinifera* cell suspension culture and purified 35-fold.[47] Some characteristics of this enzyme are presented together with those of other anthocyanin methyltransferases (Table 5). In general, the methyltransferases reported to date carry out the methylation of hydroxyl moieties in position 5' and/or 3', and have stringent requirement for a specific substitution pattern in position 3. Activities are either low or nil with anthocyanidins.

Time-course experiments carried out with *Vitis vinifera* cell culture show a correlation between methyltransferase activity and the increase in peonidin-based anthocyanins.[47] After 11 days, the methyltransferase activity of cultures grown in the anthocyanin-production medium is 6-fold greater than that of cultures grown in a maintenance medium. This explains the greater accumulation of the 3'-methylated derivatives of cyanidin 3-glucoside in the *Vitis* cell culture.

Ultrastructural Features and Anthocyanin Composition

Anthocyanins are stored in the vacuoles of cells. Often anthocyanin-producing cells of plants contain intensely-pigmented spherical bodies which are located in the vacuole.[48] These structures were incorrectly named anthocyanoplasts since ultrastructural studies have demonstrated that they are inclusions (void of an outer membrane) and not an organelle.[13] In the *Vitis vinifera* cell culture, both the selection of highly productive cell lines such as No. 13.1 and the growth under anthocyanin-promoting conditions contribute to increasing the frequency of occur-

Table 5. Properties of anthocyanin methyltransferases

Source (reference)	Position of methylation	Substrate (Km): Consumed	Substrate (Km): Methyl donor	Substrate (Km): Not used	Divalent cations[c]	MW (KDa)	pH	Remarks
Petunia hybrida—flower buds (Refs. 45, 46)								
	3′, 5′	3-(*p*-coumaroyl)-rutinosido-5-glucoside derivatives of: cyanidin (22 μM), petunidin (25 μM), delphinidin	SAM[a] (50 μM)	cyanidin; cyanidin 3-glucoside; cyanidin 3-rutinoside; *p*-coumaric acid; caffeic acid[b]	10 mM Mg^{2+}(+)	45~50	7.7, 8.5~9	enzyme code *Mt1*; no inhibition by 1 mM EDTA; competitive inhibition by SAH[d].
	3′, 5′	3-(*p*-coumaroyl)-rutinosido-5-glucoside derivatives of: cyanidin (6 μM), petunidin (2 μM), delphinidin	SAM (40 μM)	cyanidin; cyanidin 3-glucoside; cyanidin 3-rutinoside; *p*-coumaric acid; caffeic acid	10 mM Mg^{2+}(+)	45~50	7.7, 8.5~9	enzyme code *Mt2*; no inhibition by 1 mM EDTA; non-competitive inhibition by SAH.
	3′, 5′	3-(*p*-coumaroyl)-rutinosido-5-glucoside derivatives of: cyanidin (8 μM), petunidin (8 μM), delphinidin	SAM (95 μM)	cyanidin; cyanidin 3-glucoside; cyanidin 3-rutinoside; *p*-coumaric acid; caffeic acid	10 mM Mg^{2+}(+)	45~50	7.7, 8.5~9	enzyme code *Mf1*; no inhibition by 1 mM EDTA; competitive inhibition by SAH.
	3′, 5′	3-(*p*-coumaroyl)-rutinosido-5-glucoside derivatives of: cyanidin (22 μM), petunidin (21 μM), delphinidin	SAM (100 μM)	cyanidin; cyanidin 3-glucoside; cyanidin 3-rutinoside; *p*-coumaric acid; caffeic acid	10 mM Mg^{2+}(+)	45~50	7.7, 8.5~9	enzyme code *Mf2*; no inhibition by 1 mM EDTA; non-competitive inhibition by SAH.
Vitis vinifera—cell suspension (Ref. 47)								
	3′	cyanidin 3-glucoside (199 μM); cyanidin (13 % of the activity with cyanidin 3-glucoside)	SAM (18 μM)	cyanidin 3-*p*-coumaroylglucoside; delphinidin 3-glucoside	5 mM Mg^{2+} (+) 5 mM Fe^{2+} (0) 5 mM Cu^{2+} (−) 5 mM Zn^{2+} (−) 5 mM Ca^{2+} (−) 5 mM Hg^{2+} (−)	80	7.75, 9.75	inhibition by 1 mM EDTA; non-competitive inhibition by SAH.

Legend: [a]SAM, S-adenosyl-L-methionine; [b]determined from a crude extract;[46] [c]divalent cations: +, activation; 0, no effect; -, inhibition; [d]SAH, S-adenosyl-L-homocysteine.

rence of these inclusions. Addition of 200 μl/l geraniol to the suspension results in the destabilization of the cell membranes, and within 30 min the anthocyanins leach out of the cells.[48] Optical microscopic observations reveal that the inclusions remain pigmented and seemingly intact.[13] The extraction of 'free' anthocyanins can be completed by repeated suspension of the cells in water and pelleting by centrifugation. Thereafter, the inclusions are broken down in 1% HCl and the anthocyanins which are 'bound' to the inclusions are recovered. HPLC analysis of this fraction reveals the predominance of peonidin 3-*p*-coumaroylglucoside while other anthocyanins are present in low levels (unpublished results). This phenomenon might be explained by the stronger binding of the acylated anthocyanin compared with the other (non acylated) anthocyanins. Although the inclusions bear some resemblance to tannin bodies, the chemical nature of the bulk of the inclusions is unknown. Regardless, it would appear that there is a correlation between the frequency of occurrence of the inclusions and the percentage of the acylated anthocyanin peonidin 3-*p*-coumaroylglucoside.

Conclusion

The substitution pattern of anthocyanins has an effect on color hue and stability.[49] Therefore, anthocyanin composition is of utmost importance when developing a process to manufacture anthocyanin-based food colorants from plant cell cultures. To date, not much is known in this area. Studies carried out on *Vitis vinifera* cell culture indicate that the selection of highly-productive cell lines and the cultivation of this material, under physical and nutritional conditions which favor the accumulation of anthocyanins, have an effect on anthocyanin composition. In the metabolic branch beginning with cyanidin 3-glucoside, the accumulation of 3'-methylated anthocyanins, especially peonidin 3-*p*-coumaroylglucoside is favored. Similarly, in the branch starting with delphinidin 3-glucoside, the 3',5'-dimethylated anthocyanin malvidin 3-glucoside is accumulated. The anthocyanidin glucosyltransferase of *Vitis* cells cannot discriminate between cyanidin and delphinidin, and hence cannot regulate the flow of metabolites in both branches. The activity of the methyltransferase which is responsible for the formation of the 3'-methylated derivatives of cyanidin 3-glucoside, correlates well with the accumulation of these anthocyanins.[47] Structures such as pigmented inclusions may play a role in the accumulation of the monoacylated anthocyanin, peonidin 3-*p*-coumaroylglucoside, which binds to the inclusions.

CROCETIN GLYCOSIDE PRODUCTION USING A SAFFRON (*Crocus sativus* L.) CELL CULTURE

Carotenoids are produced by a diverse group of organisms ranging from bacteria to higher plants. The total natural production has been estimated at over

100 million tons per annum.[50] Over 600 different carotenoids have been identified from natural sources.[51] The colors of carotenoids range from yellow to orange and red; blue, green, and purple are possible when carotenoids are complexed to proteins. While carotenoids *e.g.* β-carotene, bixin (annatto) and others, are suitable for many food color applications, most carotenoids are insoluble in water and therefore have limited application in food. Crocetin glycosides, however, are unique water-soluble carotenoids and have higher tinctorial potency as well. The LD_{50} of crocetin is very high, about 2 g/kg.[52] Crocetin glycosides are ideally suited as a replacement for the artificial colorant tartrazine *i.e.* FD&C Yellow No. 5.[50] Crocin *i.e.* all-*trans*-di-(β-D-gentiobiosyl) ester (Fig. 2) is the major pigment obtained from saffron (*Crocus sativus*) stigmas and gardenia (*Gardenia jasminoides*) fruits. Saffron is unlikely to become a major source of crocin because of its high cost (most expensive of all spices). Gardenia is approved in Japan and China but not in Europe or North America. Therefore, a process to produce hydrosoluble carotenoids using saffron cell cultures is currently under development.

Crocetin Glycosides

Using high performance liquid chromatography-mass spectrometry, Tarantalis and collaborators provided evidence for the presence in saffron of *cis* and *trans* isomers of crocetin glycosides carrying one to 5 glucose moieties.[53] Crocetin derivatives have been extracted and purified from saffron stigmas and gardenia fruits by reverse-phase HPLC and characterized by 1D and 2D ^{1}H and ^{13}C NMR and UV-visible spectroscopies and mass spectrometry.[54] The following compounds were found in saffron (Fig. 2): all-*trans*-crocetin di-(β-D-gentiobiosyl) ester *i.e.* crocin (**1a**), all-*trans*-crocetin β-D-gentiobiosyl-β-D-glucosyl ester (**1b**), all-*trans*-di-(β-D-glucosyl) ester (**1c**), all-*trans*-crocetin mono-(β-D-gentiobiosyl) ester (**1d**), 13-*cis*-crocetin β-D-gentiobiosyl-β-D-glucosyl ester (**2a**) and 13-*cis*-crocetin β-D-gentiobiosyl-β-D-glucosyl ester (**2b**) (Fig. 2). Compounds **1a**, **1b** and **1d** were also found in gardenia, in addition to all-*trans*-crocetin mono-(β-D-glucosyl) ester (**1e**).

Cell Cultures

The production of saffron by *in vitro* culture techniques has been reported. Callus cultures induced to produce "red globular" callus and "red filamentous structures" contained crocin at a lower level than in the stigmas.[55] Stigma-like structures which were formed *in vitro* on cultured young half ovaries[56] and on stigma and style tissues[57] contained small amounts of crocin. Callus cultures also had limited production of crocin, crocetin β-D-gentiobiosyl-β-D-glucosyl ester and crocetin di-(β-D-gentiobiosyl) ester.[58]

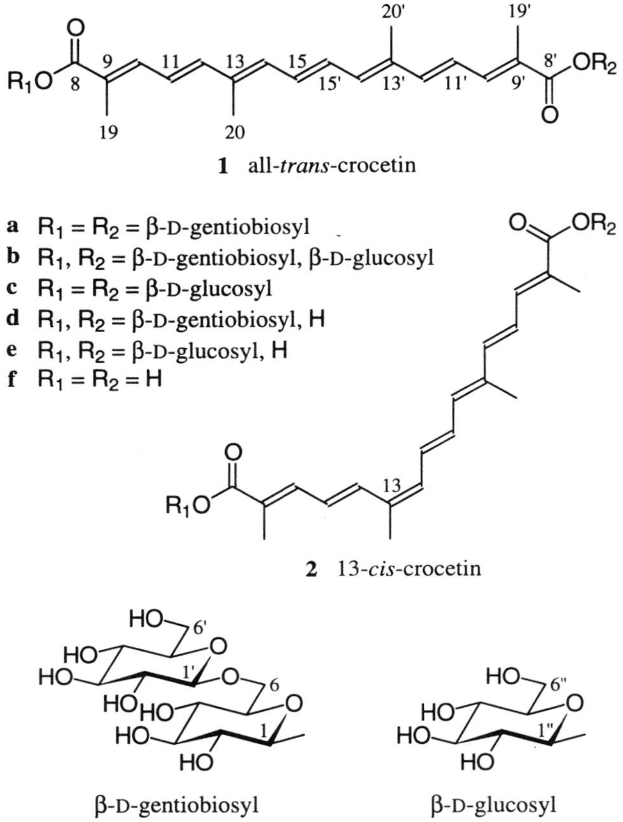

1 all-*trans*-crocetin

a $R_1 = R_2 =$ β-D-gentiobiosyl
b $R_1, R_2 =$ β-D-gentiobiosyl, β-D-glucosyl
c $R_1 = R_2 =$ β-D-glucosyl
d $R_1, R_2 =$ β-D-gentiobiosyl, H
e $R_1, R_2 =$ β-D-glucosyl, H
f $R_1 = R_2 = H$

2 13-*cis*-crocetin

β-D-gentiobiosyl β-D-glucosyl

Figure 2. Structure of crocetin derivatives.

A callus culture of saffron was established in our laboratory but attempts to select a cell line which produced crocetin glycosyl esters were unsuccessful. Indeed, through cell line selection, yellow to orange callus cultures, which did not contain the desired pigments, were obtained. The callus material turned red upon exposure to ammonia vapors, leading us to believe that the pigmentation could be attributed to chalcones or aurones. Because of our lack of success and the limited production reported elsewhere, the possibility of producing the pigments through a bioconversion process was explored.

Crocetin Bioconversion

Various methods to chemically-synthesize crocetin have been reported.[59,60] Thus, the use of crocetin as a substrate in the enzymatic synthesis of its related glycosides is conceivable. Plant cell cultures can perform a wide array of bioconversions of which glucosylation seems one of the most interesting.[61] It occurs readily in plant cells, but only with difficulty in microorganisms, and the organic chemist has problems with this reaction. The glucosylation of crocetin by a cell-free extract of saffron callus, in the presence of uridine-diphospho-glucose, was investigated.

The first difficulty encountered is with the substrate crocetin which is insoluble in water. In many instances, the enzymatic glucosylation of poorly-soluble substrates is accomplished in the presence of a water-soluble solvent such as dimethylsulfoxide. However, the glucosylation of 100 μM crocetin is inhibited by dimethylsulfoxide at concentrations which would permit the solubilization of more crocetin.[62] The inhibition is linear with increased dimethylsulfoxide concentration, with the half-reaction rate occurring at *ca* 2.7 M. Similarly, 2.8 M dimethylsulfoxide was reported to inhibit the enzymatic conversion of cardioglycosides by *Digitalis lanata* cell extracts.[63] To circumvent this problem, crocetin was solubilized by encapsulation into maltosyl-β-cyclodextrin.[62]

Cyclodextrins are clatharing agents that can form a water soluble complex with apolar compounds. The formation of inclusion complexes of the aliphatic carotenes, norbixin and crocetin, with various cyclodextrins has been previously reported.[64] Maltosyl-β-cyclodextrin was chosen as the encapsulating agent in the enzymatic glucosylation of crocetin because it is more soluble in water than the unbranched α-, β- or γ-cyclodextrins thus enabling the solubilization of a higher concentration of crocetin. At crocetin to maltosyl-β-cyclodextrin molar ratios of 1:3 and 1:6 the glycosylation can be carried out with up to 1,750 and 2,500 μM crocetin respectively. A time-course experiment provided some evidence that two distinct reactions are involved in the synthesis of all-*trans*-crocin from all-*trans*-crocetin: (i) the step-wise addition of a glucose moiety to a free carboxyl function of crocetin and (ii) the 1→6 addition of a glucose moiety to a glucose ester function.[65]

The bioconversion of crocetin can be carried out by the saffron cell culture in the presence of glucose (unpublished results). Crocetin, which is taken up by the cells, is readily converted mainly into two new major crocetin glycosides which are stored in the vacuole. Diode array reverse-phase HPLC analysis of the pigments indicates that they are all-*trans* crocetin glycosides and more polar than crocin. There are few intermediates accumulated in the cells, making any attempts to stop the reaction at the crocin level impossible. With regards to the possible identity of the two new compounds which accumulate in the saffron cell culture, there are reports of crocetin glycosyl esters which possess a higher degree of glycosylation than crocin.

A minor crocetin glycoside which has three and two glucose moieties on each end has been isolated in saffron stigmas.[53] The trisaccharide moiety, which has been named "neapolitanose", has been identified as O-β-D-glucopyranosyl-

(1→2)-O-[β-D-glucopyranosyl-(1→6)]-D-glucose.[66] Elsewhere, crocetin di-(β-D-neapolitanosyl) ester has been found in *Crocus neapolitanus*.[67]

Conclusion

Although cell cultures of saffron are unable to synthesize *de novo* substantial amounts of valuable crocetin glycosides, cell-free extracts and whole cells possess the capability of adding glucose moieties to the aglycone crocetin. The addition of two glucose moieties on each carboxyl function of crocetin yielding crocin, the major hydrosoluble carotene of saffron, has been demonstrated. However, there are preliminary indications that more glucose units are added to crocin yielding more polar pigments. From a regulatory point of view, the approval of novel crocetin glycosides might present some difficulties. Toxicological data for the compounds do not exist. Bioconversion processes are sometimes subject to unwanted reactions. Such reactions might be inactivated by anti-sense RNA technology.

ACKNOWLEDGMENTS

I wish to acknowledge the major contribution of Dr. Chi Bao Do to the anthocyanin study, of Ms. Christiane Dufresne to the crocetin glycoside study and of Dr. Marie-Rose Van Calsteren to the elucidation of the structure of most compounds presented herein.

REFERENCES

1. NEWSOME, R.L. 1990. Natural and synthetic coloring agents. In: Food Additives, (A.L. Branen, P.M. Davidson, S. Salminen, eds.), Marcel Dekker, Inc., New York, Basel, pp. 327–345.
2. O'CALLAGHAN, M. 1993. Natural color systems for beverages. In: Papers Presented at the First International Symposium on Natural Colorants for Foods, Nutraceuticals, Beverages and Confectionery, (F.J. Francis, ed.), The Herald Organization, Hamden, Connecticut.
3. LEPRE, J. 1994. Natural colors, Chemical Marketing Reporter (New York), June 27, p. SR-9.
4. HUTCHINGS, J.B. 1994. Food color and appearance. Blackie Academic & Professional, London, 513 pp.
5. DÖRNENBURG, H., KNORR, D. 1996. Generation of colors and flavors in plant cell and tissue cultures. Crit. Rev. Plant Sci. 15(2): 141–168.
6. CORMIER, F., BRION, F., DO, C.B., MORESOLI, C. 1996. Development of process strategies for anthocyanin-based food colorant production using *Vitis vinifera* cell cultures. In: Plant Cell Culture Secondary Metabolism. Toward Industrial Application, (F. DiCosmo, M. Misawa, eds.), CRC Press, Boca Raton, pp. 167–185.
7. LATCHÉ, A., FALLOT, J. 1984 Obtention d'anthocyanes marquées au carbone 14 par voie biologique à l'aide de cellules de raisin cultivées *in vitro*. Compte-rendu contrat DGRST no 81C376, Toulouse, France.

8. YAMAMOTO, Y., MIZUGUCHI, R., YAMADA, Y. 1982. Selection of high and stable pigment producing strain in cultured *Euphorbia millii* cells. Theor. Appl. Genet. 61: 113–116.
9. NOZUE, M., KAWAI, J., YOSHIMATA, K. 1987. Selection of high anthocyanin-producing cell lines of sweet potato cell cultures and identification of pigments. J. Plant Physiol. 129: 81–88.
10. CORMIER, F., DO, C.B., NICOLAS, Y. 1994. Anthocyanin production in selected cell lines of grape (*Vitis vinifera* L.). In Vitro Cell. Dev. Biol. 30P: 171–173.
11. HOLTON, T.A., CORNISH, E.C. 1995. Genetics and biochemistry of anthocyanin biosynthesis. Plant Cell 7: 1071–1083.
12. MCCLURE, J.W. 1975. Physiology and functions of flavonoids. In: The Flavonoids, (J.B. Harborne, T.J. Mabry, H. Mabry, eds.), Chapman and Hall, London, pp. 970–1055.
13. CORMIER, F., DO, C.B. 1993. XXVII *Vitis vinifera* L. (Grapevine): *In vitro* production of anthocyanins. In: Biotechnology in Agriculture and Forestry, Vol. 24, Medicinal and Aromatic Plants V. (Y.P.S. Bajaj, ed.), Springer-Verlag, Berlin, Heidelberg, pp. 373–386.
14. DO, C.B., CORMIER, F. 1990. Accumulation of anthocyanins enhanced by a high osmotic potential in grape (*Vitis vinifera* L.) cell suspension. Plant Cell Reports 9: 143–146.
15. DO, C.B., CORMIER, F. 1990. Accumulation of peonidin 3-glucoside enhanced by osmotic stress in grape (*Vitis vinifera* L.) cell suspension. Plant Cell Tissue and Organ Culture 24: 49–54.
16. HIRASUNA, T.J., SCHULER, M.L., LACKNEY, V.K., SPANSWICK, R.M. 1991. Enhanced anthocyanin production in grape cell cultures. Plant Sci. 78: 107–120.
17. RAJENDRAN, L., RAVISHANKAR, G.A., VENKATARAMAN, L.V., PRATHIBA, K.R. 1992. Anthocyanin production in *Daucus carota* as influenced by nutrient stress and osmoticum. Biotechnol. Lett. 14: 707–712.
18. SAKAMOTO, K., IIDA, K., SAWAMURA, K., HAJIRO, K., ASADA, Y., YOSHIKAWA, T., FURUYA, T. 1993. Effects of nutrients on anathocyanin production in cultured cells of *Aralia cordata*. Phytochemistry. 33(2): 357–360.
19. THOLAKALABAVI, A., ZWIAZEK, J.J., THORPE, T.A. 1994. Effect of mannitol and glucose-induced osmotic stress on growth, water relations, and solute composition of cell suspension cultures of poplar (*Populus deltoides* var *occidentalis*) in relation to anthocyanin accumulation. In Vitro Cell. Dev. Biol. 30P: 164–170.
20. SUZUKI, M. 1995. Enhancement of anthocyanin accumulation by high osmotic stress and low pH in grape cells (*Vitis* hybrids). J. Plant Physiol. 147: 152–155.
21. SATO, K., NAKAYAMA, M., JUN-ICHI S. 1996. Culturing conditions affecting the production of anthocyanin in suspended cell cultures of strawberry. Plant Sc. 113: 91–98.
22. DO, C.B., CORMIER, F. 1991. Effects of low nitrate and high sugar concentration on anthocyanin content and composition of grape (*Vitis vinifera* L.) cell suspension. Plant Cell Reports 9: 500–504.
23. KNOBLOCH, K.-H., BAST, G., BERLIN, J. 1982. Medium- and light-induced formation of serpentine and anthocyanins in cell suspension cultures of *Catharanthus roseus*. Phytochemistry. 21: 591–594.
24. YAMAKAWA, T., KATO, S., ISHIDA, K., KOMADA, T., MINODA, Y. 1983. Production of anthocyanins by *Vitis* cells in suspension culture. Agric. Biol. Chem. 47: 2185–2191.
25. PARK, H.H., KANG, S.K., LEE, J.H., CHOI, J.Y., LEE, Y.S., KWON, I.B., YU, J.H. 1989. Production of anthocyanins by *Vitis* hybrid cell culture. Kor. J. Appl. Microbiol. Bioeng. 17: 257–262.
26. YAMAMOTO, Y., KINOSHITA, Y., WATANABE, S., YAMADA, Y. 1989. Anthocyanin production in suspension cultures of high-producing cells of *Euphorbia millii*. Agric. Biol. Chem. 53: 417–423.
27. ASADA, Y., SAKAMOTO, K., FURUYA, T. 1994. A minor anthocyanin from cultured cells of *Aralia cordata*. Phytochemistry. 35 (6): 1471–1473.

28. HELLER, W., FORKMANN, G. 1988. Biosynthesis. In: The Flavonoids: Advances in Research Since 1980, (J.B. Harborne, ed.), Chapman and Hall, London, pp. 1–20.
29. GRISEBACH, H. 1982. Biosynthesis of anthocyanins. In: Anthocyanins as Food Colors, (P. Markakis, ed.), Academic Press, London, pp. 69–92.
30. DO, C.B., CORMIER, F., NICOLAS, Y. 1995. Isolation and characterisation of a UDP-glucose:cyanidin 3-O-glucosyltransferase from grape cell suspension cultures (*Vitis vinifera* L.). Plant Sci. 112: 43–51.
31. GLÄβGEN, W.E., SEITZ, H.U. 1992. Acylation of anthocyanins with hydroxycinnamic acids via 1-O-acylglucosides by protein preparation from cell cultures of *Daucus carota* L.. Planta 186: 582–585.
32. SALEH, N.A.M., POULTON, J.E., GRISEBACH, H. 1976. UDP-glucose cyanidin 3-O-glucosyltransferase from red cabbage seedlings. Phytochemistry. 15: 1865–1868.
33. PETERSEN, M., SEITZ, H.U. 1986. UDP-glucose: cyanidin 3-O-glucosyltransferase in anthocyanin-containing cell cultures from *Daucus carota* L.. Plant Physiol. 125: 383–390.
34. ROSE, A., GLÄβGEN, W.E., SEITZ, H.U. 1996. Purification and characterization of glycosyltransferases involved in anthocyanin biosynthesis in cell-suspension cultures of *Daucus carota* L. Planta 198: 397–403.
35. TANAKA, Y., YONEKURA, K., FUKUCHI-MIZUTANI, M., FUKUI, Y., FUJIWARA, H., ASHIKARI, T., KUSUMI, T. 1996. Molecular and biochemical characterization of three anthocyanin synthetic enzymes from *Gentiana triflora*. Plant Cell Physiol. 37(5): 711–716.
36. SALEH, N.A.M., FRITSCH, H., GRISEBACH, H. 1976. UDP-glucose:cyanidin 3-O-glucosyltransferase from cell cultures of *Haplopappus gracilis*. Planta 133: 41–45.
37. HRAZDINA, G. 1988. Purification and properties of a UDP-glucose: flavonoid 3-O-glucosyltransferase from *Hippeastrum* petals. Biochem. Biophys. Acta 955: 301–309.
38. TEUSCH, M. 1986. Uridine 5'-diphosphate-xylose: anthocyanidin 3-O-glucosyltransferase from petals of *Matthiola incana* R. Br.. Planta 169: 559–563.
39. TEUSCH, M., FORKMANN, G., SEYFFERT, W. 1986. Genetic control of UDP-glucose: anthocyanin 5-O-glucosyltransferase from flowers of *Matthiola incana* R. Br., Planta 168: 586–591.
40. KHO, K.F.F., KAMSTEEG, J., VAN BREDERODE, J. 1978. Identification, properties and genetic control of UDP-glucose: cyanidin 3-O-glucosyltransferase in *Petunia hybrida*. Z. Pflanzenphysiol. 88: 449–464.
41. JONSSON, L.V.M.; AARSMAN, M.E.G., VAN DIEPEN, J., SMITH, N., SCHRAM, A.W. 1984. Properties and genetic control of anthocyanin 5-O-glucosyltransferase in flowers of *Petunia hybrida*. Planta 160: 341–347.
42. KAMSTEEG, J., VAN BREDERODE, J., VAN NIGTEVECHT, G. 1978. Identification and properties of UDP-glucosyltransferase isolated from petals of the red campion (*Silene dioica*). Biochem. Genet. 16: 1045–1058.
43. KAMSTEEG, J., VAN BREDERODE, J., VAN NIGTEVECHT, G. 1978. Identification, properties and genetic control of UDP-glucose: cyanidin-3-rhamnosyl-(1→6)-glucoside-5-O-glucosyltransferase isolated from petals of the red campion (*Silene dioica*). Biochem. Genet. 16: 1059–1071.
44. KAMSTEEG, J., VAN BREDERODE, J., VAN NIGTEVECHT, G. 1980. Identification, properties and genetic control of a UDP-L-rhamnose: anthocyanidin 3-O-glucoside, 6''-O-rhamnosyltransferase isolated from petals of the red campion (*Silene dioica*). Z. Naturforsch. 35c: 249–257.
45. JONSSON, L.M.V., AARSMAN, M.E.G., POULTON, J.E., SCHRAM, A.W. 1984. Properties and genetic control of four methyltransferases involved in methylation of anthocyanins in flowers of *Petunia hybrida*. Planta 160: 174–179.
46. JONSSON, L.M.V., AARSMAN, M.E.G., SCHRAM, A.W., BENNIK, G.J.H. 1982. Methylation of anthocyanins by cell-free extracts of flower buds of *Petunia hybrida*. Phytochemistry. 21(10): 2457–2459.

47. BAILLY, C., CORMIER, F., DO, C.B. 1997. Characterization and activities of S-adenosyl-L-methionine: cyanidin 3-glucoside 3'-O-methyltransferase in relationship to anthocyanin accumulation in *Vitis vinifera* cell suspension cultures. Plant Sci. in press.
48. CORMIER, F., DO, C.B., MORESOLI, C., ARCHAMBAULT, J., CHAVARIE, C., CHAOUKI, F. 1992. Anthocyanin release from grape (*Vitis vinifera* L.) cell suspension. Biotechnol. Letters 14(11): 1029–1034.
49. FRANCIS, F.J. 1989. Food colorants: anthocyanins. Critical Rev. Food Sci. Nut. 28(4): 273–314.
50. TIMBERLAKE, C.F., HENRY, B.S. 1986. Plant pigments as natural food colors. Endeavour, New Series 10(1): 31–36.
51. AUSICH, R.L. 1994. Production of carotenoids by recombinant DNA technology. Pure Appl. Chem. 66(5): 1057–1062.
52. GAINER, J.L., WALLS, D.A., JONES, J.R. 1976. The effect of crocetin on skin papillomas and Rous sarcoma. Oncology 33: 222–224.
53. TARANTALIS, P.A., TSOUPRAS, G., POLISSIOU, M. 1995. Determination of saffron (*Crocus sativus* L.) components in crude plant extract using high-performance liquid chromatography-UV-visible photodiode-array detection-mass spectrometry. J. Chrom. A. 699: 107–118.
54. VAN CALSTEREN, M.-R., BISSONNETTE, M.C., CORMIER, F., DUFRESNE, C., ICHI, T., LEBLANC, Y., PERRAULT, D., ROEWER, I. 1996. Spectroscopic characterization of crocetin derivatives from *Crocus sativus* and *Gardenia jasminoides*. J. Agric. Food Chem. in press.
55. RAVISHANKAR, V.S., VENKATARAMAN, L.V. 1990. Induction of crocin, picrocrocin, and safranal synthesis in callus cultures of saffron - *Crocus sativus* L. Biotechnol. Appl. Biochem. 12: 336–340.
56. HIMENO, H., SANO, K. 1987. Synthesis of crocin, picrocrocin and safranal by saffron stigma-like structures proliferated *in vitro*. Agric. Biol. Chem. 51(9): 2395–2400.
57. KOYAMA, A., OHMORI, Y., FUJIOKA, N., MIYAGAWA, H., YAMASAKI, K., KOHDA, H. 1987. Formation of stigma-like structures and pigments in cultured tissues of *Crocus sativus*. Shoyakugaku Zasshi 41(3): 226–229.
58. HORI, H., ENOMOTO, K., NAKAYA, M. 1988. Induction of callus from pistils of *Crocus sativus* L. and production of color compounds in the callus. Plant Tissue Cult. Lett. 5(2): 72–77.
59. QUINKERT, G., SCHMIEDER, K.R., DÜRNER, G., HACHE, K., STEGK, A., BARTON, D.H.R. 1977. Light-induced reaction, XII. A convenient synthesis of dimethylcrocetin. Chem. Ber. 110: 3582–3614.
60. BIENAYMÉ, H. 1994. A convergent synthesis of symmetrical carotenoids using the Heck reaction. Tetrahedron Lett. 35(37): 6867–6868.
61. PRAS, N. 1992. Bioconversion of naturally occurring precursors and related synthetic compounds using plant cell cultures. J. Biotechnol. 26: 29–62.
62. CORMIER, F., DUFRESNE, C., DORION, S. 1995. Enhanced crocetin glucosylation by means of maltosyl-β-cyclodextrin encapsulation. Biotechnol. Tech. 9(8): 553–556.
63. KREIS, W., HOELZ, H., MAY, U., REINHARD, E. 1990. Storage of cardenolides in *Digitalis lanata* cells. Effect of dimethyl sulfoxide (DMSO) on cardenolide uptake and release. Plant Cell Tissue and Organ Culture 20: 191–199.
64. HASHIMOTO, H., TANAKA, T., OKEMOTO, H. 1993. Proc. First Int. Symp. Nat. Colorants for Food, Nutraceutical, Beverages and Confectionery, Amherst, MA, Nov. 7–11, 1993. p. 12
65. DUFRESNE, C., CORMIER, F., DORION, S. 1996. *In vitro* formation of crocetin glucosyl esters by *Crocus sativus* L. callus extract. Planta Med. in press.
66. PFISTER, S., MEYER, P., STECK, A., PFANDER, H. 1996. Isolation and structure elucidation of carotenoid-glycosyl esters in gardenia fruits (*Gardenia jasminoides* Ellis) and saffron (*Crocus sativus* Linne). J. Agric. Food Chem. 44(9): 2612–2615.

67. RYCHENER, M., BIGLER, P., PFANDER, H. 1984. Isolation and structure elucidation of neapolitanose, a new trisaccharide from the stigmas of garden crocusses (*Crocus neapolitanus* var.). Helv. Chim. Acta, 67, Fasc. 2 (45): 386–391.

Chapter Ten

TRANSGENIC MANIPULATION OF EDIBLE OILSEEDS

Toni Voelker

Calgene Inc.
1920 Fifth Street
Davis, California 95616

INTRODUCTION

Plants deposit fatty acids in triglycerides for high-density carbon and energy storage. The triglycerides accumulate predominantly in seeds, where the oil can make up a large component of the total weight of the tissue, from several to about 65 percent.[1] Vegetable oils are of major economic importance, representing the source for practically all plant-derived fats in our diet. The fatty acid triglyceride composition of a given oil determines its physical, chemical, and nutritional value. For example, acyl chain length and desaturation determine melting characteristics and other functional properties, such as nutritional values, (see Refs.2,3 for reviews). To date, the fatty acid composition of commercial oil seeds falls into only a handful of categories (Fig. 1). Temperate crops produce highly unsaturated C18 oils with different degrees of unsaturation (70–94%). Palmitate (16:0) (in this fatty acid nomenclature, the first number indicates the length of the fatty acyl chain, the second, the number of double bonds) represents

Functionality of Food Phytochemicals
edited by Johns and Romeo, Plenum Press, New York, 1997

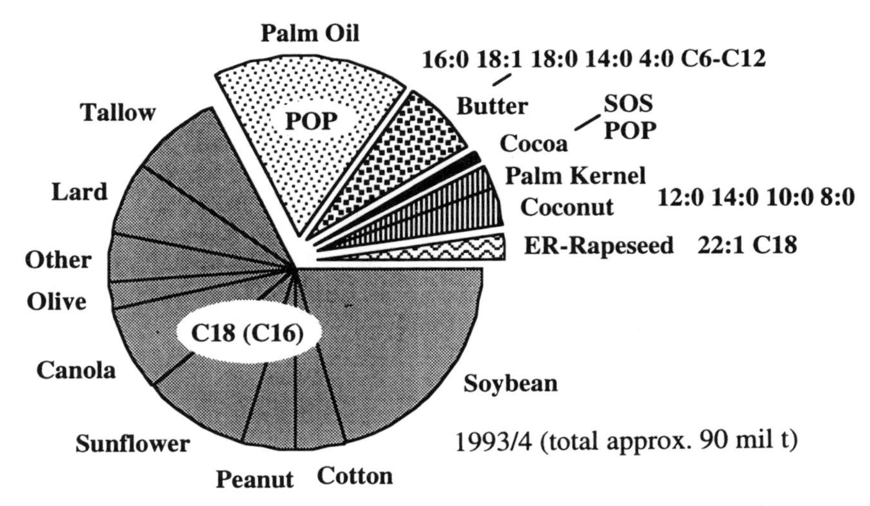

Figure 1. World production of fats and oils. The annual production is broken up by the respective crops. Oils with relatively similar composition are grouped. The predominant fatty acids, triglycerides are indicated. POP stands for a triglyceride with palmitate, oleate, palmitate as acyl groups, SOS stands for a triglyceride with stearate, oleate, stearate as acyl groups.

most of the saturated fatty acid component in these oils. All are liquid at room temperature and have to be hydrogenated in order to be useful for solid fat applications. Hydrogenation is currently becoming a non-desired oil modification.

Commercial medium-chain fatty acid (C8-C14) oils are exclusively of tropical origin, with coconut and palm kernel being the predominant sources. Lauric acid (12:0, 40–50%) is the major fatty acid in these oils.[1] In the US, these oils are mostly used in non-food applications, but they are found also in confectionery and non-dairy creamers.[4] Oils with rather unique compositions are palm oil (40% palmitate, 40% oleate, 18:1) and cocoa butter (high stearate (18:0), palmitate), which are used in shortenings, for frying applications, and in confectionery. This rather limited set of fatty acid compositions in commercially important vegetable oils contrasts with an extreme variability found in seed oils of different plant species, where fatty acids may vary with respect to chain lengths, degree of saturation, and presence of fatty acid modifying groups.[5] Since variations of the plant fatty acid biosynthesis pathway for the production of seed oil are so frequently found in nature, it was speculated that engineering of the common pathway by addition or suppression of a few enzymes should be feasible. In contrast to classical breeding, which is empirical in nature and mainly restricted to elimination of gene functions or the introduction of genes from close relatives, genetic engineering should make possible a rational redesign of a given seed oil using genes from any organism. Therefore, detailed knowledge of the biosynthetic pathway of plant lipids is necessary in order to define target genes crucial for the engineering of desired traits. Transferring genes to the target species then should allow the introduction of a novel enzyme, leading (if compatible with the new host) to a new trait, for example, deposition of a fatty acid previously not present in the plant.[6] Indeed, the first round of engineering, which will be summarized in this report, yielded a diverse crop of modified oils.

BIOSYNTHESIS OF PLANT OILS

Fortunately, in the past two decades, the general framework of plant lipid biosynthesis has been elucidated, and many genes have been cloned, (see Refs. 7,8 for reviews). Like all other plant carbon compounds, fatty acids are ultimately derived from photo assimilate. The *de novo* fatty acid biosynthesis is located in a special compartment of plant cells, the plastids. There, a set of seven enzymes, together comprising the fatty acid synthase, assembles the carbon chains (Fig. 2).

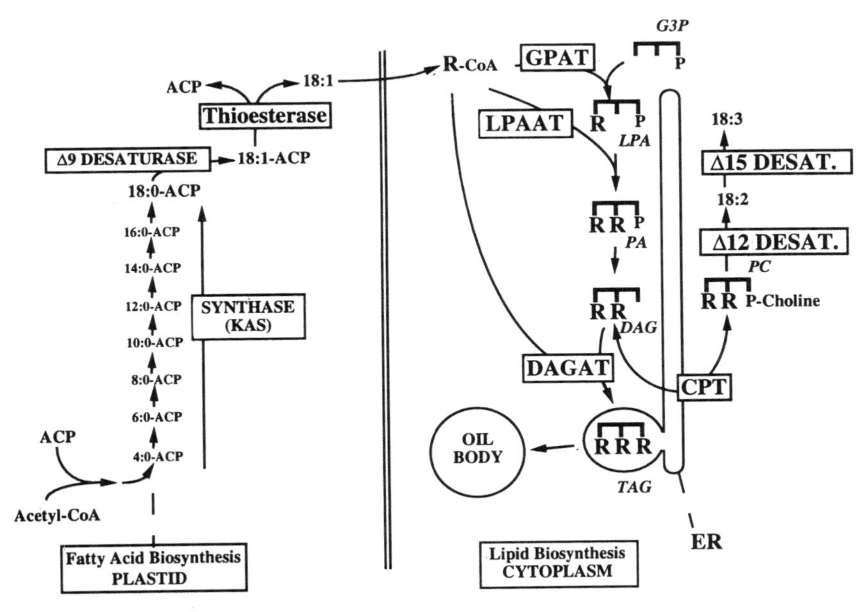

Figure 2. Overview of plant oil biosynthesis. Schematic, only the major steps and enzymes shown. ACP = acyl carrier protein; CPT = choline phosphotransferase; GPAT = glycerol 3-phosphate acyltransferase; LPAAT = lysophosphatidic acid acyltransferase; DAGAT = diacylglycerol acyltransferase; LPA = lysophosphatidic acid; PA = phosphatidic acid; DAG = diacylglycerol; ER = endoplasmatic reticulum; R = acyl residue; PC = phosphatidyl choline; TAG = triacylglycerol.

In a first step, a 2-carbon primer is attached to a protein, the acyl-carrier protein (ACP). Subsequently, the primer is extended by repeated condensation of C2 components to its chain, *i.e.* the acyl-ACP grows with each step by 2 carbons. Each cycle is initiated by a β-ketoacyl synthase (KAS), followed by a sequence of reduction and dehydration reactions. In plants, some of the chain elongation is terminated at the C16 stage, but the majority continues to 18:0-ACP. Subsequently, a Δ9 desaturase (Δ9: double bond between carbon 9 and 10, when counting from the carboxyl group) forms 18:1-ACP. The fatty acid synthase reactions are terminated by specialized thioesterases which release the acyl chains from their proteinaceous primer, the ACP, thereby producing free fatty acids. For subsequent modification and assembly into triglycerides, the free fatty acids exit the plastid, become reesterified to CoA, and enter the cytoplasmic glycerolipid biosynthesis pathway. There, a set of membrane-bound enzymes attach the three acyl chains to the glycerol backbone in sequence (Fig. 2). In the same compartment, 18:0, bound to a membrane lipid, phosphatidyl choline (PC), can be desaturated further. This leads to the formation of polyunsaturated fatty acids. Usually, double bonds are first introduced by specific desaturases at Δ12 and subsequently at Δ15. In certain plants, *e.g.*, rapeseed, specialized elongases can further increase chain length. It is clear that the glycerolipid pathway is used for the synthesis of membrane lipids as well as triglycerides. The key enzyme for the channeling into the storage triglycerides is the diacylglycerol acyltransferase (DAGAT). Polyunsaturated fatty acyl residues, after synthesis in phosphatidyl choline membrane lipids, can be shuttled into triglycerides. Finally, the completed triglycerides are deposited as oil in storage compartments, the oil bodies (Fig. 2). Oil bodies are basically triglyceride droplets surrounded by a single-layer of phospholipids.[8]

PRINCIPLES OF ENGINEERING OIL COMPOSITION

In order to modify a given plant oil by genetic engineering, it is necessary to have the target enzyme genes available (in most cases as cDNAs) and have the means to introduce them into the genome of the crop species, together with a promoter which confers seed specific expression. In most cases reported to date, promoters of seed storage protein genes from plants like rapeseed, soybean, etc., have been shown to be most useful, since in many oilseeds, lipid and seed storage protein synthesis takes place in the same cells and at the same developmental stage.[9,10] After splicing the reading frame of the target enzyme gene to the appropriate promoter, the chimeric gene has to be transferred to the host genome. This can be accomplished either by cocultivation of the target tissue with *Agrobacterium tumefaciens* containing the appropriate vectors, or by direct bombardment with DNA-coated microprojectiles. After transformation of the plant cells (insertion of the novel genes into the plant genome), complete plants are regenerated from the cultured tissue. To facilitate the selection of transformed tissues after cocultivation, usually a selective marker (*e.g.*, a resistance gene against a certain antibiotic) is cotransformed with the target gene.[11] To date, canola varieties of rapeseed, *Arabidopsis*, and soybean have been used most frequently for the modification of seed oil composition. Since the general pathway structures and cell biology of all oil-producing tissues are the same, it can be expected that the composition of all oil crops can be modified by genetic engineering, if desired.

In principle there are two approaches for manipulating lipid biosynthesis through genetic engineering. One can either genetically repress or overexpress the gene for a certain enzyme already active in the crop plant, or one can introduce a new enzyme with a novel specificity. In the following, I will first give a short overview of what has been achieved to date, and then will report in more detail on the path which led to high laurate canola, the first genetically engineered modified oil commercially available.

MODIFICATION OF UNSATURATION

Currently, in the USA, most of the vegetable oil used for food applications is hydrogenated, either to reduce polyunsaturates for increased oxidative stability, or to make the oil solid for use as a feed stock in edible spreads and margarines (elevate saturates). In the technical hydrogenation of plant oils, so-called *trans*-fatty acids are produced in large amounts. *Trans*-fatty acids have been regarded as risk factors for coronary heart disease.[12] Therefore, the prime targets of genetic engineering of unsaturation have been the repression of the respective desaturases in the oil crop itself. To date, enzyme repression strategies have been executed with all of the three major desaturases (Fig. 2) in soybean, *Arabidopsis*, and rapeseed, which demonstrated that the manipulation of unsaturation is possible with this approach and effective in a variety of oil seeds from different plant families. The principal rationale is illustrated in Fig. 3.

The pathway is seen as carbon flux through all respective intermediates, and the outcome at each step is determined by the competition of the respective enzymes for the particular intermediate. This is shown for 18:0-ACP. For example, in canola, the 18:0-ACP $\Delta 9$ desaturase reaction is much more efficient than the 18:0-ACP thioesterase, resulting in a 99%-efficient channeling to 18:1-ACP (Figs. 3, 4). Repression of 18:0-ACP desaturation by genetic engineering ($\Delta 9$ desaturase, Fig. 2), as demonstrated first by Knutzon, Thompson et al.[13] in rapeseed, led to high stearate canola (Fig. 4). Activities of the $\Delta 12$ and $\Delta 15$ desaturases, which are responsible for polyunsaturation (Fig. 2), were also genetically suppressed leading to low linolenic or high oleic varieties (Fig. 4). The most dramatic reduction of polyunsaturates is demonstrated with the high-oleic soybean (to generate this phenotype, the $\Delta 12$ desaturase reduction [14] was

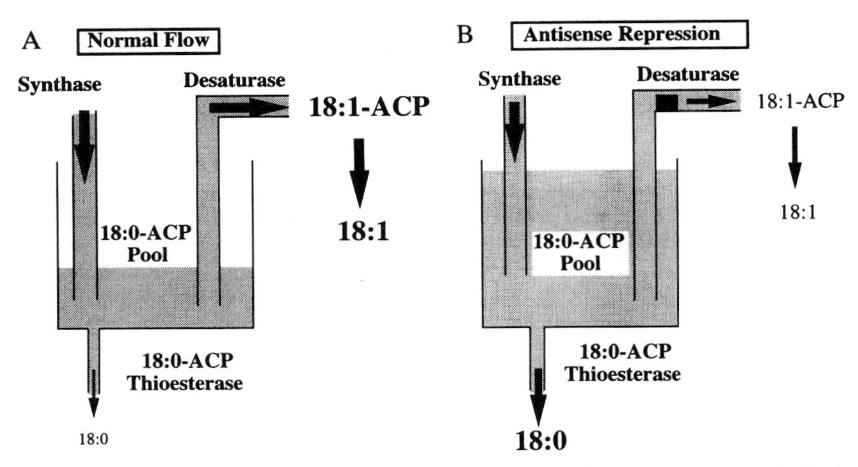

Figure 3. Flux model of pathway engineering using enzyme repression. Enzymes are symbolized as arrows, thickness of arrows indicates respective fluxes. The acyl-ACP pool concentrations are symbolized as depth of the "pool".

engineered by so-called co-suppression).[15] To date, all of these novel traits have been tested in multigeneration field trials, and the varieties are in advanced states of product development.

There is considerable interest in the elevation of g-linolenic acid (18:1 Δ 6, 9,12) in our diet. This fatty acid is an important precursor of polyunsaturated long-chain fatty acids and is essential for humans. In plants, this fatty acid is found in the oils of evening primrose and borage. Recently, using random sequencing of borage seed cDNAs, a Δ6 desaturase was found and expressed in *Arabidopsis* seeds. The genetic engineering resulted in the accumulation of g-linolenic acid, normally absent from *Arabidopsis* oil, up to 20% of the total (Terry Thomas, Texas A&M; oral presentation at the 12th International Symposium on Plant Lipids, Toronto, July 7–12, 1996).

MANIPULATION OF CHAIN LENGTH

Oils having 8:0, 10:0, 12:0, or 14:0 fatty acids are commonly referred to as medium-chain oils, with the commercially dominant sources coconut and palm kernel (with similar, laurate-dominated, composition).[1] Considerable interest in the possibility of the production of these fatty acids in temperate crops has emerged, especially for a temperate laurate crop. Currently, lauric oils are harvested exclusively from palm kernel and coconut seeds, and, in some instances, the supply has been insufficient in the past. Laurate is used predomi-

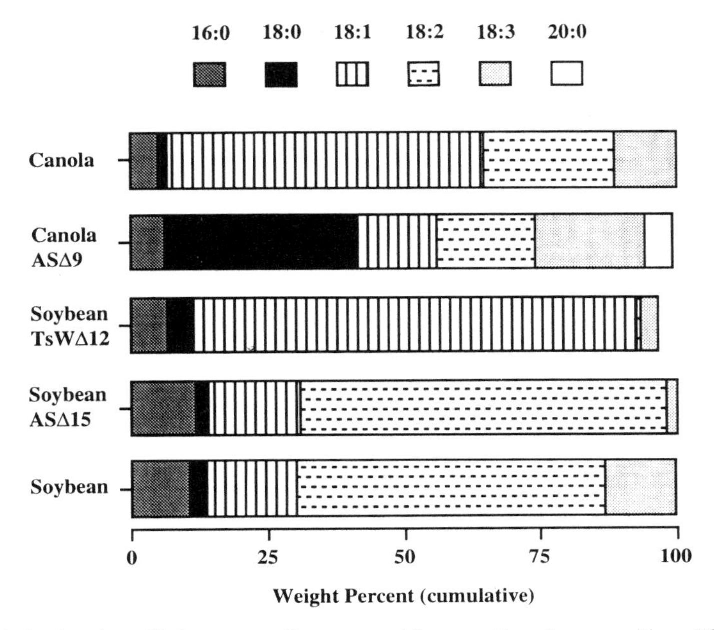

Figure 4. Engineering with desaturases. Bars represent the respective oils compositions. 16:0, palmitic acid; 18:0, stearic acid; 18:1, oleic acid; 18:2, linoleic acid; 18:3, linolenic acid; 20:0, arachidic acid. TsW, transwitch; AS, antisense; Δ9,12,15, delta-9, 12, 15 desaturases; CanolaAS9 from ref. 13 others from ref. 14.

nantly in detergent production, and annual US lauric oils imports approach one billion lbs/yr. The current strategy used for this metabolic engineering involves thioesterases for premature chain termination. As the acyl chain is elongated in the *de novo* fatty acid synthase (Fig. 2), pools of even chain-length acyl-ACP intermediates are formed. These pools of intermediates should be accessible for chain-length engineering.

The model proposes the introduction of a medium-chain-ACP hydrolyzing thioesterase, which would then compete with the elongation reaction (Fig. 5). Free medium chains would be produced, ready for plastid export and triglyceride synthesis. The first successful engineering of medium-chain fatty acid production was reported in *Arabidopsis* and rapeseed.[16] To achieve this goal, a thioesterase specific for 12:0-ACP and to a lesser degree for 14:0-ACP was isolated from the developing seeds of the California bay tree (laurate represents 60% of its seed oil). Subsequently, its cDNA was cloned and then expressed in canola under a seed storage protein promoter. Figure 6, top two bars, shows the impact of this engineering on canola oil. To date, thioesterases with a wide variety of specificities from different plants have been isolated. Figure 6, bottom bars, illustrates the versatility of this engineering.

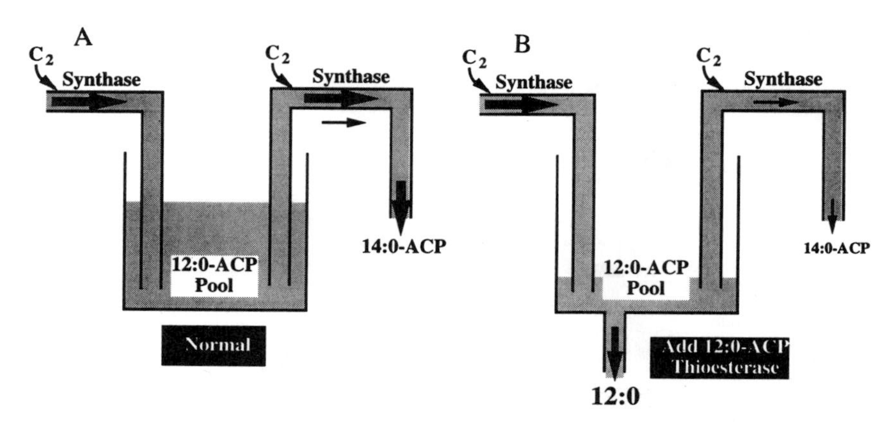

Figure 5. Flux model of pathway engineering using a novel enzyme. Enzymes are symbolized as arrows, thickness of arrows indicates respective fluxes. The acyl-ACP pool concentrations are symbolized as depth of the "pool".

REARRANGING TRIGLYCERIDES

The physical and nutritional properties of a given oil are not only defined by the nature of the fatty acids, but are also determined by the position of the respective fatty acids in the glycerol backbone.[2] For example, cocoa butter has the special melting characteristics which make chocolate stable in our hand, but let it instantly melt in the mouth. If one would randomize those residues, the melting curve would change. This property is conferred by its special triglyceride structure, namely saturate-oleate-saturate, with the saturates being palmitate and stearate.[1] In addition, there is evidence in the literature that saturated fatty acids are less cholesterogenic when on position one or three on the triglycerol backbone then in the middle.

Due to the biochemistry of plant lipid biosynthesis, triglyceride positions are usually not filled randomly. For example, in palm oil, palmitate occupies only *sn1* and *sn3* positions; in cocoa butter, palmitate and stearate are excluded from position *sn2*.[1] This exclusion of saturated fatty acids from *sn2* of the glycerol backbone is generally the rule for most plant species investigated,[21] but there are exceptions. For example, the tropical lauric oils (coconut, palm kernel) have up to 95% saturated fatty acids.[1] Obviously, these plants must have acyl transferases with modified acyl preferences. In summary, not only have plants developed the ability to synthesize a wide range of fatty acids, they have also become capable of precisely directing the triglyceride position of a given fatty acid.

This specificity of position 2 acylation is conferred by an enzyme labeled lysophosphatidic acid acyl transferase (LPAAT, see Fig. 2 for its position in the lipid biosynthesis pathway). Indeed, when high lauric canola oil was analyzed,

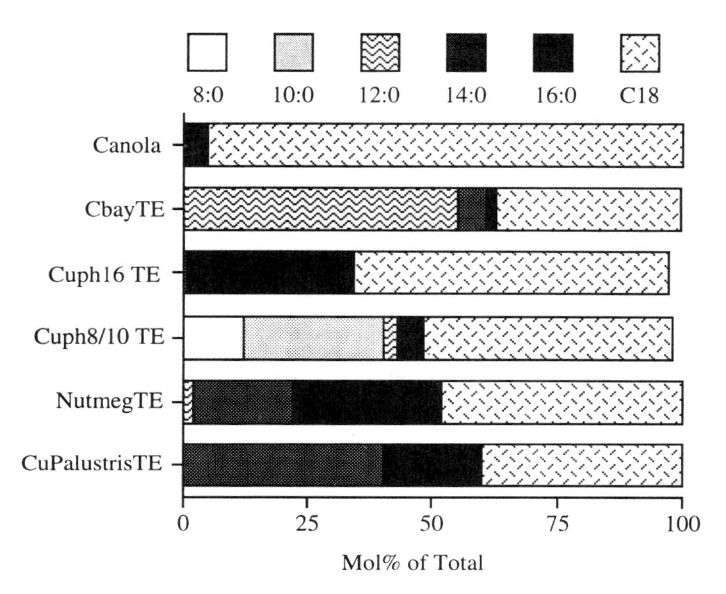

Figure 6. Engineering with thioesterases. Bars represent the respective oils compositions. Top bar is unmodified canola oil. CbayTE from ref. 17, Cuph16 TE from ref. 18, Cuph8/10 TE from ref. 19, NutmegTE from ref. 20, and CuPalustrisTE from Voelker, unpublished.

the medium-chains were found to be excluded from the middle position of the triglycerides. Since all saturated fatty acids together are about 60 mol% of the total oil composition, a structured triglyceride, similar to cocoa butter or palm oil, was created with shorter saturated fatty acids, predominantly C12. In Figure 7, the triglyceride compositions of coconut oil and high laurate canola are compared, to show the drastically different nature of these high-lauric oils. More recently, coconut LPAAT has been purified and its cDNA isolated. When expressed in high-laurate canola, LPAAT drastically increases laurate at position 2 (Fig. 7). This should allow us to make even higher laurate levels, making super-high laurate canola a competitive supplier for lauric acid.

LAURATE CANOLA: THE PATH FROM PROPOSAL TO MARKETPLACE

This program is rooted in 1984 when the price for coconut oil skyrocketed. At this time, several groups in the US began to search for the creation of alternate laurate crops. At Calgene and other places, biochemists attempted to elucidate a mechanism and find an enzyme which could be responsible for the production of this fatty acid. Success came as late as 1988 when Calgene researchers found

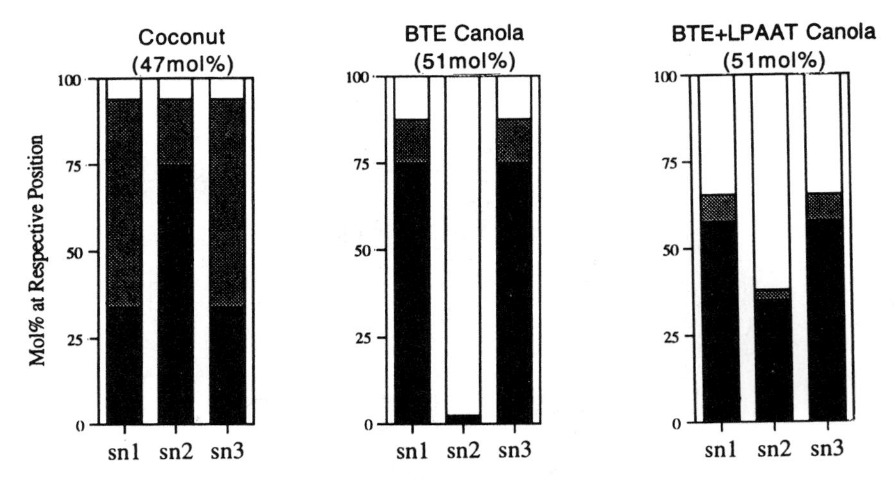

Figure 7. Triglyceride structures of lauric oils. Compared are: conventional coconut oil 1 (left); canola oil engineered with the California bay thioesterase 17 (BTE, middle); canola oil engineered with BTE and a coconut LPAAT, Knutzon and Davies, unpublished (BTE+LPAAT, right). The three columns represent the three positions on the glycerol molecule, sn1, 3, are assumed as equal. Bar coding: black, laurate; gray, other saturates; open, unsaturated acyl chains.

evidence for a laurate-ACP preferring thioesterase in seeds of California bay trees. Protein purification took until 1990 when enough bay thioesterase was isolated to allow protein sequencing. Based on protein sequence, the bay thioesterase cDNA ("gene") was isolated and expressed under a seed storage promoter in canola.[16] The first laurate-producing canola seeds were harvested in summer 1991. Since then, many different activities have been conducted in parallel in order to ensure the creation of a novel oil crop. About three years of breeding (two seasons/year) were needed to bring the yields up to normal canola levels. Once a desirable canola plant has been identified, seed increase is very fast. Since one plant can produce up to several thousand seeds, it needs only a few generations (2 years) to produce adequate amounts of seeds needed for commercial production from a one-plant selection (one pound of seed can be increased to 500,000 pounds in one year).

The generation of novel crops via plant biotechnology is currently tightly supervised by the respective regulatory agencies, Environmental Protection Agency (EPA; only for pesticide resistant plants), US Department of Agriculture (USDA) and the Food and Drug Administration (FDA). There are several aspects of concern, namely the growth of the novel engineered crop itself and the properties of the new, potentially changed products derived from the plant (Ref. 22 summarizes these aspects with the FlavrSavr Tomato). For the generation of crops with modified vegetable oils, two agencies are concerned. USDA regulates the outdoor release and cultivation of the novel crops. Major concern here is the introduction of genes into a crop plant which might cause environmental problems (weediness, outcrossing potential of the crop plants, potential of the new genes to cause harm in the environment after transfer to another organism, etc.). Calgene first applied for laurate canola field trials in 1991 and had the first trial in summer 1992. Since then, more than 10 field trials have been conducted as part of the breeding program and to create data for a deregulation application for the USDA. Additionally, more than a dozen field trials have been conducted in Canada, Finland and Great Britain. In no case did the laurate canola behave differently from conventional canola. Therefore, after testing this variety over 6 seasons, Calgene applied to the USDA for deregulation. The exempt status was granted in October 1994. Since then, the laurate canola can be grown and shipped like any other canola variety.

A second federal agency, the FDA, regulates the release of genetically engineered plant products for food. In the case of oil seed crops, FDA is concerned about changes in the meal (the solid seed residue remaining after oil extraction), which is used in animal formulations, and the actual oil composition. The meal contains the genes and enzymes introduced via plant transformation, and it is important to assess potential for increased toxicity, allergenicity, etc.. Of concern with respect to the actual oil is the total impact on the average diet; for example whether the use of the new oil will significantly increase the total intake of saturated fatty acids. The FDA's general policy guidelines for plant biotechnology products are available.[23] For the deregulation of laurate canola for food (and its meal for feed), Calgene accumulated a safety data package. For example, we demonstrated that the bay thioesterase protein is not detectable in the oil, and that its levels in the meal are very low. All important nutrient factors commonly found in canola oil and the meal were unchanged. We also demonstrated that the bay thioesterase is digested by common gut enzymes. Basically, aside from the fatty acid composition, we found that, in all measured aspects, laurate canola is not substantially different from conventional canola. The USDA/FDA deregulation of Calgene's marker gene (resistance to the antibiotic kanamycin) had been already achieved earlier (1994), as a co-application with the FlavrSavr tomato documents, and was not part of the high laurate application. Since lauric acid is already considered "GRAS" (generally considered as safe), the FDA, after reviewing the data, deregulated laurate canola in the spring 1995. In summary, since spring 1995, Calgene's laurate canola has been treated in the US as any other canola variety, in respect to agronomics and uses, including for food (Canada deregulation, spring 1996). During the last year, Calgene has established a production system where it supplies planting seed to the farmer, then buys back the crop. Subsequently, the seeds are crushed, and oil is extraction by contracted crushers. In summer 1995, the first commercial crop was harvested in Georgia (1.2 mill pound of oil). A larger crop was planted for the summer 1996 season. A more detailed report on the laurate canola crop production and its application is published elsewhere.[24]

The laurate project was started predominantly to create a temperate crop for lauric acid, a major feed stock for detergents, and also for potential food applications. As mentioned earlier, the resulting triglyceride (Fig. 7, middle) is novel, and therefore might have novel functionality. To date, we have conducted a physical characterization of this oil, and have performed a limited amount of application research. Basically, the laurate canola oil can be modified to resemble fats used in confectionery manufacturing (*e.g.*, laurate canola chocolate has a very good melting characteristics, flavor release, and mouthfeel). It has been demonstrated that this oil can be used to produce simulated dairy products (low-fat coffee whiteners, dips, imitation cheeses).[24] Although the results are preliminary, it is clear that a variety of functional properties can be provided through the use of laurate canola in food products. A second case can be found with the high-oleic soybean: not only were, for many applications, the undesired polyunsaturates reduced to a few percent, but the oil was found to be exceptionally heat stable. In frying applications, it compares well with heavily hydrogenated soybean oil (A. Kinney, Dupont, personal communication).

SUMMARY, OUTLOOK

With the first products of single-gene manipulations already at the market place, plant biotechnology is at the threshold of making its impact on the market of vegetable oil feed stocks.[25] It has been shown that through directed genetic manipulations near-total control over the saturation status of a given oil can be achieved. Also, acyl chain-length distributions of a given oil can be manipulated to a large extent by the introduction of a novel thioesterase. As the laurate canola story indicates, often the new product may be useful in different applications from what was envisioned at the outset of the project. Also, as history with established technologies has shown, novel feed stocks are usually not readily accepted by the market, and the generator must drive its implementation by giving samples away, support application research, etc.[26] The near future will show whether these novel oils (summary of current varieties in Refs. 27,28) can maintain their promise and carve a niche in the current oils marketplace. Of utmost importance is not only the generation of the desired oil compositions, but also the ability to produce these new feed stocks at a reasonable price. With the rapid increase in knowledge concerning every aspect of plant fatty acid biosynthesis and the refinement of the involved technologies, it is likely that in the next several decades, nearly complete control over triglyceride fatty acid structure and composition can be achieved. This will provide the food industry with a new set of feed stocks for creating food systems having desirable attributes for the consumer.[29]

REFERENCES

1. PADLEY, F.B., GUNSTONE, F.D., HARWOOD, J.L. 1994. Occurrence and characteristics of oil and fats. In: The lipid handbook, (F.D. Gunstone, J.L. Harwood, F.B. Padley, eds.), Chapman and Hall, London, pp. 49–170.
2. HEGENBART, S. 1992 (11). Taming the Tempest: Health Developments in Fats and Oils. Food Product Design, pp. 27–43.
3. WILLETT, W.C. 1994. Diet and health: What should we eat? Science 264: 532–537.
4. WEISS, T.J. 1982. Food Oils and Their Uses. Avi Publishing, Westport, CN, 310 pp .
5. BATTEY, J.F., SCHMID, K.M., OHLROGGE, J.B. 1989. Genetic Engineering for plant oils: potential and limitations. TIBTECH 7:122–126.
6. SLABAS, A.R., SIMON, J.W., ELBOROUGH, K.M. 1995. Information needed to create new oilseed crops. Inform 6:159–166.
7. SOMERVILLE, C., BROWSE, J. 1991. Plant lipids: Metabolism, mutants, and membranes. Science 252:80–87.
8. OHLROGGE, J., BROWSE, J. 1995. Lipid biosynthesis. The Plant Cell 7:957–970.
9. KRIDL, J.C., MCCARTER, D.W., ROSE, R.E., SCHERER, D.E., KNUTZON, D.S., RADKE, S.E., KNAUF, V.C. 1991. Isolation and characterization of an expressed napin gene from *Brassica rapa*. Seed Science Research 1:209–219.
10. HÖGLUND, A.-S., RÖDIN, J., LARSSON, E., RASK, L. 1992. Distribution of napin and cruciferin in developing rapeseed embryos. Plant Physiology 98:509–515.
11. KLEE, H., HORSCH, R., ROGERS, S. 1987. *Agrobacterium*-mediated plant transformation and its further applications to plant biology. Ann. Rev. Plant Physiology 38:467–486.
12. ALLEN, A.H. 1995. Translating the mixed signals on *trans* fat. Food Prod. Design 5 (Nov.): 30–49.
13. KNUTZON, D.S., THOMPSON, G.A., RADKE, S.E., JOHNSON, W.B., KNAUF, V.C., KRIDL, J.C. 1992b. Modification of *Brassica* seed oil by antisense expression of a stearoyl-acyl carrier protein desaturase gene. Proc. Natl. Acad. Sci. 89:2624–2628.
14. FADER, G.M., KINNEY, A.J., HITZ, W.D. 1995. Using biotechnology to reduce unwanted traits. Inform 6:167–169.
15. MATZKE, M., MATZKE, A.J.M. 1993. Genomic imprinting in plants: Parental effects and trans-inactivtion phenomena. Ann. Rev. Plant Phys. Plant Molec. Biol. 44:53–76.
16. VOELKER, T.A., WORRELL, A.C., ANDERSON, L., BLEIBAUM, J., FAN, C., HAWKINS, D., RADKE, S., DAVIES, H.M. 1992. Fatty acid biosynthesis redirected to medium chains in transgenic oilseed plants. Science 257:72–74.
17. VOELKER, T.A., HAYES, T.R., CRANMER, A.C., DAVIES, H.M. 1996. Genetic engineering of a quantitative trait: Metabolic and genetic parameters influencing the accumulation of laurate in rapeseed. The Plant Journal 9:229–241.
18. JONES, A., DAVIES, H.M., VOELKER, T.A. 1995. Palmitoyl-acyl carrier protein (ACP) thioesterase and the evolutionary origin of plant acyl-ACP thioesterases. The Plant Cell 7:359–371.
19. DEHESH, K., JONES, A., KNUTZON, D.S., VOELKER, T.A. 1996. Production of high levels of 8:0 and 10:0 fatty acids in transgenic canola by over-expression of *Ch FatB2*, a thioesterase cDNA from *Cuphea hookeriana*. The Plant Journal 9:167–172.
20. VOELKER, T.A., JONES, A., CRANMER, A.M., DAVIES, H.M., KNUTZON, D.S. 1997. Broad-range and binary-range acyl-ACP thioesterases suggest an alternative mechanism for medium-chain production in seeds. Plant. Phys. in press.
21. BROWSE, J., SOMERVILLE, C. 1991. Glycerolipid synthesis: Biochemistry and regulation. Ann. Rev. Plant Phys. Plant Molec. Biol. 42:467–506.
22. REDENBAUGH, K., HIATT, W., MARTINEAU, B., EMLAY, D. 1994. Regulatory assessment of the FLAVR SAVR tomato. Trends in Food Science & Technology 5(4):105–110.

23. KESSLER, D.A. 1992. Statement of Policy: Foods Derived From Now Plant Varieties. Federal Register 57 (104):22984–23005.
24. DELVECCHIO, A.J. 1996. High-laurate canola. Inform 7:230–242.
25. ANONYMOUS. 1995. Transgenic oilseed harvests to begin in May. Inform 6:152–157.
26. MANGOLD, H.K. 1994. "Einfache Triacylglycerine" in Fetten und Ölen aus den Samen von Wildpflanzen: Wertvolle Rohstoffe für die chemische Industrie. Fat Sci. Technol. 1:23–27.
27. SLABAS, A., ROBERTS, P., OSMESHER, J. 1982. Characterisation of fatty acid synthesis in a cell free system from developing oil seed rape. In: Biochemistry and Metabolism of Plant Lipids, (J.F.G.M. Wintermans, P.J.C. Kuiper, eds.), Elsevier, Amsterdam, pp. 251–256.
28. MURPHY, D.J. 1996. Engineering oil production in rapeseed and other oil crops. TIBTECH 14:206–213.
29. KINNEY, A.J. 1996. Designer oils for better nutrition. Nature Biotech. 14:946.

Chapter Eleven

QUANTITATIVE MICROSCOPIC APPROACHES TO CARBOHYDRATE CHARACTERIZATION AND DISTRIBUTION IN CEREAL GRAINS

R. Gary Fulcher,[1] S. Shea Miller,[2] and R. Roger Ruan[1,3]

[1] Department of Food Science and Nutrition
University of Minnesota
St. Paul, Minnesota 55108
[2] Center for Food and Animal Research
Agri-Food and Agriculture Canada
Ottawa, Ontario, K1A OC6
[3] Department of Biosystems and Agricultural Engineering
University of Minnesota
St. Paul, Minnesota 55108

INTRODUCTION

Development and utilization of improved cereal varieties for domestic and international use depends upon the ability of both breeders and processors to: (a)

Functionality of Food Phytochemicals
edited by Johns and Romeo, Plenum Press, New York, 1997

identify potential sources of new or improved traits; (b) incorporate these traits into agronomically acceptable cultivars; and (c) exploit these characteristics in improving traditional products or developing new ones. To do so the breeder must identify and measure desirable traits in a large number of potential cultivars in a relatively rapid and simple manner, and the processor must have access to rapid and precise methods for defining grain "quality", the combination of chemical and structural attributes which defines the utility of grains in processing conditions. In most major cereals, including wheat, rice, barley, oats, maize, and sorghum, our ability to identify and characterize these traits in phytochemical terms is quite variable depending on the history of the crop, but in all cases it is rudimentary and somewhat empirical. This is not surprising in view of the large number of molecular species which interact to contribute to the overall biochemistry of the grains. Consequently, the most important tool for identifying new cereal varieties is pilot or micro-scale processing, (*e.g.* pilot milling, malting, baking or extrusion) in which relatively large numbers of samples can be analyzed for their suitability for use in food systems. Although the majority of the storage reserves in cereal grains are polysaccharides (*e.g.* starch, pentosans, β-glucans) or carbohydrate-linked complexes (*e.g.* phenolic glycosides, lignin), with few exceptions (*e.g.*, β-glucan determination[1]) individual chemical traits are either too costly or cumbersome to measure routinely in large numbers, or are ill-defined and inappropriate for daily use.

The essential elements in the analysis of grain quality are *simplicity*, *speed*, and *low cost*. The procedures used by the plant breeder must be rapid if they are to be applied to the hundreds or thousands of genetic lines from which suitable parental material will be selected. In turn, this requirement dictates that the procedures developed for screening genetic traits must be simple and inexpensive: costly, tedious protocols are normally not appropriate for routine analysis, nor are they often rapid and easy to use, except in the most exceptional circumstances. The processor also faces the need to perform routine analyses of thousands of samples of both raw and processed material each year in order to ensure that product specifications are met.

In this context, it bears repeating that the *functionality* of cells and tissues is a direct reflection of their *structural characteristics*, an observation based on cereals and first made over a hundred years ago.[2] For several years, we have exploited this association in evaluating a selected number of new screening methods for rapid and inexpensive characterization of quality traits. The procedures are based primarily on two quantitative imaging techniques, *scanning microspectrophotometry* and *digital image analysis (DIA)*. Both methods have become increasingly popular in the food and grain industries, and show considerable promise for routine and precise evaluation of a wide range of quality parameters in grains and grain products. We have concentrated primarily on grain carbohydrates, and a few simple applications of quantitative imaging are described in the following sections to show the potential of these techniques:

characterizing grain constituents *in situ*; mapping the distribution of molecular components in the grain; and rapid measurement of key quality traits in different cultivars. Recently, we have added magnetic resonance imaging (MRI) techniques for improved definition of key physiological processes such as hydration and germination.

STRUCTURE/FUNCTION RELATIONSHIPS IN GRAINS

Every biochemical constituent in a cereal grain is synthesized and stored in discrete, identifiable structures in specific areas of the kernel. This is especially true of the many different carbohydrates in the grain, and in order to fully understand the complex interactions among these various constituents during development, germination and processing, it is essential to understand the structure and organization of the grain. Figure 1 shows the relationships among the different cell and tissue types in a typical barley kernel. In most respects, the architecture of the barley grain closely parallels that of other cereals.[3,4]

In all cereals, the outer grain tissues (including the hulls, if present) contain high concentrations of phenolic compounds, including lignin, and esterified phenolic acids such as ferulic and coumaric acids.[5–8] Regardless of their exact composition, the hulls, pericarp, testa, and nucellus are major contributors to a range of important and diverse physiological and processing characteristics including: (a) permeability of the grain to water and nutrients during germination; (b) resistance to a wide range of potentially damaging pests such as *Sitophilus zeamais*[9] and *Fusarium graminearum*;[10] (c) high levels of insoluble dietary fiber which are characteristic of cereal brans; (d) flavor, color, and texture of finished products; and (e) antioxidant effects in products. Genetic variation in the composition and thickness of these layers contributes significantly to variation in most of these characteristics. The characteristics of the outer layers have been described in detail for barley[11] and oats.[3,4]

Perhaps one of the most important tissues in the typical cereal grain is the one to three cell thick aleurone layer (Fig. 2) which completely surrounds the starchy endosperm and part of the germ. It contains a wide range of storage materials, including lipids, phytin, minerals, protein, and phenolic compounds, and it is a primary source of the hydrolytic enzymes which degrade the starchy endosperm during germination.[12] It is here that the fibrous cell walls (in concert with those of the associated pericarp, nucellus, and testa) exert some level of resistance to invading soil microorganisms. They also presumably play a role in permeability to hydration and rate of release of enzymes. The thickness, number of cell layers, phenolic acid content, and relationship to other tissues all play an important part in the functioning of the aleurone layer. However, measurement of variation due to genetic influences on several of these characteristics traditionally has been tedious, slow and imprecise.

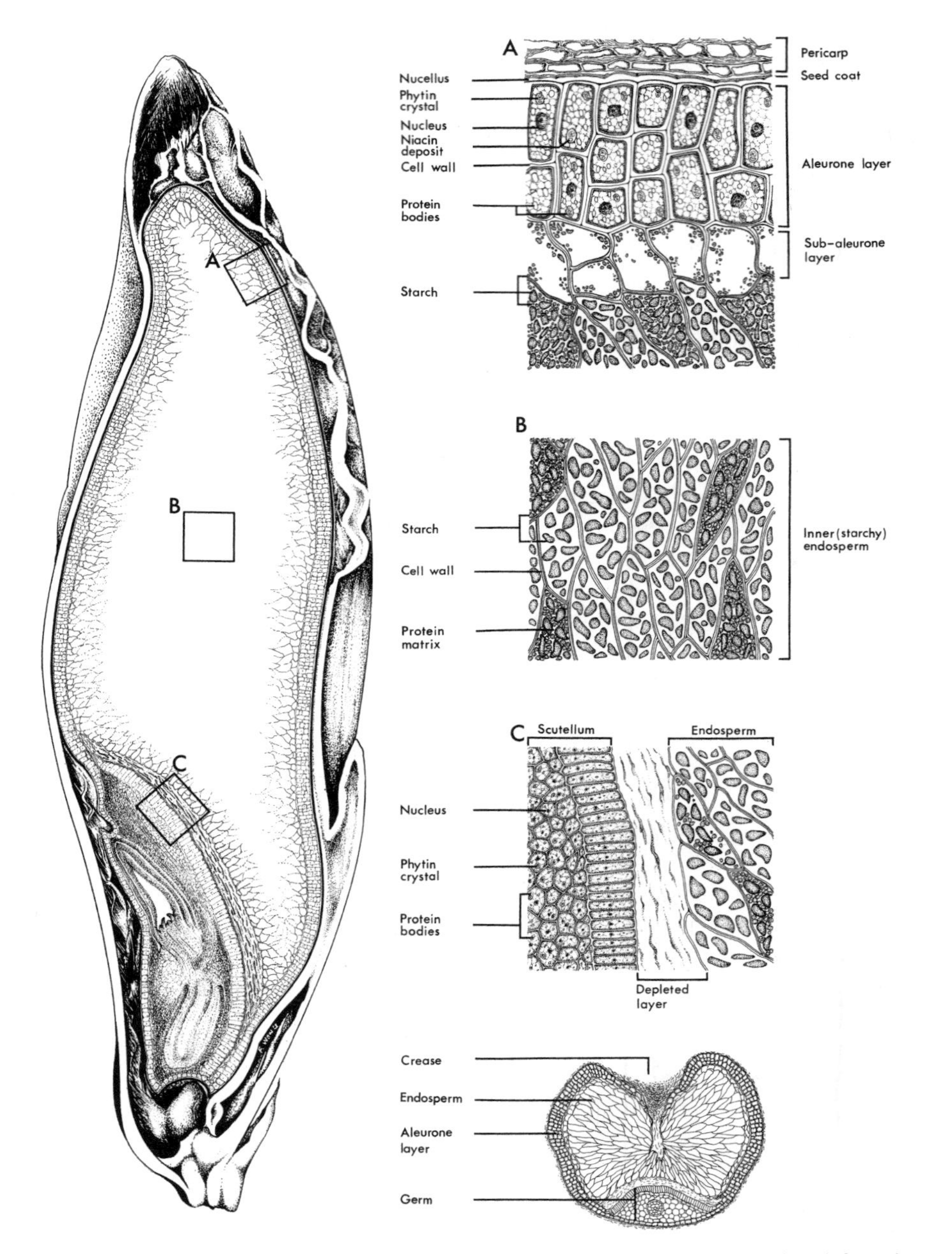

Figure 1. Diagrammatic representation of a barley kernel, showing longitudinal section (left) and enlarged selected regions of the grain at A, B and C.

The starchy endosperm is the primary source of the starch (50–70% of the grain dry weight) that is eventually converted to fermentable sugars during germination. It is also the primary non-cellulosic polysaccharide in industry, and products range from adhesives and thickeners to paper and textile sizing agents. Starch is typically comprised of high molecular weight amylose and amylopectin molecules in a wide range of ratios depending on genetic and environmental origin, and the polymers are stored in discreet granules with unique architectures in each species. The polymers are not uniformly distributed across each granule,[13] and in some grains (*e.g.* rye, wheat, and barley), starch granules also occur in two or more distinct size classes.[14,15] There is some evidence that each contributes differently to product quality. Typical wheat starch granules are shown in microscopic section in Figure 3, which illustrates the distinct bimodality of wheat starch populations. If product quality is dependent upon simple differences in granule size ratios, then it follows that rapid and precise measurements of this trait would be useful in predicting the performance of raw materials during processing; imaging techniques have proven to be very useful in this context, although they are not yet widely applied in the cereal industry.

Although starch is the most abundant polysaccharide in cereals, a major determinant of both malting and feed quality in barley, baking performance in wheat and rye flour, and viscosity of oat products is the complex of high molecular weight polymers which comprise the majority of the cell wall sur-

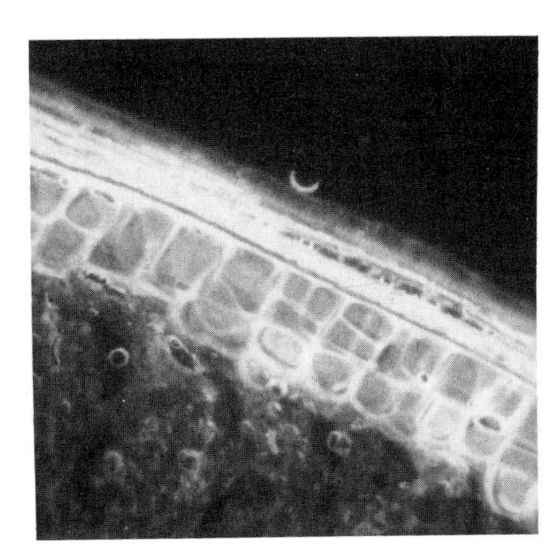

Figure 2. A fluorescence micrograph of a portion of a barley grain showing the bright (white) fluorescence of phenolic acid rich cell wall components in the aleurone layer. The lower portion of the micrograph primarily shows starchy endosperm.

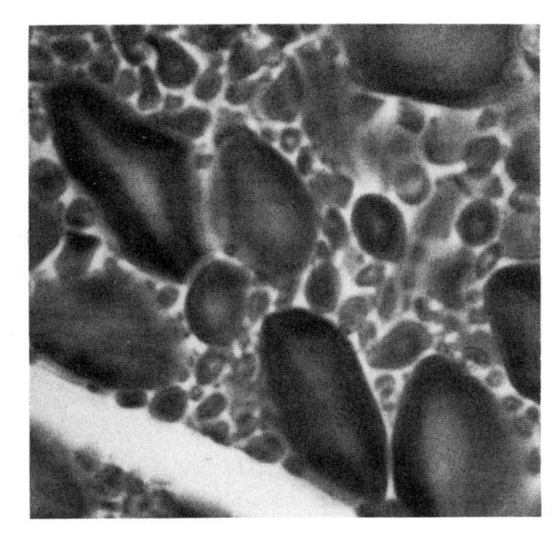

Figure 3. Bright field micrograph of a portion of the starchy endosperm of wheat after staining with iodine/potassium iodide. The starch granules exist as two distinct populations, the large A granules, and much smaller B granules.

rounding each starch-filled endosperm cell (Fig. 4). The cell wall contributes perhaps 1–5% of the endosperm dry weight, yet it provides a significant barrier to movement of enzymes secreted into the starchy endosperm by the aleurone layer or scutellum. The wall's high viscosity inhibits filtering during brewing, is a major source of both soluble and insoluble dietary fiber for humans and livestock, and is a source of high viscosity polymers for several new industrial and medical applications. In most common cereals, the endosperm cell wall contains pentosans (arabinoxylans) with small amounts of galactose and mannose, mixed linkage (1,3)(1,4)-β-D-glucan,[6] as well as small amounts of phenolic acids and protein.[16–18] Each cereal type characteristically synthesizes and stores these polymers in distinctive ratios,[19] and within each type, there is tremendous heterogeneity in their distribution across the endosperm, depending on the cultivar. Wheat, for example, contains pentosans as the major endosperm polysaccharide (>80%) with only minor amounts of β-glucan.[19] Barley and oats, in contrast, synthesize over 80% of their endosperm cell wall as β-glucans of varying solubilities and molecular weights, with only a small amount of pentosan.[17] Rye includes significant quantities of both pentosan and β-glucans.[20] Although these polymers represent only a small amount of the grain dry weight, they are extremely hydrophilic. This is of significance to the normal germination

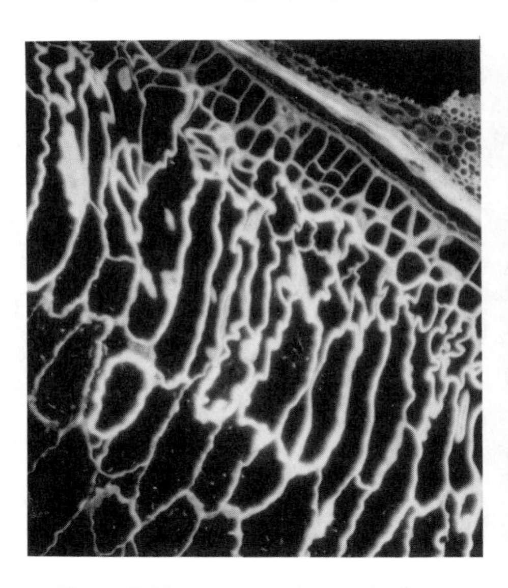

Figure 4. Fluorescence micrograph of a cross section of a mature barley kernel after staining with Calcofluor to emphasize β-glucans. Note the tremendous variation in cell wall thickness in different regions of the tissue.

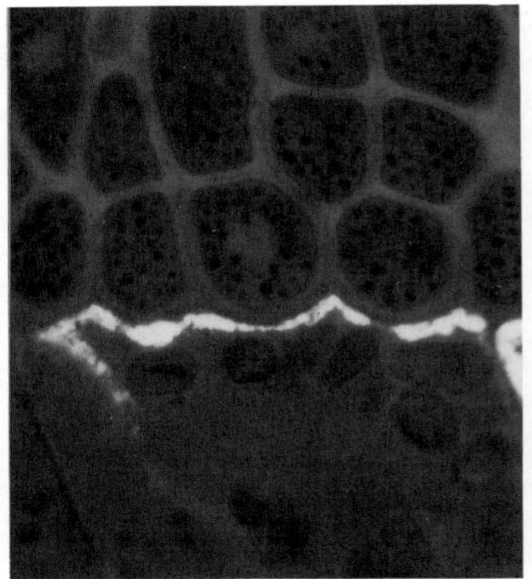

Figure 5. Fluorescence micrograph of a barley kernel cross section after treatment with aniline blue. The bright layer is tentatively identified as (1-3)-β-D glucan, which appears to form a barrier between the aleurone layer (top) and the starchy endosperm.

process, but it also means that they contribute significantly to a wide range of hydration and dehydration effects in typical food or beverage processing systems. Thus, they are of critical economic importance. Endosperm cell walls are also important barriers to penetration of enzymes through the endosperm during germination, a fact that has concerned brewers for decades.[12]

Because most endosperm cell walls contain significant quantities of mixed-linkage β-glucans, it is not surprising that we expend major efforts on their analysis, genetic modification, hydrolysis and industrial properties. The polymers are easily observed by fluorescence microscopy[21] (Figs. 4,5), and they are readily extracted and measured by enzymatic[1] or flow injection methods.[22,23] There is some suggestion from microscopic analyses that other materials such as β-(1-3)-D-glucans[24–26] (Fig. 5), phenolic esters, and proteins[17] are deposited in strategic regions of the endosperm cell wall. Detailed evaluation of structure/function relationships in the cereal kernel show clearly that many of the most important processing and quality attributes within the grain are also organized in a highly polarized or localized fashion; this is readily apparent in micrographs of cell wall phenolics (Fig. 2), starch granule populations (Fig. 3), and (1-3)(1-4)-β-D-glucan-rich cell walls (Fig. 4).

Regardless of the relative contribution of individual cell traits to grain quality, it follows that they must be measured quickly and efficiently in order to provide useful genetic screening or quality control opportunities. Two distinctly different techniques, *scanning microspectrophotometry* and *digital image analysis*, are especially useful for starch, cell wall, and polysaccharide analysis. They may be used with bright field, fluorescence, or other forms of light microscopy, and they may be used separately or in combination. In either case, the measurement of relatively small amounts of cellular carbohydrate is relatively simple, and often exceeds the limit of detection of many other analytical systems. A third technique, magnetic resonance imaging (MRI), also allows direct examination of hydration events in grains, particularly as they relate to carbohydrate-rich structures.

SCANNING MICROSPECTROPHOTOMETRY

Because of the intimate association between structure and function, we have chosen to use quantitative microscopic imaging techniques to identify, characterize, and map the distribution of many plant polysaccharides and associated compounds such as phenolic esters and proteins. Quantitative light microscopy can be used to detect naturally occurring substances in plant cells and tissues and raw or processed foods, either by direct analysis of their fluorescence or absorbance, or by using applied fluorochromes or diachromes. Where practical, fluorescence probes are preferred for maximum sensitivity and specificity. Current sensors are able to detect relatively faint fluorescence, and improved specificity is apparent when fluorescence techniques are employed. Reagents which are detected in traditional bright field applications (diachromes) are also extremely useful, however, and microscope photometry adapts readily to either mode. In addition, optical systems capable of focusing and detecting either transmitted or reflected light signals are also commonly available. Measurement may involve spectral characterization, mapping of components, or kinetic analysis of changes in absorption or fluorescence intensities. Provided that the investigator adheres to the basic principles of spectral analysis,[27,28] excellent definitions of constituent properties are possible, with high levels of sensitivity rarely matched by other analytical techniques. Equally important, food scientists now may exploit the opportunities provided by a rapidly expanding list of highly specific cellular probes which are often readily adapted to food applications. For example, a detailed list[29] of specific probes for a wide range of biological components, from enzymes to lectins to carbohydrates, proteins and lipids, among many others, is updated routinely, and an extensive list of fluorescent probes for food analysis also has been developed.[30] These lists are constantly expanding, and the range of applications is virtually unlimited. Extensive details have been published elsewhere[21] along with a complete review of fluorescence techniques for food analysis. [31] New applications are under constant development and many more will arrive as the challenges of food chemistry accelerate. Because fluorescence offers such dramatic advantages in sensitivity and selectivity, however, the following remarks will focus primarily on fluorescence detection only.

Advantages of Fluorescence Analysis

Because fluorescent objects are essentially self-luminous and viewed against a dark background, both resolution and sensitivity are optimized. An object can be smaller than the limit of resolution of the light microscope and still be detected as a luminous source. In addition, the high contrast characteristic of fluorescence analysis offers increased ability to differentiate the diverse mixtures of biological compounds normally encountered in food materials. Reflectance optics also improves sensitivity, precision, and selectivity quite dramatically in comparison to older transmission techniques.[21]

An essential component of microscopic photometry is the adaptation of microcomputers for operation of the wide range of filters, monochromators, and detectors which are necessary for routine use in food and biological research programs. Continuing improvements in microcomputer architecture and software and parallel developments in optical systems ensure that these technologies will provide unprecedented ease, speed, and precision of microchemical characterization of foods. Many software programs are now available for rapid accumulation of large amounts of spectral data using desk-top computers, and data are transferred easily to other analytical programs for statistical evaluation (*e.g.*

neural networks, principal component analysis, or simple regression analysis). Multivariate calibration and analysis of fluorescence data, in particular, is now a well-developed tool and provides interpretative opportunities previously unavailable to occasional users of fluorescence photometric instruments.[32] Commercial software programs such as *Unscrambler* (CAMO A/S, UUC, Trondheim, Norway) also have been used widely in the food industry for routine analysis of multiple spectra. They provide opportunities for differentiating background light, stray scattered light, and other sources of nonspecific energy which contribute to the detected signal. Use of such interpretative programs is highly recommended if comparative and continuing calibration is desired.

Instrumentation

Fluorescence microspectrophotometers consist of several essential components and a number of additional items which may be useful in selected applications. The primary component, the light microscope, is the core of the system and for many uses needs only to be a high quality research microscope, including fluorite or similar objectives/condensers which optimize fluorescence. These optics are available from most major microscope manufacturers. If ultraviolet analyses are required, the instrument must include UV-transparent quartz optics throughout the detection system. This adds considerable expense, and its use is relatively uncommon. It is, however, of some use in special applications relating to natural fluorescence of proteins and nucleic acids, and it is especially useful for the measurement of structures which contain significant concentrations of phenolic esters or lignin. The latter are found in most commercial fibers of plant origin, and this approach has been used in both absorbance and fluorescence mode to differentiate food fibers derived from a range of grains.[33]

An appropriate instrument combines a scanning stage, ultraviolet/visible and near infrared (UV/Visible/NIR) optics, and systems for both incident and transmitted light illumination, and is described in detail elsewhere.[21] Briefly, however, the instrument with most flexibility and utility includes a scanning stage with 0.25 μm matrix step-scanning capability, a photomultiplier and PbS detector for UV/VIS and NIR detection respectively, and appropriate monochromators to allow illumination in all spectral regions. For fluorescence analysis, a minimum of two monochromators is necessary to allow characterization of both excitation and emission wavelengths. Illumination is provided by mercury (HBO) illuminators for routine fluorescence work, xenon (XBO) illuminators for short wavelength (UV) analysis, and/or halogen illuminators for bright field and long wavelength fluorescence analysis. For highest sensitivity and efficiency in fluorescence detection, an epi-illuminating system and infinity-corrected optics are common components of modem fluorescence microscopes and provide excellent sensitivity.

Fluorescence microspectrophotometry typically provides chemical information in three modes: *spectral characterization*, *constituent scanning* or *mapping* in specimens, and *kinetic measurements* of enzyme systems or photobleaching. All three approaches assist in defining chemical composition and properties *in situ*, and one or all may be incorporated into modern instruments. Software control of monochrometers allows precise analysis of absorption and/or fluorescence emission characteristics in plant tissues and foods, and routine detailed spectral analysis of large numbers of food elements (*e.g.* cells, fibers, fat droplets, protein bodies, crystals, etc.) is accomplished easily.

Spectral Scanning

Microscopic *spectral scanning* of materials is a useful method for determining composition and relative concentrations *in situ*. Briefly, the microscopic object is illuminated with the range of wavelengths appropriate for the specimen in question, and both the excitation (or absorption) wavelengths and emission wavelengths are obtained by sequential scanning and intensity detection through the appropriate wavelengths. Most cereal grains (and many of their products) contain high concentrations of ferulic and coumaric acids, as well as a range of higher molecular weight phenolic polymers (*e.g.* Lignin) in some cell walls. These compounds are major determinants of grain and product quality, and because they are fluorescent and may be examined directly using fluorescence optics without other chemical enhancements, they are prime candidates for quantitative microscopy. Figure 6 shows the relative fluorescence intensities of individual cell wall components in different portions of the outer regions of the wheat kernel, and includes both excitation and emission properties of these naturally-occurring phenolic compounds.

This observation has become an important contributor to the development of rapid, automatic scanning of outer tissues of grains (primarily bran tissues) which contribute both strong color and taste characteristics to grain products such as wheat flour (see section 5 following). The ability to measure both the concentration and distribution of such components is paramount to quality control in mills and bakeries, and to raw material definition.[33–35]

Similar approaches are also available for both UV and NIR absorption measurements. The former provides a potential useful method for characterizing and identifying food fiber sources based on lignin absorption spectra. For example, fibers from diverse seed and non-seed sources can be characterized and differentiated simply on the basis of UV absorption properties. Microspectrophotometry has been used to differentiate and characterize several common insoluble fibers used in bakery products.[36]

Similarly, near infrared (NIR) analysis is a useful method for characterizing strong IR absorbing substances, such as fats and water in foods and raw materials. We have used the scanning microspectrophotometer in NIR mode to assess a wide

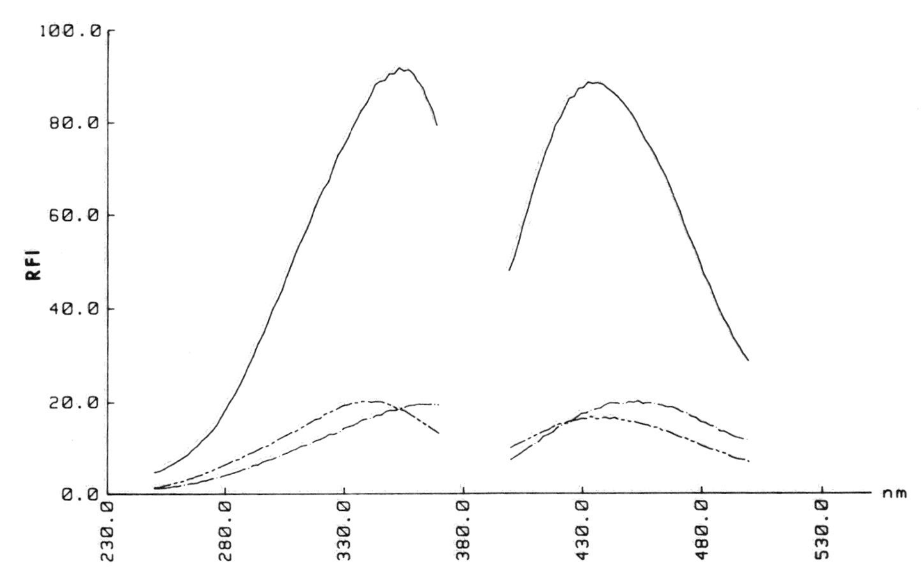

Figure 6. Excitation (left) and emission (right) spectra at the optimum for aleurone tissue showing intensity differences between aleurone, endosperm, and pericarp tissues. The emission monochromator was set at 460 nm for excitation scans and the excitation monochromator was set at 350 nm for emission scans. RFI = relative fluorescence intensity. Under these conditions, the high intensity of the ferulic acid rich aleurone cell walls is readily apparent. Each scan was obtained from a single cell wall approximately 20–75 micrometers thick.

range of components in grains and foods, including water content in grains and grain products, and solubilized sugar droplets on the surfaces of stored bakery products. Once again, the primary advantage of this approach is that very small amounts of material can be characterized chemically, and in a short time.

Specimen Scanning

Once a particular food constituent has been characterized and/or identified by spectral analysis, it becomes a relatively simple matter to quantify the material further by mapping its distribution or by quantifying the material by associated imaging procedures. Mapping involves the measurement of absorption or fluorescence intensities at fixed intervals across a specimen. This approach requires the instrument to be equipped with a scanning stage; a number of these are available commercially and range in capability from very fine and relatively slow matrix step scans (0.25 μm intervals) to scans at 10–2,000 μm intervals. The method permits either overlapping scans, such that continuous images of fluo-

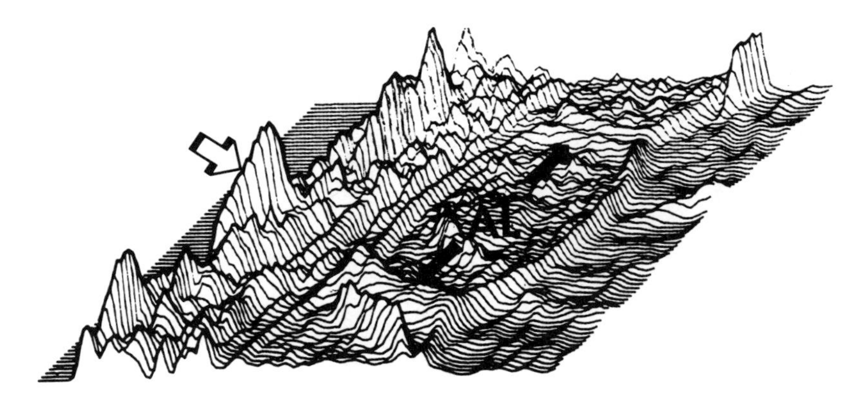

Figure 7. A fixed wavelength matrix scan of a portion of wheat bran. The section was scanned at approximately 450nm and the fluorescence emission was monitored at 520nm in approximately 2 um x 2 um scan steps. Peak height increases with fluorescence intensity. A single aleurone cell (AL) is clearly visible, as are the intensely fluorescent pericarp cell walls (open arrow).

rescence or absorbance values are created, or it permits scans which are essentially sampling procedures providing statistical evaluations of the total fluorescence or absorbance across a specimen. Figure 7 provides an excellent example of the former in which high resolution scans have allowed detailed analysis of the distribution of the phenolic acids in individual cell walls of a wheat kernel.

In combination with detailed extractive and HPLC procedures, this approach has allowed accumulation of both compositional and distribution data relating to cell wall structure and organization, especially in relation to ferulic acid content and its association with kernel hardness and insect resistance.[5,17,34] Specimen scanning procedures are also useful on a larger, semi-micro scale. We routinely use the scanning microspectrophotometric approach in fluorescence mode to evaluate distribution of functionally important constituents in raw materials such as grains. For example, Figure 8 shows the result of scanning a complete cross section of a wheat kernel with fluorescence filters set at 365 nm excitation and 450 nm emission. Using relatively crude scan step intervals (~100 x 100 μm) one has a simple procedure to exploit natural fluorescence to map phenolic compounds (primarily ferulic and coumaric acids) in grains.

More recently, the microspectrophotometer has been used extensively to map the distribution of mixed-linkage β-glucans in oat and barley endosperm cell walls, again, in relation to differences in processing and nutritional quality. Because β-glucans have been implicated in serum cholesterol reduction and in modifying carbohydrate metabolism[37–39] they have been the subject of considerable interest for several years. More importantly, it has become apparent that these high MW polymers vary considerably in their distribution patterns in

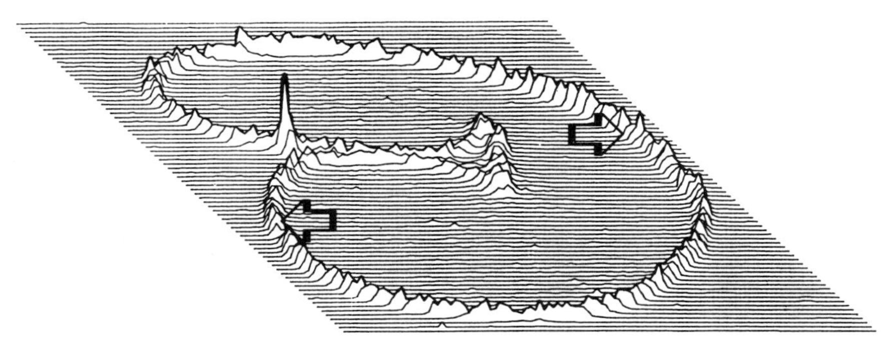

Figure 8. Lower magnification equivalent to Figure 7. In this case, a complete cross section of a wheat kernel has been mapped for fluorescence intensity, and the high levels of phenolic compounds are clearly illustrated in the outer regions of the specimen (arrows).

barley and oats, and this observation has been exploited commercially in identifying desirable cultivars of oats for food processing applications. In addition to apparent differences in cell wall composition among a range of oat cultivars, for example,[17,40] it is also apparent that the morphological organization of the cell walls also differs.[41] These studies of β-glucan distribution have been facilitated by the fact that the polymers can be detected quite specifically with a microscope using Calcofluor, an intensely fluorescent compound with tremendous affinity for β-glucans. An example of the mapping capability of the scanning microspectrophotometer is shown (Fig. 9).

Kinetic Changes

Instruments of the types described may also be used effectively to evaluate the kinetics of time-dependent changes in foods, be they enzymatic or other reactive changes. The computerized controllers and data-acquisition capabilities

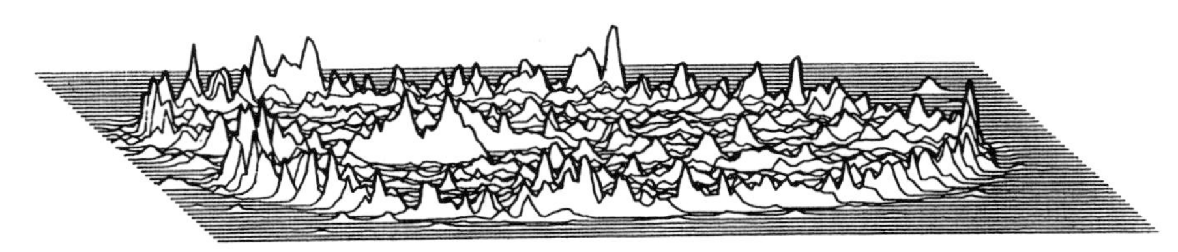

Figure 9. Cross section of a barely kernel stained with Calcofluor and scanned to map distributions of β-glucans in the grain. The highest concentrations of β-glucans correspond to the regions with maximum peak heights.

of these instruments allow precise shutter control and measurement of absorbance or fluorescence changes at millisecond intervals. This is particularly useful for analysis of fluorescence decay rates and measurement of enzymatic activity *in situ*. A number of enzyme substrates are available commercially[29] which produce fluorescent reaction products after hydrolysis by appropriate enzymes. The kinetic approach is a relatively under-used capability of computerized microspectrophotometers, but one which has considerable capability for comparing activities in individual cells or cellular components. We have used similar systems to measure esterase activity in single barley aleurone cells during germination.[12]

DIGITAL IMAGE ANALYSIS (DIA)

Digital image analysis, or *computer vision*, is the enterprise of automating and integrating a wide range of processes and representations used for visual perceptions. It includes, in part, many techniques that are useful by themselves, such as image processing and statistical pattern classification. A typical image analysis system includes an image acquisition device consisting of light source, image sensor (usually a video camera), digitizer, and computer with image array processor. The system acquires, processes, and displays images.

The image acquisition device scans and digitizes a continuous scene and breaks it into an array of digital intensity values called pixels, and puts the pixel array into the computer image memory. The computer processes the pixels in the image memory, and the display device converts the processed pixels into a spatially organized and (usually enhanced) image.

The most important element in image analysis is software. Image analysis systems usually offer standard routine software for image processing including: (a) enhancement or modification of the image to improve its appearance or highlight information; (b) measurement of image elements; (c) classification or matching of image elements; and (d) recognition of items or features in the image. Image analysis is a fascinating technology which is beginning to have an important impact on plant and food processing industries. Applications of image analysis in the food industry can be classified into the following areas:[42]

a. dimensional measurements, such as cell size and shape, starch granule sizes, product sizing, and apple bruise evaluation, volume measurements and grading, plant embryo and tissue culture analysis;
b. number counting for particle distributions and classification of blood cells, chromosome, plant callus, and starch granules;
c. texture measurement of product surfaces (*e.g.* fruit and bakery products);
d. quality assessment, such as color and maturity evaluation, bruise detection on fruits, and disease detection;

e. orientation and position, such as guidance of robots for orange and apple harvesting and grapevine pruning.

Applications in cereal grain analysis have been mainly in size and shape measurements based on the silhouettes, or binary images, and color for varietal and grade classifications.[43–47] When combined with microscopy, DIA is especially useful for measuring shapes, sizes, and distributions of cellular and sub-cellular plant constituents, and most applications in the grain industry have related to cell size and shape[12,48] and to starch granule size, shape, and frequency.[14–15,48–50]

In cereal grains particularly, starch granule size may be influenced quite significantly by both genetic and environmental factors, and, in turn, the starch properties are important components of quality. Figure 3 shows a typical distribution of "A" and "B" granules in a cereal endosperm. While the smaller of the two starch granule populations in barley (the so-called B granules) contribute approximately 90% of the total *number* in a typical starchy endosperm, the less frequent but much larger A granules provide between 92 and 96 % of the total *volume* of the starch.[49] Moreover, there is significant genetic variation among 14 barley genotypes, a factor which might be expected to influence starch related product quality factors such as malt or feed value. A similar situation occurs in wheat, and it is becoming clear that slight differences in the ratios of large to small granules can have an apparent effect on gluten performance.[50] Such observations may relate to the fact that a slight change in the ratio of A to B granules dramatically changes the surface areas of the starch in barley and wheat products. Furthermore, there appear to be three distinct populations of granules (including a much smaller "C" population) in both wheat[15] and barley.[49]

While the differences in size between A and B granules in wheat and barley are quite obvious to even the unskilled microscopist, it is the addition of DIA to the microscope that has allowed precise and reproducible measurements of starch granule populations and other carbohydrate components of plant cells. Image analysis allows measurement of tens of thousands of individual objects, and simultaneously extracts shape, texture and other features. However, until recently, DIA systems in plant and food research laboratories have been quite expensive and the domain of specially trained individuals, *i.e.* have not been widely adapted for routine and rapid analysis of quality attributes of raw or processed plant materials in the food industry. In part, this has also been due to the fact that most DIA systems have not been automated or are robust enough to tolerate the demands of processing environments. In many instances, common DIA systems have produced a wealth of detail on very large populations of cellular constituents, but, in most cases, this has not evolved further to routine food analysis. Similarly, the scanning microspectrophotometer also provides an obvious range of opportunities for detailed evaluation of the intimate association between composition and structure in raw and processed foods. However, such instruments tend to be rather expensive and are rarely useful for routine, on-line or at-line analysis during processing.

AUTOMATED HYBRID SYSTEMS

Recently, hybrid systems which combine relatively simple spectral measurements with digital analysis of microscopic images have evolved to the extent that they are capable of providing rapid analysis of microscopic structures in foods (and hence their chemistry) with minimal specimen preparation. In addition to the rapid evolution of microprocessors and the ease with which large data fields can be processed, recent improvements in fluorescence standardization, inexpensive scanning stages, and high speed image processing boards permit numerous measurements, which previously had been possible only with specialized equipment, to be obtained much more inexpensively and rapidly.

One such instrument, the I440F Power Scope (Maztech MicroVision Ltd., Ottawa, Canada), exploits the principle that any food component clearly identified with a microscope can be quantified accurately and rapidly using automated, filter-mediated digital image analysis. The I440F uses one of several common illumination systems including: (a) incident light for fluorescence analysis of natural constituents or added fluorochromes; (b) transmitted light illumination for diachromes and detection of food components with measurable absorbencies; and (c) combinations of the above. The instrument is modular, relatively easy to operate, and provides measurements of several hundred fields of view in a few minutes. Although the I440F continues to evolve as applications are more clearly identified, it is becoming apparent that the approach embodied in the instrument, namely the coincident rapid measurement of structure and chemistry, is both a novel and rewarding one. Although designed initially for high speed food component analysis, the instrument should be equally capable of routine measurements of tissue culture or other cell systems.

For each programmed measurement module provided by the I440F, the instrument records images at fixed wavelengths from a large number of microscopic fields of view (300–500 fields in 3–5 minutes). The number of fields is established according to the standard deviation and reproducibility of the data from each module, and this value is determined by the user. The instrument employs a standard commercial fluorescence / bright field microscope fitted with several modifications for routine analysis.

Two types of samples are in common use by the 1440F and each includes customized software and sample holders calibrated for individual instruments. The sampling methods include use of: (a) dry, powdered samples (approximating those in use in common near infrared reflectance analyzers); and (b) wet monolayers which are essentially samples placed on a large scale microscope slide and a cover glass. A holder for dry powdered samples (*e.g.* flour, starch, bran, fiber) consists of a glass window bonded to a shallow rigid frame to form a rectangular container. The sample is placed in the container, leveled with a scraper, and covered with a metal backing designed to provide uniform compression. Samples are mounted on a fast scanning stage, illuminated through appropriate filter systems to selectively highlight specific objects, and several hundred microscopic fields of view are obtained through the glass window and analyzed in a few minutes. Each field is obtained and measured in a fraction of a second. Although there are a number of important food components which are naturally fluorescent (*e.g.* cereal brans, lignified materials such as pea, soy and cotton fiber, and even proteins and pigments), detection of many requires application of specific fluorochromes or diachromes. Therefore, quantitative analysis using microscopic imaging also requires judicious use of sensitive dyes or stains suitable for visualization and rapid measurement. The dyes must be stable, non-toxic, easily obtained, and of consistent quality. A number are available, and even food grade colors can be employed for rapid measurements, provided they exhibit specificity for particular components.

To measure dyed microscopic components rapidly, instruments such as the 1440F include a "wet" sample holder designed to allow examination of a relatively large field of view (~2.5 x 4.5 inches) of uniformly distributed particles with little evaporation. This holder also consists of a glass window bonded to a shallow metal frame. A trough is etched around the edge of the window to provide a seal for the liquid once a cover glass has been added. The samples (usually specimens suspended in liquid fluorescent or absorbing dye) are spread over the window surface and a cover glass added to provide a "monolayer" of sample. The advantages of this system are that a wide range of dyes can be used for different chemical constituents, more than one dye can be used at a time (and measured sequentially), and the specimen can be examined with either epi-illumination or transmitted illumination. Common assays include measurement of the percent of damaged starch granules in a flour or starch preparation and frequency distributions of particulates in liquids. New protocols are under constant development, but the primary advantage of the instrument is that it is essentially an automatic measuring microscope with computer control of light intensity and filter systems, extremely rapid auto-focusing of images, and no condensers or diaphragms to adjust. It provides data and views of biological materials which often are difficult or impossible with conventional microscopic or chemical techniques. The data shown in Figure 10 represent an automated 3–5 minute analysis of ~10^3 starch granules in barley (Fig. 10a) and oat (Fig. 10b) samples.

MAGNETIC RESONANCE IMAGING (MRI)

Although the methods identified in the preceding paragraphs are all well-adapted to identification of many cellular and sub-cellular plant components,

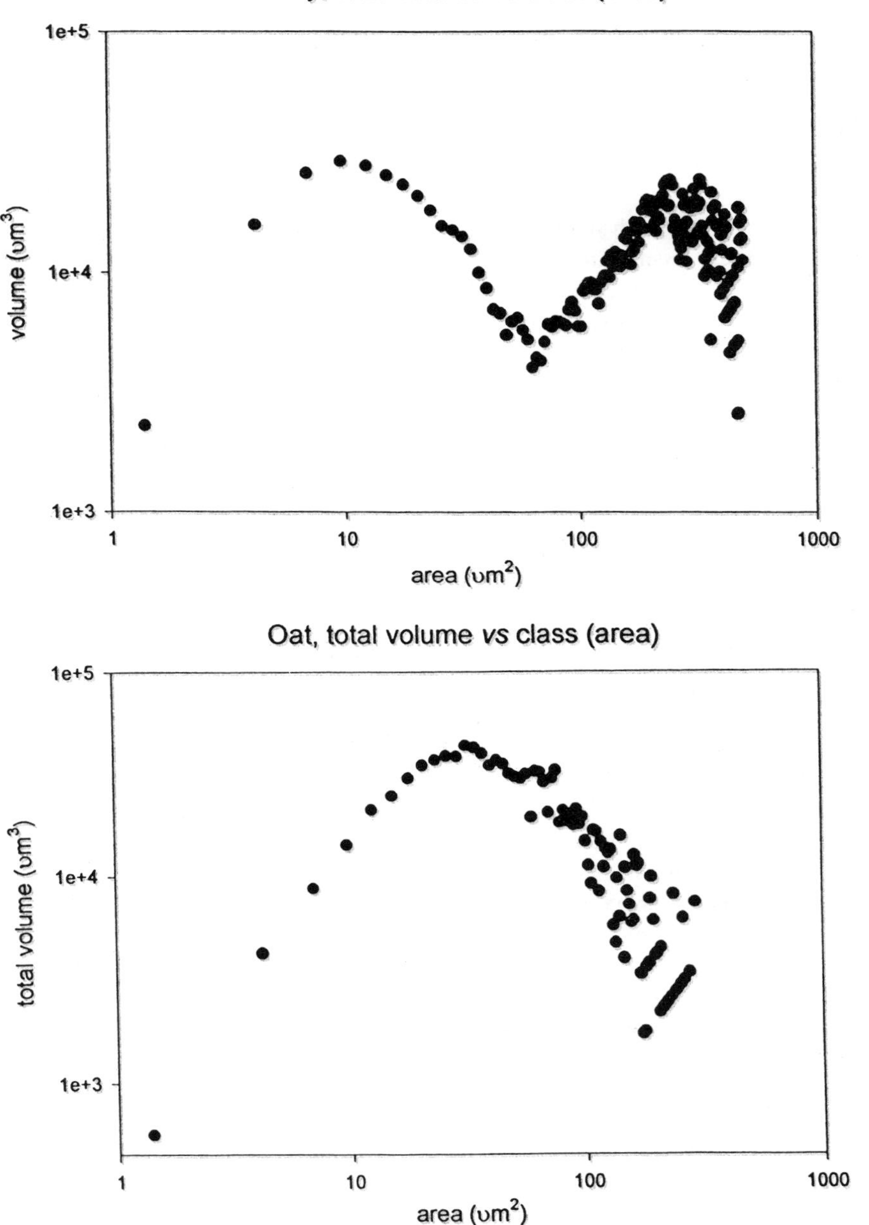

Figure 10. Frequency distributions vs. size of barley (top) and oat (bottom) starch granules as determined using the I440F automated microscopic image analysis system. Each population represents 5–10 thousand granules, and assays are complete within a few minutes. The bimodality of the barley granules is evident.

including storage and cell wall carbohydrates, they all suffer from the need to grind, section, or otherwise disrupt the specimen. Furthermore, they do not allow continuous observation of biochemical changes in specimens, nor can they provide real time images of fundamental changes in structure which accompany hydration events.

Consequently, alternative methods are constantly sought for *in situ* analysis of materials in an effort to better understand the associations among components both before and after processing. Although not especially suited to direct carbohydrate analysis, magnetic resonance imaging has proven to be a useful adjunct to plant and food structure evaluation, especially in relation to the process by which water is absorbed and redistributed in cells and tissues. Because carbohydrates are particularly important in controlling the hydration patterns of germinating grains and in plant-derived foods, there is special interest in mapping water concentrations in products during processing.

MRI allows *nondestructive and noninvasive* observation of moisture and temperature distribution and structure of an object. This suggests that MRI can be used for the continuous study of hydration during steeping or tempering, two preliminary grain processing steps which are common to the malting and flour milling industries, respectively, provided that the hydration step can be implemented inside the MRI.

MRI, also referred to as nuclear magnetic resonance (NMR) imaging, was developed in 1973[51] and is based on the principles of NMR spectroscopy. Compared to basic NMR spectroscopic methods, which provide an average signal from all excited nuclei within a sample, MRI techniques involve the manipulation of several magnetic field gradients oriented at right angles to each other, resulting in spatial encoding of signals and further information about the position of the excited nuclei within the object. For example, a linear field gradient is superimposed on the main magnetic field, and the resultant Larmor frequency (ω) will depend on position along the gradient direction, *e.g.*, y axis:

$$\omega\ (y) = g\ (Bo + yGy)$$

where Gy is the linear gradient (Gauss/cm) applied along the y axis, and Bo represents a homogeneous external magnetic field.

By making the active nuclei in different volume elements of a specimen resonate at different frequencies, MRI can *non-destructively and non-invasively* measure the density of the active nuclei as a function of spatial coordinates. Thus, MRI produces images of heterogeneous systems based on the NMR properties of the bulk fraction, *e.g.*, water or lipid distributed within the sample.

By adjusting the several parameters in an NMR imaging pulse sequence, many characteristics can be obtained, including proton density (ρ), spin-lattice relaxation time (T1), spin-spin relaxation time (T2), and the diffusion coefficient (D). The proton density directly reflects the apparent water content, while the relaxation times are correlated with the local structure and the fraction of bound water or water mobility. Diffusion also reflects the local water mobility and temperature.

The total MRI signal is a function of the imaging pulse sequence, probe characteristics and other hardware. Magnetic field gradients are generated by additional sets of coils inside the main magnet of the NMR spectrometer. By producing gradients along the three main axes (x, y, z), two- and three-dimensional images can be obtained. MRI probes are used to apply pulses to the sample (transmit mode), and to receive the signals emitted by the sample nuclei in the magnetic field (receiver mode). Pulse sequencing is the basis of NMR imaging. It indicates the order and timing of the application of radio frequency and the gradient fields as well as signal acquisition. Using these relatively simple techniques, two-dimensional (2D) image slices can be obtained rapidly for quantitative analysis of moisture profiles in steeping grain. The moisture distribution in a single kernel of barley after 12 hours of hydration is shown (Fig. 11).[52]

Typically, we use a 4.7 Tesla, 200 MHz /330 mm Spectroscopy Imaging System (Spectroscopy Imaging System Co., USA), located at the Center for Magnetic Resonance Research at the University of Minnesota, for most MRI analyses of foods and food systems. A small and high sensitivity NMR imaging probe and a gradient coil have been designed and constructed specifically for grain applications. The imaging probe has an internal diameter of 1.0 cm, and the wire used for the probe is approximately 0.1 cm diameter and mounted on the sample holder surface to minimize the image probe internal diameter and avoid movement during image data acquisition. The distance between adjacent coils is 0.1 cm, which is equal to the width of the wires. The length of the specially constructed image probe is 2.0 cm, which is long enough to enclose a barley or wheat kernel but short enough to provide desirable signal intensity. Typically, a kernel is placed in a sealed water-filled pipette tip and the entire sample is inserted into the image probe, which is held in the center of the probe. The gradient coil designed for these analyses has a 10.8 cm internal diameter with maximum field gradient strength of 40 G/cm, and it is double-layered to allow temperature stability.

Using this system, we have evaluated a wide range of food systems, with particular emphasis on the effects of added carbohydrates to products and processes. The MRI approach is somewhat expensive, but in many cases provides the only method for non-destructive and non-invasive examination of water and lipids, in particular, as well as their interaction with food hydrocolloids.

The process by which serial (2D) images of moisture profiles, as observed in a single grain during steeping, can be reassembled for three-dimensional (3D) mapping of water in grains has also been described in detail.[53,54] Briefly, multislice techniques consist of three steps: 1) the excitation of nuclei in the plane of detection; 2) spatial encoding of the signal from nuclei in the excited plane; and

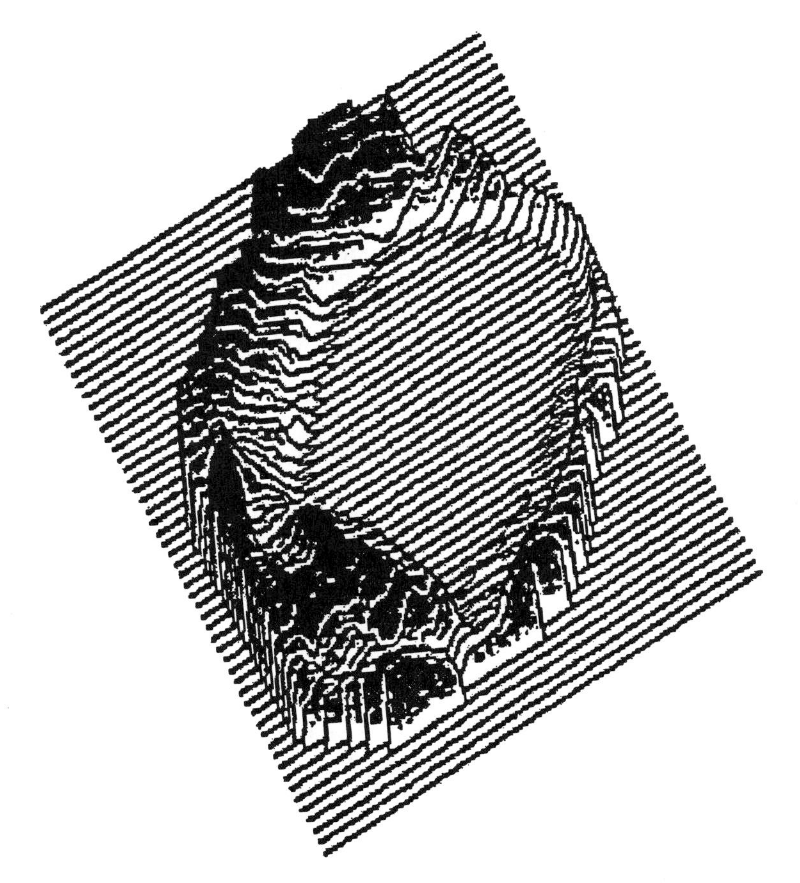

Figure 11. MRI image of a mature barley kernel in longitudinal section showing hydration patterns after 8 hours of imbibition. The lower left dark region of the image contains the germ tissues. Peak height relates directly to moisture content. The center of the endosperm contains little moisture at this stage.

3) detection of the NMR signal. This method is rapid, and all the data can be collected at the same time for a single imaging experiment.

MRI is an excellent technique for studying ingredients, processes, structures, and functional relationships in food systems. It has been used in a number of studies to determine the moisture profiles of high moisture food samples such as fruits and vegetables. McCarthy et al.[55] obtained images of peaches during freezing, and Chen et al.[56] employed MRI to obtain two dimensional NMR proton density images of various fruits and vegetables for evaluation of their internal quality. Some quality factors, including fruit abnormalities such as bruises, worm damage, dry regions and void spaces, can also be observed. Breakdown processes which relate to cell wall degradation in apples and pears have also been monitored using MRI.

Increased spatial resolution of MRI has improved the ability to acquire noninvasive images of internal structure in food materials and the ability to monitor changes in internal structure caused by processing. Eccles et al.[57] applied microscopic NMR imaging to study the circulation of water within developing wheat grain using a diffusion-velocity sensitive sequence. However, the use of MRI to obtain moisture profiles and to observe structural changes of low moisture materials such as mature wheat, barley and other grains are rather rare, although a recent report described a process for evaluation of moisture migration during drying of soybeans.[58]

SUMMARY

Complete details of the several different imaging systems discussed in this chapter are available elsewhere. It should be obvious that digital image analysis, microspectrophotometry, and/or magnetic resonance imaging provide unprecedented opportunities for direct examination of the complex association between structure and function in plant systems. The systems are not particularly new, but the rapid development of computer controllers has brought the routine use of these instruments into more laboratories. Each system has advantages and includes precise physical measurements (DIA), sub-cellular chemical characterization (microspectrophotometry), and non-destructive examination of water movement in plant tissues (MRI). They are particularly suitable for examining carbohydrate and related cellular constituents.

REFERENCES

1. MCCLEARY, B.V., GLENNIE-HOLMES, M. 1985. Enzymic quantification of (1,3)(1,4)-β-D-glucan in barley and malt. J. Inst. Brew. 91:285–295.
2. FULCHER, R.G. 1986. Morphological and chemical organization of the oat kernel. In: Oats: Chemistry and Technology. (F.H. Webster, ed.), American Association of Cereal Chemists, Inc. Minnesota. pp. 47–74.
3. HABERLANDT, G. 1890 Die Kleberschicht des Gras-Endosperms als Diastase ausscheidendes Drusengewebe. Bei Deutch Bot. Ges. 8:40–48.
4. FULCHER, R.G., MILLER, S.S. 1993. Structure of oat bran and distribution of dietay fiber components. In: Oat Bran. (P.J. Wood, ed.), American Association of Cereal Chemists, St. Paul, MN. pp. 1–24.
5. FULCHER, R.G., O'BRIEN, T.P., LEE, J.W. 1972. Studies on the aleurone layer. I. Conventional and fluorescence microscopy of the cell wall with emphasis on phenol-carbohydrate complexes in wheat. Aust. J. Biol. Sci. 25:23–34.
6. FINCHER, G.B. 1975. Morphology and chemical composition of barley endosperm cell walls. J. Inst. Brew. 81:116–122.

7. PUSSAYANAWIN, V. 1986. High Performance Liquid Chromatographic Studies of Ferulic Acid in Flour Milling Fractions. Ph.D. Dissertation. Kansas State University. 274 pp.
8. SEN, A., BERGVINSON, D., MILLER, S.S., ATKINSON, J., FULCHER, R.G., ARNASON, J.T. 1994. Distribution and microchemcial detection of phenolic acids, flavonoids and phenolic acid amides in maize kernels. J. Agric. Food Chem. 42:1879–1883.
9. ARNASON, J.T., GALE, J., DE BEYSSAC, C., SEN, A., MILLER, S.S., PHILOGEN, B.J.R., LAMBERT, J.D.H., FULCHER, R.G., SERRATOS, A., MIHM, J. 1992. Role of phenolics in resistance of maize grain to the stored products insects *Sitophilus zeamais* and *Prostephanus truncatus*. J. Stored Prod. 28:229–236.
10. REID, L.M., MATHER, D.E., ARNASON, J.T., HAMILTON, R.I., BOLTON, A.T. 1992. Changes in phenolic constituents of maize silk infected with *Fusarium graminearum*. Can.J.Bot. 70:1697–1702.
11. FREEMAN, P.L., PALMER, G.H. 1984. The structure of the pericarp and testa of barley. J. Inst. Brew. 90:88–94.
12. FULCHER, R.G., DENEKA, T., MILLER, S.S. 1992. Structure/function relationships in barley quality: Analysis by microscopy and quantitative imaging. In: Proc. Sixth International Barley Genetics Symposium, (L. Munck ed.), Helsingborg, Sweden. pp. 711–724.
13. WASILUK, K.R., FULCHER, R.G., JONES, R.J., GENGENBACH, B.G. 1994. Characterization of starch granules in maize using microspectrophotometry. Starch 46(10):369–373.
14. PEDERSEN, L.H. 1987. Development of Screening Methods for Evaluation of Starch Structure and Synthesis in Barley. Ph.D. Thesis, Chemistry Dept., Royal Veterinary and Agricultural University, Copenhagen, Denmark. 171 pp.
15. BECHTEL, D.B., ZAYAS, I., KALEIKAU, L., POMERANZ, Y. 1990. Size distribution of wheat starch granules during endosperm development. Cereal Chem. 67(1):59–63.
16. HOSENEY, R.C., FAUBION, J.M. 1981. A mechanism for the oxidative gelation of wheat flour solubel pentosans. Cereal Chem. 58(5):421–428.
17. MILLER, S.S., FULCHER, R.G., SEN, A., ARNASON, J.T. 1995. Oat endosperm cell walls: I. Isolation, composition, and comparison with other tissues. Cereal Chemsitry 72(5):421–427.
18. FINCHER, G.B. 1976. Ferulic acid in barley cell walls: A fluorescence study. J. Inst. Brew. 82:3347–349.
19. FINCHER, G.B., STONE, B.A. 1986. Cell walls and their components in cereal grain technology. In: Advances in Cereal Science and Technology, Vol 8. (Y. Pomeranz, ed.), American Association of Cereal Chemists, St. Paul, MN. pp. 207–295.
20. VINKX , C.J.A., DELCOUR, J.A. 1996. Rye (*Secale cereale L.*) arabinoxylans: A critical review. J. Cereal Sci. 24(1):1–14.
21. FULCHER, R.G., IRVING, D.W., DE FRANCISCO, A. 1989. Fluorescence microscopy: Applications in food analysis. In: Fluorescence Analysis in Foods. (L. Munck, ed.) Longman Scientific and Technical, UK. pp. 59–109.
22. JORGENSEN, K.G. 1988. Quantification of high molecular weight (1,3)(1,4)-β-D-glucan using Calcofluor complex formation and flow injection analysis. I. Analytical principle and its standardization. Carlsberg Res. Commun. 53:277–285.
23. JORGENSEN, K.G., AASTRUP, S. 1988. Quantification of high molecular weight (1,3)(1,4)-β-D-glucan using Calcofluor complex formation and flow injection analysis. II. Determination of total beta-glucan content of barley and malt. Carlsberg Res. Commun. 53:287–296.
24. FULCHER, R.G., SETTERFIELD, G., MCCULLY, M.E., WOOD, P.J. 1977. Observations on the aleurone layer. II. Fluorescence microscopy of the aleurone/subaleurone junction with emphasis on possible β-(1,3)-glucan deposits in barley. Aust. J. Plant Physiol. 4:917–928.
25. BACIC, A., STONE, B.A. 1981. Isolation and ultrastructure of aleurone cell walls from wheat and barley. Aust. J. Plant Physiol. 8:453–474.
26. MACGREGOR, A.W., BALLANCE, G.M., DUSHNICKY, L. 1989. Fluorescence microscopy studies on (1,3)-β-D-glucan in barley endosperm. Fd. Microstructure 8:235–244.

27. PILLER, H. 1977. Microscope Photometry. Springer Verlag, 253 pp.
28. DHILLON, S.S., MIKSCHE, J.P., CECICH, R.A. 1983. Microspectrometric applications in plant science research. In: New Frontiers in Food Microstructure. American Association of Cereal Chemists, (D.B. Bechtel, ed.), St. Paul, MN, pp. 27–74.
29. HAUGLAND R.P. 1994. Molecular Probes: Handbook of Fluorescent Probes and Research Chemicals. Molecular Probes Inc., Eugene, OR, 420 pp.
30. DE FRANCISCO, A. 1989. Fluorochromes: wavelengths, recipes and applications. Appendix In: Fluorescence Analysis in Foods. (L. Munck, ed.), Longman Scientific and Technical, UK, pp. 268–282.
31. MUNCK, L. (Ed). 1989. Fluorescence Analysis in Foods. Longman Scientific and Technical, UK, 289 pp.
32. PEDERSEN, B., MARTENS, H. 1989. Multivariate calibration of fluorescence data. In: Fluorescence Analysis in Foods. (L. Munck, ed.), Longman Scientific and Technical, UK, pp. 215–267.
33. FULCHER, R.G., COLLINGWOOD, K. 1987. Quantitative microscopy in barley research. Proc. Australian Barley Technical Symposium,. (M. Glennie-Holmes, H. Taylor, ed.), NSW Dept. of Agriculture, Wagga Wagga, NSW. pp. 222–228.
34. IRVING, D.W., FULCHER, R.G., BEAN, M.M., SAUNDERS, R.M. 1989. Differentiation of wheat based on fluorescence, hardness, and protein. Cereal Chem. 66(6):471–477.
35. SYMONS, S.J. DEXTER, J.E. 1993. Relationship of flour aleurone fluorescence to flour refinement for some Canadian hard common wheat classes. Cereal Chem. 70:90–95.
36. ROONEY, M.K., FULCHER, R.G. 1992. Differentiation of insoluble fibers and fiber mixtures using high resolution scanning absorption microspectrophotometry. J. Food Sci. 57(5):1246–47, 1257.
37. WOOD, P.J., ANDERSON, J.W., BRAATEN, J.T., CAVE, N.A., SCOTT, F.W., VACHON, C. 1989. Physiological effects of β-D-glucan rich fractions from oats. Cereal Foods World 34:878–882.
38. WOOD, P.J., BRAATEN, J.T., SCOTT, F.T., RIEDEL, D., POSTE, L.M. 1990. Comparison of viscous properties of oat and guar gum and the effects of these and oat bran on glycemic index. J. Agric. Food Chem. 38:753–757.
39. DAVIDSON, M.H., DUGAN, L.D., BURNS, J.H., BOVA, J., STORY, K., DRENNAN, K.B. 1991. The hypocholesterolemic effects of β-glucan in oat meal and bran. A dose controlled study. JAMA 265:1833–1839.
40. MILLER, S.S., FULCHER, R.G. 1995. Oat endosperm cell walls: II Hot-water solubilization and enzymatic digestion of the wall. Cereal Chem. 72(5):428–432.
41. MILLER, S.S., FULCHER, R.G. 1994. Distribution of (1,3)(1,4)-β-D-glucans in kernels of oats and barley using microspectrophotometry. Cereal Chem. 71(1):64–68.
42. RUAN, R., NING, A., BRUSEWITZ, G.H. 1989. Imaging processing techniques for food engineering applications. American Society of Agricultural Engineers. Paper No. 896614.
43. SYMONS, S.J., FULCHER, R.G. 1988. Determination of wheat kernel morphological variation by digital image analysis. I. Variation in Eastern Canadian milling quality wheats. J. Cereal Sci. 8:211–218.
44. SYMONS, S.J., FULCHER, R.G. 1988. Determination of wheat kernel morphological variation by digital image analysis: II. Variation in cultivars of soft white winter wheats. J. Cereal Sci. 8:219–229.
45. KEEFE, P.D. 1990. Observations concerning shape variations in wheat grains. Seed Sci. and Technol. 18:629–640.
46. ZAYAS, I., POMERANZ, Y., LAI, R.S. 1985. Discrimination between Arthur and Arkan wheats by image analysis. Cereal Chem. 62:478–480.
47. PIETRZAK, L.N., FULCHER, R.G. 1995. Polymorphism of oat kernel size and shape in several Canadian oat cultivars. Can. J. Plant Sci. 75:105–109.

48. FULCHER, R.G., FAUBION, J.M., RUAN, R., MILLER, S.S. 1994. Quantitative microscopy in carbohydrate analysis. Carbohydrate Polymers 25:285–293.
49. OLIVEIRA, A.B., RASMUSSON, D.C., FULCHER, R.G. 1994. Genetic aspects of starch granule traits in barley. Crop Sci. 34:1176–1180.
50. HABERER, K.M. 1994. Evaluation of Starch Quality in Relation to Mixing Characteristics of Minnesota Grown Wheat Varieties. M.S. Thesis. University of Minnesota. 159 pp.
51. LAUTERBUR, P.C. 1973. Image formation by induced local interactions: examples employing NMR. Nature 242:190–191.
52. MCENTYRE, E. 1995. Evaluation of the Physical Properties of Barley in Relation to Hydration. M.S. Thesis. University of Minnesota. 145 pp.
53. RUAN, R., LITCHFIELD, J.B. 1992. Determination of water distribution and mobility inside corn kernels during steeping using magnetic resonance microscopy. Cereal Chem. 69(1):13–17.
54. MANSFIELD, P., MAUDSLEY, A.A. 1976. Planar spin imagin by NMR. J. Phys. C:9 L409–411.
55. MCCARTHY, M.J., KAUTEN, R. 1990. Magnetic resonance imaging applications in food research. Trends in Food Sci. and Technol. Dec. 143.
56. CHEN, P., MCCARTHY, M.J., KAUTEN, R. 1989. NMR for internal quality evaluation of fruits and vegetables. Trans. of ASAE 32(5):1747–1753.
57. ECCLES, C.D., CHALLAGHAN, C.F., JENNER, C.F. 1988. Measurement of the self-diffusion coefficient of water as a function of position in wheat using nuclear magnetic resonance imaging. Biophys. J. 53(1):77–81.
58. ZENG, X.S. 1994. Study of Soybean Seedcoat Cracking During Drying Using Magnetic Resonance Imaging. M.S. Thesis. University of Minnesota. 106 pp.

INDEX